Springer Series in Materials Science

Volume 358

The Springer Series in Materials Science covers the complete spectrum of materials research and technology, including fundamental principles, physical properties, materials theory and design. Recognizing the increasing importance of materials science in future device technologies, the book titles in this series reflect the state-of-the-art in understanding and controlling the structure and properties of all important classes of materials.

Igor Krivtsun · Leonid Golovko · Serhii Fomichov · Oleksii Kaglyak · Yevgenia Chvertko

Laser and Hybrid Technologies for Surface Engineering and Welding

Springer

Igor Krivtsun
E. O. Paton Electric Welding Institute
National Academy of Sciences of Ukraine
Kyiv, Ukraine

Serhii Fomichov
Institute of Materials Science and Welding
IIW/EWF Approved Training Body
National Technical University of Ukraine
"Igor Sikorsky Kyiv Polytechnic Institute"
Kyiv, Ukraine

Yevgenia Chvertko
Institute of Materials Science and Welding
IIW/EWF Approved Training Body
National Technical University of Ukraine
"Igor Sikorsky Kyiv Polytechnic Institute"
Kyiv, Ukraine

Leonid Golovko
Institute of Materials Science and Welding
National Technical University of Ukraine
"Igor Sikorsky Kyiv Polytechnic Institute"
Kyiv, Ukraine

Oleksii Kaglyak
Institute of Materials Science and Welding
National Technical University of Ukraine
"Igor Sikorsky Kyiv Polytechnic Institute"
Kyiv, Ukraine

ISSN 0933-033X ISSN 2196-2812 (electronic)
Springer Series in Materials Science
ISBN 978-3-032-15037-0 ISBN 978-3-032-15038-7 (eBook)
https://doi.org/10.1007/978-3-032-15038-7

This Springer imprint is published by the registered company Springer Nature Switzerland AG
The registered company address is: Gewerbestrasse 11, 6330 Cham, Switzerland

Preface

The aim of the book is to reveal the full range of technological possibilities for using laser radiation to form the desired properties and improve the quality of working surfaces of parts and structures.

This book summarizes the experience of the scientific schools at the Igor Sikorsky Kyiv Polytechnic Institute and the Paton Electric Welding Institute in developing laser technologies. It covers mathematical models of thermal and hybrid plasma-laser processes; patterns of microstructure formation and mechanical properties for major metal groups; and technology for laser surface engineering.

A distinctive feature of the book is its comprehensive examination of the scientific foundations of laser technologies, namely:

- the physical processes involved in laser radiation penetration and ways to increase absorption efficiency;
- the features of the physical interaction of the laser beam with plasma in the cathode, column, and anode regions and develops mathematical models of hybrid processes;
- calculations of thermal fields during heating with pulsed and continuous laser radiation;
- the formation and control of microstructure features and properties, such as wear resistance, distribution of residual surface stresses, corrosion resistance, fatigue strength, and heat resistance in surface layers during the laser processing of major metal groups.

The book opens up opportunities for the implementation and improvement of the areas of laser surface engineering presented by the authors:

- Increasing the rigidity of thin sheet metal products by creating structural stiffening ribs with a laser to compensate for working stresses and reinforce the product by creating areas of increased strength and hardness.
- Shaping sheet metal products involving giving the product the desired shape by applying a specific pattern of laser tracks to the surface, creating a temperature gradient, bulging, and polymorphic transformations.

- Combined laser-thermal deformation strengthening of metal products by combining ultra-fast local laser heating with mechanical plastic deformation using a roller. This process suppresses the development of recrystallization processes, increases the density and directionality of dislocations, forms branched sub-boundaries, refines grains, increases microhardness, and increases the depth of the hardened layer. It also forms compressive residual stresses.
- Hardening cutting tools taking into account the influence of laser beam travelling speed, distance between the laser beam axis and cutting edge, heating spot radius, wedge sharpening angle, and its effect on the wedge temperature field and hardening microstructure formation.
- Combined plasma-laser coating consisting of two sequential processes: plasma spraying for controlled surface alloying, followed by laser thermal cycling to distribute the doping elements and ensure favorable microstructure and mechanical properties of the surface layer.
- Laser thermal deformation synthesis of diamond-containing composite coatings involving laser melting of the product surface with the simultaneous addition of a metal powder binder and artificial diamonds. This is followed by the plastic deformation of the composite layer during cooling from 650 °C to 500 °C.
- Laser powder surfacing, which has the ability to control the maximum heating temperature, heating and cooling rates of the surface layer, lifetime, and shape of the surfacing pool by adjusting the heat flux density distribution function in the focusing zone.
- Laser casting and projection casting technologies for producing bimetals. Laser casting technology involves melting a rectangular section of the surface layer of the bimetal base with a laser beam while simultaneously feeding the molten functional component of the bimetal from a foundry induction ladle onto the melted surface. Projection casting technology consists of two operations: forming a macro-relief on the surface of the bimetal base by laser cutting, milling, or plastic deformation, and pouring the molten functional component of the bimetal from a foundry induction ladle over the entire length of the workpiece with a surface macro-relief.

The book is intended for master's students, doctoral students, university professors, and enterprise specialists involved in the development and implementation of laser metal processing technologies.

Kyiv, Ukraine

Igor Krivtsun
Leonid Golovko
Serhii Fomichov
Oleksii Kaglyak
Yevgenia Chvertko

Competing Interests The authors have no competing interests to declare that are relevant to the content of this manuscript.

Introduction

Laser radiation forms the foundation of the most rapidly advancing technologies in surface engineering, welding, and metal cutting.

The efficiency and competitiveness of modern production depend on the extent to which high technologies are used. These technologies primarily include the latest material processing methods developed based on the latest fundamental scientific achievements. Lasers and laser radiation are among the greatest achievements of human scientific activity.

Laser radiation is a universal tool for imparting desired properties to metals and shaping products. Lasers are already used in manufacturing and perforating oil and gas wells, in metallurgy, and in medium and heavy engineering. This is due to their ability to concentrate energy more than any other source, allowing energy to be introduced into materials quickly, precisely, and locally.

Despite short-term periodic declines, laser technologies have steadily risen over a short period of development.

The impetus for the development of laser technologies stems from the introduction of new equipment with high output radiation power, high efficiency, high beam quality, and small dimensions. Recently, there has been a surge in developments associated with new LED and fiber lasers that have high output radiation power (up to 50 kW) and high efficiency (50–60%). These lasers emit radiation with wavelengths of 1.06–0.96 μm, which is well absorbed by metals. This radiation can be transported relatively easily over long distances and delivered to hard-to-reach places using optical fiber light guides.

Excimer lasers, which generate ultraviolet radiation, are of considerable interest. Their short wavelengths and nanosecond radiation pulses allow many issues related to nanotechnology, micro-object creation, and medicine to be addressed.

With rare exceptions, laser material processing is purely thermal and characterized by a classic set of parameters: heating temperature and distribution, heating time and rate, and cooling rate. Laser radiation enables the precise control of these parameters, driving new technological developments.

The theoretical basis for controlling laser material processing and industrial applications is formed by a triad:

(a) description of the physical essence of laser and hybrid laser-plasma processes, patterns of laser radiation absorption by the materials being processed;
(b) mathematical modeling of thermal, hydro-gas-dynamic processes, stress-strain state of materials under various conditions of laser surface treatment with pulsed, continuous, and scanning radiation;
(c) research of structural and phase transformations and their influence on the physical and mechanical characteristics of surface layers and metal products.

Laser processing technologies are based on theoretical foundations, among which laser surface engineering technologies are the leading direction: increasing the rigidity of thin-sheet metal structures; contactless forming of products; thermo-deformation strengthening of metal products, in which laser hardening is combined with mechanical plastic deformation; thermal cycling of coatings; surfacing; surface cleaning; nitriding.

Solving these pressing problems is a promising direction for the development of modern science and technology.

Contents

About the Authors

Prof. Academician Igor Krivtsun, D.Sc. is the director of the E. O. Paton Electric Welding Institute. He is an eminent Ukrainian physicist specializing in gas discharge physics and welding process theory.

His research focuses on low-temperature technological plasmas—such as welding arcs, plasma jets, and various gas discharges—and their interaction with materials during arc, plasma, laser, and hybrid welding, surfacing, and spraying processes. His pioneering work in hybrid laser-arc welding and metal processing has earned international recognition.

Igor Krivtsun developed foundational design and calculation principles for laser-arc plasma torches used in advanced technologies. Under his leadership, several innovative hybrid processes were created, including laser–microplasma welding of thin metals, laser–plasma powder surfacing, and spraying, as well as laser–plasma deposition of diamond and diamond-like coatings. His team also engineered integrated laser–arc plasma torches, unmatched globally.

He actively contributes to the development of efficient welding and coating methods for the production and repair of modern military equipment.

In addition to research, Igor Krivtsun is engaged in education and scientific leadership. He has supervised three Candidates and two Doctors of Sciences and continues mentoring Ph.D. and doctoral researchers.

He is a full member of the American Welding Society (AWS) and ASM International.

Prof. Leonid Golovko, D.Sc. is a professor at the Department of Laser Engineering and Physical Technologies at the National Technical University of Ukraine "Igor Sikorsky Kyiv Polytechnic Institute".

Leonid Golovko is one of the founders of the scientific school of laser technologies at Kyiv Polytechnic Institute under the leadership of Prof. Volodymyr Kovalenko in the early 1970s.

Leonid Golovko headed the Department of Laser Engineering and Physical Technologies for about 20 years. As a result of his scientific activity, 356 scientific works have been published, including 15 monographs, and more than 150 Ukrainian patents have been obtained. Three monographs have been republished in the USA, China, and Bulgaria.

His field of scientific research is the study of physicochemical processes and phenomena in materials under the influence of concentrated energy flows (laser radiation, plasma, ultrasound, electric discharge) with the aim of creating new materials and highly efficient technologies for their processing, as well as the latest technological equipment. The scope of scientific interests includes: studying the properties of laser radiation depending on the type of working environment, type of pumping, resonator design, transformation methods, and their influence on the processes of laser radiation absorption by different environments, their heating, melting, evaporation, sublimation, etc. The most important component of this research is the study of structural and phase transformations that occur in materials under the action of powerful laser radiation, processes of anomalous mass transfer initiated by pulsed laser heating, and the study and development of processes of laser heating of biological media.

A significant area of focus is the development of new processes for non-contact laser forming of metal spatial structures, research and development of processes to increase their rigidity by creating special metallurgical stiffeners, research and development of laser surface alloying and surfacing processes, which are the basis for the restoration of worn parts, the application of functional coatings, and the growth of three-dimensional products. The developed and implemented processes of laser thermal cycling of plasma, detonation, and galvanic coatings have demonstrated high efficiency, increasing their adhesion strength to the base by 4–5 times. New processes have been developed for laser sintering of diamond and cubonite-based composites on the cutting surfaces of disc and drum-type tools designed for high-performance machining of large-sized products in mechanical engineering, aircraft and shipbuilding.

The developed laser technologies and equipment have been implemented at leading industrial enterprises in Ukraine (Kyiv, Kharkiv, Zaporizhzhia) and at the Vishard company (Poland), with more than 30 laser processing production sites created.

Prof. Serhii Fomichov, D.Sc. is a full professor at the Institute of Materials Science and Welding at the National Technical University of Ukraine "Igor Sikorsky Kyiv Polytechnic Institute (KPI)"; the director of IIW/EWF Approved Training Body; the chairman of the Safeguarding Impartiality Committee of Bureau Veritas Ukraine. He received two Doctor of Science degrees: in the specialty "Chemical Strength of Materials" (dissertation topic "Prediction of Resistance of Welded Pipelines and Pressure Vessels Against Corrosion Cracking") and in the specialty "Welding and Related Processes" (dissertation topic "Quality Management in Welding Fabrication"). He worked as a dean of the Welding Faculty and as a head of Department of Surface Engineering of Igor Sikorsky KPI from 2002 to 2020.

He has long experience working on development of surface engineering technologies using nano-sized powder materials. He was participating in projects for improvement of wear-resistance of parts of equipment for mining industry and transport engineering.

Most of all he loves his daughters Alina and Julia. Serhii is an instructor alpine skiing and mountaineering.

Dr. Oleksii Kaglyak is a head of Laser System and Advance Technology department at the Institute of Materials Science and Welding at the National Technical University of Ukraine "Igor Sikorsky Kyiv Polytechnic Institute".

The main areas of scientific activity are related to the study of laser technology especially laser surface. Focused on laser forming of shield materials, laser sintering of instrumental composites with diamonds or cubic boron nitride, laser assisted formation bimetallic material.

Oleksii Kaglyak has published 40 scientific works, including 1 monograph and 6 patents.

Dr. Yevgenia Chvertko received her B.S., M.S., and Ph.D. in welding and related technologies from the National Technical University of Ukraine "Igor Sikorsky Kyiv Polytechnic Institute" (Igor Sikorsky KPI), Ukraine, in 1999, 2001, and 2012, respectively. She worked as an assistant at the Welding Faculty of Igor Sikorsky KPI from 2001 to 2012 and as an associate professor from 2012 to 2020.

From 2020 to present, she works as an associate professor of the E. O. Paton Institute of Materials Science and Welding and is engaged at activities of the International Institute of Welding and European Federation for Welding, Joining and Cutting in Ukraine. In 2021–2024, she was a member of the Board of Directors of the International Institute of Welding.

Currently, Dr. Chvertko is a member of the International Authorization Board of the International Institute of Welding.

Chapter 1
Physical Phenomenon of Laser, Arc, Plasma and Hybrid Laser-Plasma Processes

Abstract The physical phenomenon of a laser beam is described, and a functional diagram of a laser is provided. The physical processes involved in the penetration of laser radiation and ways to increase the efficiency of laser radiation absorption are considered. Diagrams are provided and the physics of processes in the cathode region, column, and anode region of an electric arc are described. The features of a plasma jet and a schematic diagram of direct and indirect plasma torches are considered. The features of the physical interaction of the laser beam with plasma in the cathode region, column, and anode region are described, and a diagram of hybrid laser-plasma processes is provided. Mathematical models of hybrid laser-plasma processes are developed for coaxial and lateral positioning of the laser beam relative to of a tubular tungsten cathode. The manifestation of the synergistic effect in the interaction of the laser beam with plasma is substantiated.

1.1 Physical Phenomenon of Laser Processes

Albert Einstein postulates remain fundamental to understanding laser generation. Since then, numerous publications have addressed the physics of lasers, expanding this field to include new types of lasers [1–4].

Light is electromagnetic radiation emitted by a heated or excited substance and consists of photons.

A **photon** is an elementary electroneutral particle that is characterized by a particle-wave dualism, i.e., simultaneous possession of properties of:

- *electromagnetic waves*—diffraction and interference if the characteristic dimensions of the interference are comparable to the wavelength of the photon;
- *substance*—a photon has a certain energy and momentum, which are characteristic of bodies of a certain mass; gravity acts on a photon as on a body of a certain mass.

Light is perceived by the eye and makes the environment visible.

I. Krivtsun et al., *Laser and Hybrid Technologies for Surface Engineering and Welding*, Springer Series in Materials Science 358, https://doi.org/10.1007/978-3-032-15038-7_1

Fig. 1.1 Scheme of photon emission by light sources

The light has a narrow range from 380 nm (violet) to 780 nm (red) in the general spectrum of electromagnetic waves (from 10^{-8} nm—γ-Ray to 10 km—long radio waves).

Light sources emit photons of different electric and magnetic field intensities, frequencies (wavelengths), amplitudes, and thus energy and momentum (Fig. 1.1). Photons that are *organized* in a certain way acquire extraordinary properties—laser radiation.

A **laser** (**L**ight **A**mplification by **S**timulated **E**mission and **R**adiation) is a light quantum generator that converts the pumping energy (light, electrical, thermal, chemical, etc.) into the energy of electromagnetic waves (**laser beam**) that have ***mandatory*** properties (Fig. 1.2):

(1) ***coherence*** (Fig. 1.2a)—(coincidence in time) a constant phase difference in time of laser beam waves, which gives a total amplitude when waves of the same frequency are added;
(2) ***monochromaticity*** (Fig. 1.2b)—the uniformity of the laser beam wavelengths;
(3) ***polarization*** (Fig. 1.2c)—a fixed orientation of the electric and magnetic field intensity vectors in space relative to the direction of laser beam propagation;
(4) ***collimation*** (Fig. 1.2d)—a small angle of divergence of the laser beam.

Lasers use a wavelength range of 0.33–10.6 μm (mainly the visible light range with partial overlap of the ultraviolet and infrared ranges).

The properties of the laser beam result in an extraordinary energy density. The ratio of energy concentration per unit area in welding sources—"electric arc: plasma: laser beam" is respectively "1: 10: 170".

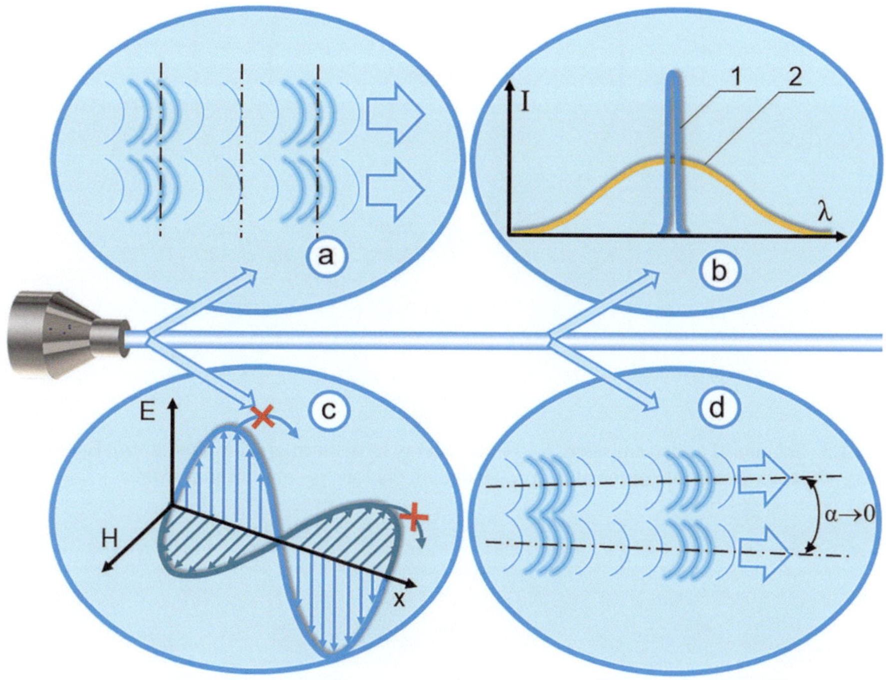

Fig. 1.2 Properties of a laser beam: **a**—coherence; **b**—monochromaticity; **c**—polarization; **d**—collimation (I—light intensity; λ—wavelength; E—electric field strength; H—magnetic field strength; α—laser beam divergence angle; 1—laser beam wavelength range; 2—sunlight wavelength range)

The generation of a laser beam is based on **forced emission** (Fig. 1.3)—the formation of a new photon γ_g when an atom transitions from a higher E_1 to a lower E_0 energy level under the influence of an inducing photon γ_i. The energy of the inducing photon is equal to the energy difference $(E_1 - E_0)$ of the energy levels of the atom. The resulting photon has the same wavelength, phase, polarization, and direction of propagation as the inducing photon. The inducing photon is not absorbed. Both photons are coherent (the produced photon is an "exact copy" of the inducing photon).

This is how a laser beam differs from spontaneous radiation, in which the emitted photons have random propagation directions, polarization, and phase. A laser beam can be *continuous*, with constant power, or *pulsed*, reaching an extremely high peak power.

The ***functional diagram*** of lasers (Fig. 1.4) includes three main components:

(1) **Active** (working) **medium** 1 is a substance in which photons 2 are formed, propagated, and amplified by forced radiation.

According to the aggregate state of the active medium, lasers are divided into solid-state, gas, liquid, and semiconductor; according to the time of laser beam generation, they are continuous and pulsed. In solid-state lasers, the active medium is optical

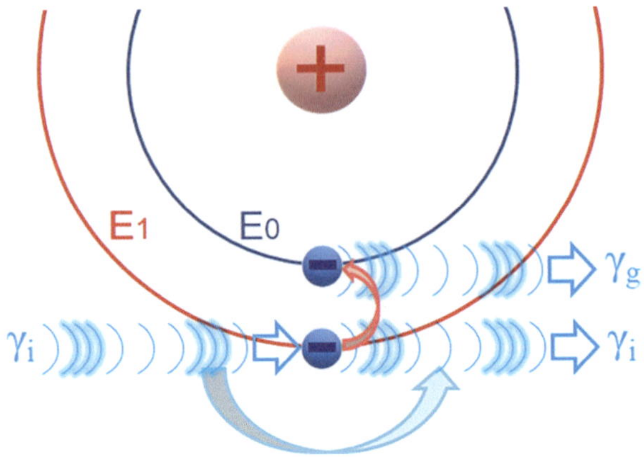

Fig. 1.3 Scheme of forced emission: E_0—lower energy level (normal state of the atom); E_1—higher energy level (excited state of the atom); γ_i—inducing photon; γ_g—produced photon

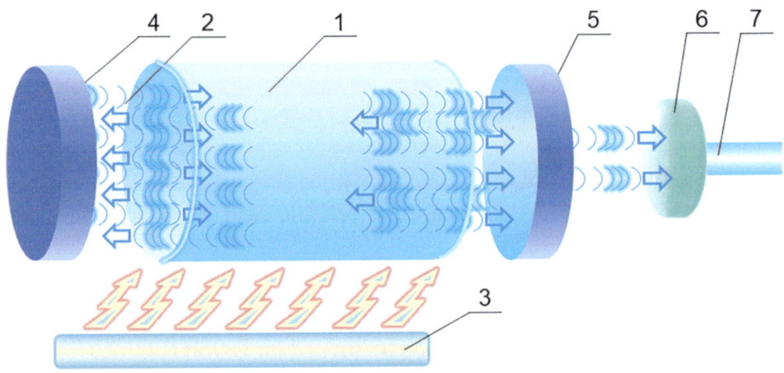

Fig. 1.4 Functional diagram of a laser: 1—active medium; 2—photons; 3—pumping system; 4—reflecting mirror of the optical resonator; 5—translucent mirror of the optical resonator; 6—focusing lens; 7—laser beam

glass with impurities of neodymium and other rare-earth elements; in gas lasers, it is carbon dioxide or a helium-neon mixture.

(2) **Pumping system** 3 is an external energy source for excitation of the active medium, transition of atoms to a higher energy state and creation of conditions for forced radiation.

An external energy source can be:

- gas discharge lamp (optical pumping)—in solid-state lasers;
- another low-power laser is used in solid-state lasers;
- electric discharge—in gas and liquid lasers;
- current through a p–n junction—in semiconductor lasers.

(3) An **optical resonator** is a device consisting of several elements that repeatedly reflect the initial electromagnetic waves in the active medium and enhance the light flux due to:

- lengthening the path along which new electromagnetic waves are formed by the forced radiation;
- formation of resonant standing optical waves.

A simple optical resonator consists of two mirrors—coaxial, parallel, and facing each other. One mirror is conventional 4. The second mirror is translucent 5—part of the light flux is reflected as from a mirror, and the other part of the light flux passes through a lens and forms a laser beam 7.

1.2 Increasing the Efficiency of Laser Radiation Absorption

Laser radiation is divided into *two parts* when it interacts with the material:

(1) part of it is ***reflected*** from its surface;
(2) part of it ***penetrates*** the surface layer and is ***absorbed into*** it.

The *efficiency* of laser processing of materials is determined *by the absorbed* radiation.

Physical processes during the penetration of laser radiation

In metals the penetration of laser radiation *occurs*:

(a) absorption of quanta by transferring energy to free *electrons*, which increases the temperature of the electron gas;
(b) heat transfer from free electrons to the crystal lattice;
(c) heat transfer due to *electronic* thermal conductivity.

Laser radiation penetrates the skin layer with a thickness of 10^{-5}–10^{-4} mm. The energy of free electrons increases, and the temperature of the electron gas rises.

The temperature equalization in the laser exposure zone takes place over a period of about 10^{-11} s in *stages*:

- the temperature of the crystal lattice remains unchanged for some time due to the large difference in the masses of the electron and ion;
- electron–ion interaction takes effect, resulting in a decrease in the temperature of the electron gas and an increase in the lattice temperature.
- The temperatures of the electron gas and the lattice are equalized.

The heat source generated by laser irradiation in metals should be considered a surface source.

The *intensity* of laser radiation decreases as it penetrates deeper into the metal according to the Bouguer-Lambert law [5]:

$$I(z) = I_0 \cdot A \cdot \exp(-\alpha z) \tag{1.1}$$

where I_0 is the intensity of laser radiation on the surface of the material;
$A = 1 - R$ is the absorption capacity of the material;
R is the reflection coefficient;
$\alpha = 1/\delta$ is the radiation absorption coefficient of the medium;
δ is the depth of radiation attenuation (penetration) in the direction from the surface; *z*—coordinate from the surface into the material.

In the optical wavelength range metals have:

- reflection coefficient R = 70–95%;
- absorption coefficient $\alpha = 10^5$–10^6 cm^{-1}.

In dielectrics (organic compounds), when laser radiation penetrates, quanta *are absorbed* by transferring energy *to intermolecular vibrations*, which causes the material to heat up.

The absorption of laser radiation in dielectrics and most semiconductors occurs at the attenuation thickness δ and depends on the wavelength.

For strong absorbers, $\alpha = 10^6$ cm^{-1}, $\delta = 10^{-6}$ cm; for relatively weak absorbers, $\alpha = 10$ cm^{-1}, $\delta = 10^{-1}$ cm.

Dielectrics do not contain free electrons that impede the penetration of laser radiation, so the thickness of the attenuation can significantly exceed the thickness of the skin layer in metals. The heat source generated by laser irradiation in dielectrics should be considered a volumetric source.

The amount of laser **energy** absorbed by the material depends on the:

- radiation wavelengths;
- electrical conductivity of the material being processed;
- thermal and physical properties and initial temperature of the material being processed;
- roughness of the irradiated surface;
- the speed of movement of the laser beam relative to the surface to be treated;
- the presence of a technological coating, its properties and thickness;
- power, radiation polarization, focus spot size, and irradiation duration;
- the angle of incidence of the laser beam;
- chemical and phase composition of the material being processed.

According to the ***wavelength***, laser radiation is divided into three ranges depending on the absorption capacity (Fig. 1.5) [5, 6]:

(1) Laser radiation in the ***ultraviolet*** (short-wave—less than 0.4 μm) region of the spectrum ***is well absorbed*** by both dielectrics and metals. In addition, laser radiation in the ultraviolet region of the spectrum is better concentrated on a smaller area. Consequently, ***ultraviolet lasers*** are the most ***effective for surface engineering***.

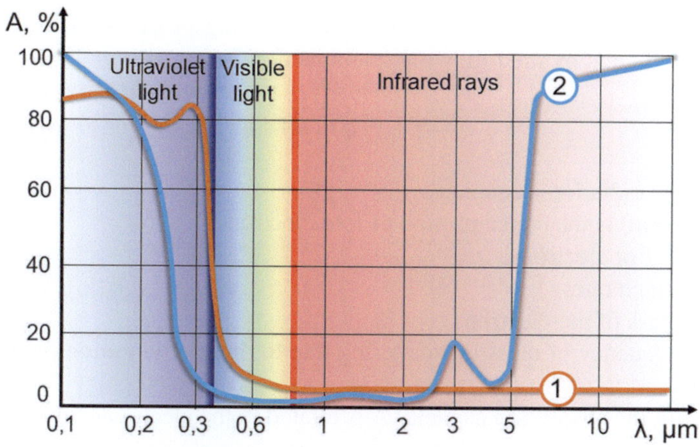

Fig. 1.5 Dependence of the absorption capacity A of an ideal surface on the wavelength λ of laser radiation: 1—metals (copper); 2—dielectrics (sapphire)

(2) Laser radiation of the ***visible*** (0.4–0.8 μm) and ***near-infrared*** (0.8–5 μm) spectral regions is ***poorly absorbed*** by both dielectrics and conductors. When using laser radiation of this range for surface treatment, special measures must be taken: increasing the temperature, providing the irradiated surface the properties of an absolutely black body, etc.

(3) Laser radiation in the ***far-infrared*** (5–10.6 μm) spectral region is *poorly absorbed by metals* and *very well by dielectrics*. The CO_2 laser radiation (λ = 10.6 μm) is poorly absorbed by all metals and alloys (no more than 15% without a technological coating).

Thus, *dielectrics* are *universal absorbers* of laser radiation. Consequently, the application of ***technological dielectric coatings*** (metal oxides, salts) is a way to ***increase the efficiency*** of laser metal processing.

Increasing the wavelength of laser radiation increases the reflection coefficient of metals (Table 1.1).

Table 1.1 Reflection coefficient of different metals depending on the wavelength λ of laser radiation (at 20 °C)

λ, μm	Laser active medium	Au	Cu	Mo	Ag	Al	Cr	Fe	Ni
0.488	Argon	0.415	0.437	0.415	0.952	0.85	–	–	0.597
0.6943	Ruby	0.93	0.831	0.498	0.961	0.82	0.555	0.575	0.676
1.06	Glass, YAG:Nd	0.981	0.901	0.582	0.961	0.733	0.57	0.65	0.741
10.6	CO_2	0.975	0.984	0.945	0.989	0.97	0.93	–	0.941

The ***electrical conductivity*** G of the material being processed is related to the reflection coefficient by the following relationship:

$$G = 1 - \sqrt{(2\omega\pi\sigma)} \tag{1.2}$$

where ω is the radiation frequency;

$\sigma = n_e e^2/(\nu m)$ is the concentration of free electrons;

n_e—number of electrons;

e—electron charge;

m is the mass of the electron;

ν is the frequency of electron scattering due to thermal vibrations of the crystal lattice.

For metals, the absorption capacity A is related to the electrical conductivity by a dependence [6]:

$$A = 112.2 \cdot \sqrt{1/\sigma_0} \tag{1.3}$$

where σ_0 is the specific electrical conductivity of the metal for direct current, Ohm · m^{-1}.

The ***temperature of the material*** affects the reflection coefficient and absorption capacity, so the *absorption properties* of the surface *change during* the irradiation process.

Figure 1.6 shows the calculated dependences of the reflection coefficient R on temperature for different laser wavelengths and different metals. The *reflection coefficient decreases* significantly with increasing metal temperature. Moreover, this dependence is more pronounced for laser radiation with a wavelength of 1.06 μm.

For engineering calculations, a linear dependence of the absorption capacity on temperature can be assumed:

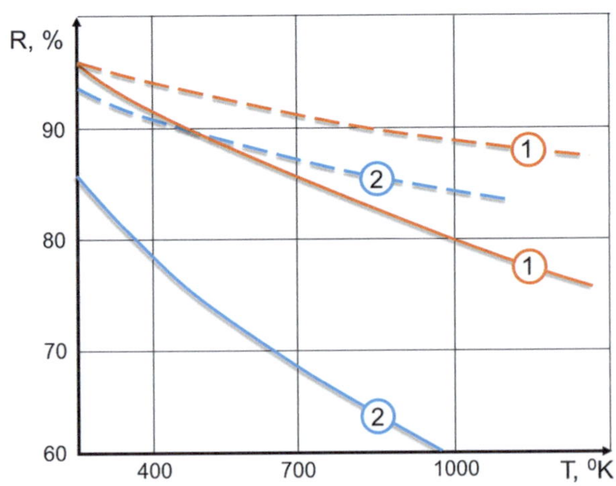

Fig. 1.6 Dependence of the reflection coefficient R on the temperature T of the metal surface: 1—copper; 2—aluminum; solid lines—wavelength λ = 1.06 μm; dashed lines—wavelength λ = 10.6 μm

$$A = A_0 + A_1(T - T_0) \tag{1.4}$$

where A_0 is the initial value of the absorption capacity;

T_0 is the initial temperature;

A_1 is the rate of change in absorption capacity with increasing temperature.

Dependence (1.4) should be used when calculating the limiting energy density of radiation required to heat the surface to the melting point. Failure to take into account dependence (1.4) leads to an overestimation of the specific limiting energy density (Fig. 1.7). The graph shows the dependence of temperature on irradiation time, calculated taking into account the linear dependence of the absorptivity on temperature, with solid lines at a constant value of the absorptivity.

The ***roughness of the*** metal ***surface*** affects the absorption of radiation.

The reflected radiation includes two components (Fig. 1.8):

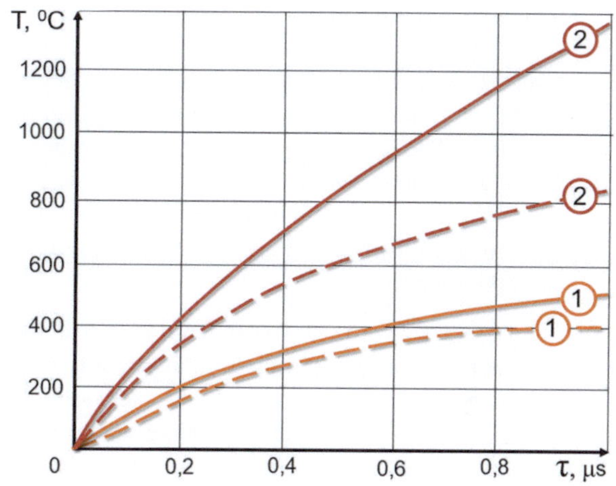

Fig. 1.7 Dependence of the surface temperature of stainless steel on the irradiation time: 1—at the laser radiation power density $W_p = 10^7$ W/cm^2; 2—at the laser radiation power density $W_p = 2 \times 10^7$ W/cm^2; solid lines—taking into account the linear dependence (1.2) of the absorptivity on temperature; dashed lines—at a constant value of the absorptivity $A = A_0$

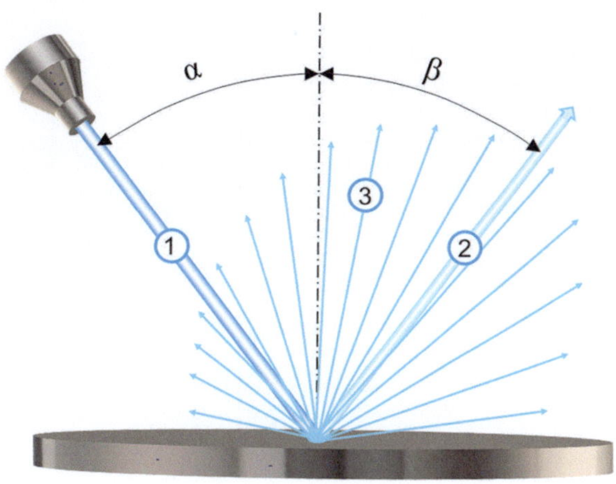

Fig. 1.8 Components of reflected radiation: 1—laser beam; 2—specular reflection; 3—diffuse reflection

(1) *Specular reflection*—occurs at an angle β equal to the angle of incidence α and lies in the plane of the incident beam and normal to the surface;
(2) *Diffuse reflection*—occurs at an angle different from the angle of incidence and does not necessarily lie in the plane of the incident beam and normal to the surface.

The cause of diffuse reflection from metals is surface roughness, which forms randomly arranged micro-irregularities at different angles to the incident beam.

The proportion of diffuse reflection is maximized if the dimensions of the surface irregularity are of the order of the wavelength (or exceed it).

If diffuse radiation hits an adjacent bump, part of it can be absorbed by the metal. Thus, three ranges of metal surface roughness can be identified by their effect on absorption capacity:

(a) Low roughness ($R_a \le 0.5$ μm) causes a large proportion of specular reflection and weak absorption of laser radiation.
(b) *Average roughness* (0.5 μm < $R_a \le 1$ μm) is commensurate with the wavelength, which causes a large fraction of diffuse reflection. At the same time, the size of the roughness contributes to the absorption of a part of the reflected diffuse radiation. The *absorption* of laser radiation becomes *higher* (Fig. 1.9).
(c) Large roughness ($R_a > 1$ μm) causes a large proportion of diffuse reflection. However, large micro-irregularities do not create conditions for absorbing the reflected diffuse radiation, it is scattered. The overall absorption of laser radiation is reduced.

With an *increase in wavelength* from $\lambda = 1.06$ μm to $\lambda = 10.6$ μm, the *effect of* metal surface *roughness* on radiation absorption *decreases.*

The absorptivity of iron and stainless-steel samples tends to increase after treatment with abrasive paper and increasing surface roughness at room temperature. However, as soon as the surface is melted by laser radiation, the absorption capacity drops to the level corresponding to that of polished samples. Therefore, for metals,

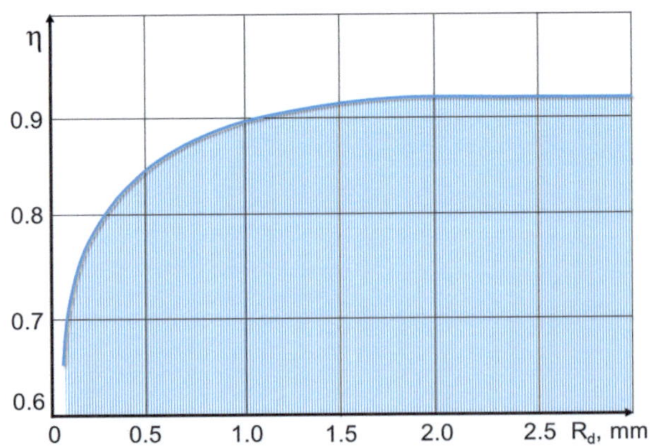

Fig. 1.9 Effect of surface roughness on the absorption capacity of tool steel ($E = 27$ J, $\tau = 7 \times 10^{-3}$ s, $d_0 = 4$ mm)

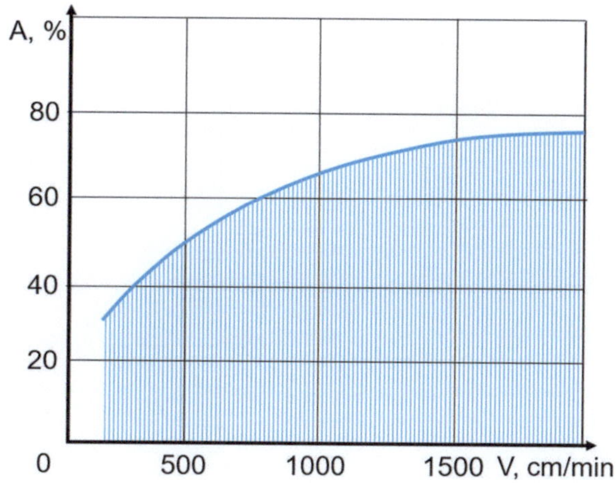

Fig. 1.10 Effect of processing speed on the absorption capacity of steel

increasing the surface *roughness is not a way to increase the absorption capacity* of laser radiation.

The ***speed of movement*** of the laser beams relative to the surface to be treated affects the efficiency of laser absorption by the material. In general, the absorption capacity *increases with the speed* of movement of the workpiece (Fig. 1.10) [5].

Technological coatings significantly increase the absorption capacity of metals and offset the effect of surface roughness on it:

- dielectric thin films;
- oxide films;
- an alcoholic colloidal solution of graphite;
- phosphate coatings;
- coating in the form of paints;
- tungsten coating.

Dielectric thin films increase the absorption capacity by reducing the electrical conductivity of the material. Fe_2S_3 and $A1_2O_3$ coatings are widely used.

The *Fe_2S_3 coating* is applied by sulfidation. The absorption capacity of stainless-steel coating increases up to 80% depending on the coating thickness. When choosing the irradiation mode, it should be borne in mind that 150 ms after the start of irradiation of stainless steel, the Fe_2S_3 coating evaporates, which leads to a decrease in the absorption capacity to the level of uncoated steel (Fig. 1.11).

The *$A1_2O_3$ coatings* are applied by anodizing. The absorption capacity of stainless-steel coating increases by 100% at a layer thickness of 3–4 μm. The absorption capacity drops sharply in the first 50 ms of laser exposure from 100 to 55%, then stabilizes at this level with further irradiation. With an increase in radiation power, the absorption capacity decreases due to the large evaporation of the $Al_2\ O_3$ layer.

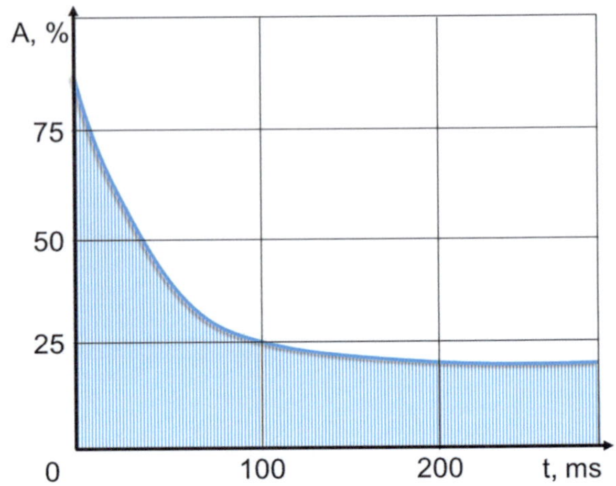

Fig. 1.11 Changes in the absorption capacity A of the Fe_2S_3 coated stainless steel surface over time under laser irradiation

Oxide films are a simple and effective tool for increasing the absorption capacity of metals. Oxide films are easily formed by heating the metal surface with focused laser radiation in air. In addition, the films are applied by chemical oxidation.

Iron oxide films Fe_3O_4 5–10 μm thick have a high absorption capacity of up to 80–90% on the surfaces of iron-carbon alloys.

Copper oxide films are also used. Coatings have a structure in the form of dense lattices of "needles", which serve as "traps" for laser radiation.

An ***alcoholic colloidal solution of graphite*** is the most effective means of increasing the absorbency. A small amount of glue is added to the solution. After evaporation of the alcohol, the graphite particles are bonded together by the adhesive and create a frame structure with cavities. This structure of the coating ensures multiple reflections of laser radiation between the surfaces of the particles and significantly increases absorption [5, 6].

Phosphate coatings are salts of orthophosphoric acid, in particular manganese phosphate $Mn_3(PO_4)_2$ and zinc phosphate $Zn_3(PO_4)_2$. Phosphate coatings are characterized by simplicity and low cost of obtaining. Phosphate coatings provide surfaces with high absorption properties due to their porosity, high heat resistance, good adhesion to the substrate and the ability to provide a uniform and adjustable layer thickness.

In addition to the technological purpose of increasing the absorption capacity, phosphate coatings are widely used to ensure the functional properties of products:

- wear resistance of friction surfaces, since phosphate coatings have a porous structure and can act as lubricant accumulators;
- corrosion resistance;
- marketable appearance.

The coatings $Mn_3(PO_4)_2$ and $Zn_3(PO_4)_2$ have approximately the same absorption capacity of 55–90%.

Coating in the Form of Paints:

(a) ***Coatings*** containing *soot pigments*—black gouache, printing ink, primer.
(b) *Non-ferrous metal oxides* in the form of *paste-like suspensions* of inorganic pigments and fillers in a dextrin-based water-based adhesive binder—yellow and white gouache, manganese phosphate (black gouache), zinc and chromium oxides, red iron oxide pigment and molybdenum micro-pigment.
(c) *Heat-resistant coatings* that are diluted with water and contain substances with high absorption capacity: sodium silicate, zinc oxide, methyl cellulose and surfactants (Table 1.2).

When applying coatings in the form of paints, a layer of absorbent particles is formed on the surface and they are bonded together. After the solvent evaporates, cavities remain between the particles. As a result, contact with the substrate is made at individual points. Coatings absorb radiation energy well and quickly heat up to a high temperature, but transfer it poorly to the metal (due to low thermal conductivity). In this case, the coating substance either melts or burns. To improve heat transfer, it is necessary that the melted coating wets the substrate well.

To prevent the coating from turning into individual droplets and reduce melt flow in the focusing spot, the viscosity is increased by adding hixotropic substances to the coating (the viscosity of heterogeneous systems is less dependent on temperature than the viscosity of glass or enamels).

The *optimal thickness* of coatings in the form of paints is 20–30 μm. Increasing the coating thickness leads to their combustion. Reducing the coating thickness increases the reflection of laser radiation.

Paint coatings are usually well diluted with soluble water which simplifies the process of applying them to the surface of products and removing them after laser processing.

Methods of applying coatings in the form of paints:

- spraying—provides the most uniform layer,
- by immersion,
- painting with a brush.

Drying time of the coating: at 80 °C—15 min., with hot air blowing—10–15 min., at room temperature—40–50 min.

After treatment, the coating can be easily washed off the surface with water.

Tungsten coatings are used to increase the absorption capacity. Tungsten coatings are applied by vacuum spraying. These coatings have a specific cellular structure

Table 1.2 Properties of heat-resistant absorbent coatings

Base	Color	Absorptivity A (at 800 °C)
Sodium silicate and methyl-cellulose	White	80
Sodium silicate	Dark brown	90
Aluminum-chrome-phosphate	Dark green	90

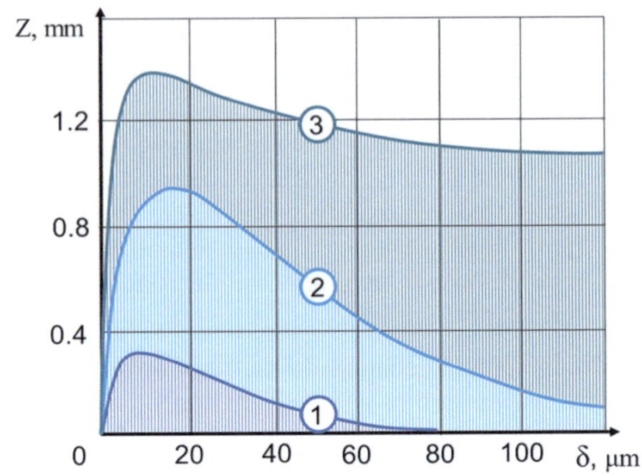

Fig. 1.12 Dependence of the depth of the heat affected zone Z on the coating thickness δ at different processing speeds: 1–0.4; 2–0.3; 3–0.2 cm/s

with a cell size of about 10 μm, which is commensurate with the wavelength of the radiation. The cells act as "traps" for radiation, so tungsten coatings have an absorption capacity of up to 96%. Despite their high efficiency, tungsten coatings are not widely used due to their high cost.

The ***thickness of coatings*** significantly affects the absorption of laser radiation (Fig. 1.12).

The coating thickness of 5–10 μm is the most effective and provides the maximum depth of the heat affected zone in a wide range of processing speeds.

As the coating thickness increases to 30 μm, the depth of the heat affected zone decreases by 30%. Further increase leads to a sharper decrease in the hardening depth, especially at high machining speeds.

The ***temperature of the coatings*** reduces the absorption capacity (in contrast to the increase in absorption capacity when the temperature of uncoated metals increases). This is due to the destruction or change in the properties of coatings when heated.

1.3 Physical Phenomena of Arc and Plasma Processes

Electric arc and plasma jet are both types of electric discharges in gases. A plasma jet differs from an electric arc in that it is more ionized and has a higher temperature.

An ***electric arc*** is a type of discharge in gases characterized by a decrease in the voltage between electrodes (up to tens of volts) as a result of an increase in the degree of ionization (density of ions and electrons) and current (up to thousands of amperes) and, as a result, a high temperature (from 3 to 20 kK).

An electric arc (Fig. 1.13) consists of a cathode region, an arc column, and an anode region.

1. ***The cathode region*** "C" is a layer about 0.1 μm thick, consisting of *free electrons* 3 and *positive ions* 4.

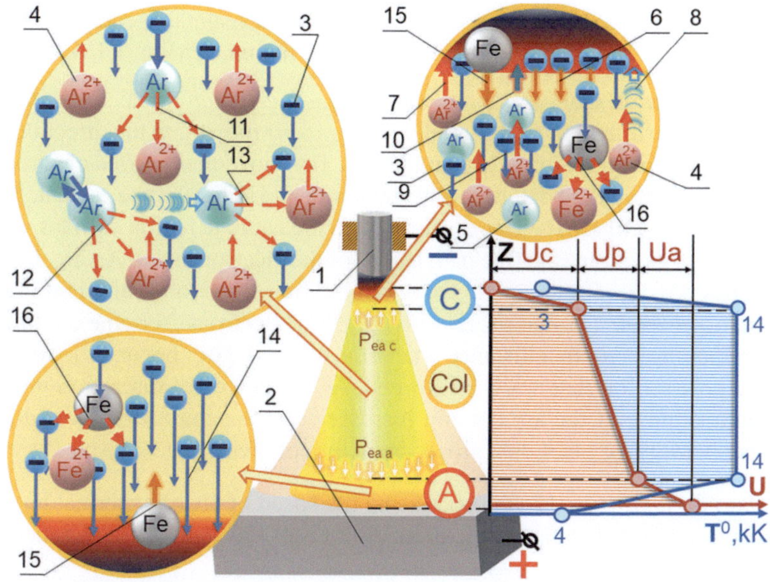

Fig. 1.13 Scheme of processes in an electric arc (C—cathode region; Col—arc column; A—anode region; $P_{ea\ c}$ and $P_{ea\ a}$—gas-dynamic pressure of the electric arc on the cathode and anode respectively): 1—electrode; 2—product; 3—free electrons; 4—positive argon ions; 5—argon atom; 6—thermoelectron emission; 7—secondary emission by positive ions; 8—photoelectron emission at the cathode; 9—neutralization of ions by free electrons; 10—inhibition of the argon atom on the cathode surface; 11—ionization by electron; 12—thermal ionization; 13—photoionization from the plasma's own light radiation;14—inhibition and heating of the anode by free electrons; 15—evaporation of metal atoms; 16—ionization of metal atoms

The *free electrons* of the cathode region come from the cathode as a result of thermoelectron emission 6, in which the work function is performed by the heating energy of the cathode. *Thermoelectron emission at the cathode* is *one* of the two main *generators of charged particles* in an electric arc. The intensity of thermoelectron emission increases with increasing cathode heating temperature.

Free electrons in the cathode region are formed to a lesser extent as a result of:

- *secondary emission* 7, in which the work function is performed due to the kinetic energy transferred by positive ions during inhibition on the cathode surface;
- *photoelectron* emission 8, in which the work function is due to the absorption of light quanta by the cathode from the arc regions.

The *positive ions* of the cathode region enter from the arc column at high speed, are neutralized 9 by free electrons and are inhibited 10 on the cathode surface with the release of a large amount of heat—this is the *main source of heat* at the cathode. The heat causes the cathode to heat up and thermoelectron emission from the cathode surface. The surface of the cathode adjacent to the cathode region is called the ***cathode***

spot. The temperature in the cathode spot area during consumable electrode welding reaches 2.5–3 kK. The voltage drop in the cathode region Uc (Fig. 1.13) is 10–20 V.

2. The ***arc column*** "Col" (Fig. 1.13) represents *free electrons* and *negative ions* moving in the direction of the anode, *positive ions* moving in the direction of the cathode, and neutral atoms and molecules. The movement of free electrons and ions in an electric field determines the electron and ionic electric current. The arc column begins where the free electron from the cathode region collides with the first neutral gas atom (the free path length of the electron ends). An arc discharge becomes possible if the kinetic energy of the free electron is sufficient to ***ionize the neutral atom*** 11—the process of detaching electrons from the neutral atom and removing them to a distance where the electron goes beyond the force field of the positive ion that has been formed. As the degree of ionization increases, the temperature of the arc column increases. Further ionization, in addition to collision with free electrons, proceeds by the following mechanisms:

   - *thermal ionization* 12—by increasing the speed of chaotic thermal motion of neutral atoms, increasing the number of collisions and increasing the energy of each collision;
   - *photoionization* 13—due to the absorption of light quanta (photons) by neutral atoms of their own plasma radiation.

The *ionization of neutral atoms in the arc column* is the *second* main *generator of charged particles* in an electric arc. The ionization intensity depends on the gas composition.

NOTE: Gases with a low ionization work (energy required for the ionization process) and, as a result, a low ionization potential are called ***plasma gases***. These include argon, as well as mixtures of argon with helium, nitrogen, or hydrogen.

Simultaneously with the ionization process, the arc column undergoes the reverse ***recombination*** process—the restoration of a neutral atom as a result of the combination of a positive ion and a free electron.

When the recombination intensity becomes equal to the ionization intensity, a steady state occurs.

Thus, the arc column is a *weakly ionized* gas (ionization degree is about 3–5 %) and, in general, *electrically neutral*—in each cross section there are equal amounts of positively and negatively charged particles at the same time. The temperature of the arc column is 6–20 kK.

The voltage drop U_p (Fig. 1.13) is proportional to the length of the arc column L_p:

$$U_p = E_p \cdot L_p \tag{1.5}$$

where $E_p \sim 2.5$ V/mm in case of arc welding of steels.

3. ***The anode region*** "A" is a layer about 1 micron thick, consisting of a stream of *free electrons* that come out of the arc column, are braked and neutralized on the

anode surface with the release of thermal energy 14—this is the *main source of heat* at the anode.

The surface of the anode bordering the anode region is called the ***anode spot***. The temperature in the anode spot region during consumable electrode welding reaches 2.5–4 kK, which is on average higher than the temperature of the cathode spot. The voltage drop in the anode region Ua (Fig. 1.13) is 3–6 V.

The ***cathode*** and ***anode spots*** on the electrodes of most metals are heated to their boiling points and above. The metal of the electrodes ***evaporates*** 15. As a result, in the layers of the *cathode and anode regions* adjacent to the electrode surface, *metal atoms* 16 (metal vapors) and *gas atoms* (oxygen and hydrogen) participate in the ionization processes, which are desorbed from the metal.

As a result of the ***directional movement*** of particles with mass at high speeds (about 120–200 m/s), the electric arc exerts ***gas-dynamic pressure*** on the electrodes ($P_{ea\,c}$ and $P_{ea\,a}$ in Fig. 1.13). The gas-dynamic pressure of the electric arc on the surface of the welding pool is the main factor in displacing the molten metal and determining the penetration depth and shape of the weld. In addition to the gas-dynamic component, the total arc pressure includes magnetostatic pressure. ***Magnetostatic pressure*** occurs under the action of the Lorentz force on moving charged particles in a stationary magnetic field of welding current. However, the magnetostatic pressure is about 7 times less than the gas-dynamic pressure.

In general, in an electric arc, the positive electrode generates 43% of the total heat, the negative electrode 36%, and the arc column 21%.

The ***heat transfer*** in the welding arc is due to:

- *thermal conductivity* as the main source;
- *radiation*—to a large extent;
- convective heat transfer is insignificant;
- energy dissipation is insignificant.

In the anode spot, the depth of metal heating is greater than ***in the cathode spot*** due to the generation of more heat and the high penetration capacity of electrons. For this reason, a distinction is made between welding and surfacing at direct current:

(1) ***Direct polarity arc***—the *negative* of the power source is connected to the *electrode* and *the positive to the workpiece*. A direct-polarity electric arc is widely used ***in welding***, as it ***increases the penetration depth*** of the parts being welded.

(2) ***Reverse polarity arc***—the *positive* pole of the power source is connected to *the electrode* and the *negative* one to *the workpiece*. The reverse polarity arc is widely used in *surfacing* applications because it provides:

   - reduced penetration of the base metal, which means a ***decrease in the proportion of base metal in the weld***;
   - increase in the volume of heating of the electrode wire or strip, hence an ***increase in the surfacing productivity***.

To strike an electric arc, the primary ionization of the gas medium between the electrodes is required. The process begins with a short circuit between the electrodes. The cathode is heated by the contact resistance. With a further increase in the distance between the electrodes, a spark discharge occurs. The number of charged particles increases. The arc process begins.

The voltage drop in the electric arc U_{ea} (Fig. 1.13) is equal:

$$U_{ea} = U_c + U_p + U_a \tag{1.6}$$

Plasma jet is an electric arc with a higher ***degree of ionization*** and a higher ***temperature*** compared to a free-burning arc, at the same current and voltage due to:

(1) *Compression* (volume reduction) of the electric arc column by reducing the diameter of the plasma-forming nozzle of a water-cooled plasmatron. Compression increases the density of charged particles and neutral atoms. The number of collisions between the particles increases, which increases the degree of ionization and the temperature.
(2) Injection of a *plasma-forming gas* (usually *argon*, less often helium or air) into the arc, which has a low ionization energy. The plasma-forming gas is ionized and heated. In addition, as a result of heating the plasma-forming gas, the pressure increases by several orders of magnitude, which leads to further heating and a significant increase in the rate of plasma flow from the nozzle.

A distinctive feature of plasma is its electrical neutrality—in any volume of plasma, the sum of charges of positive and negative particles is the same.

In welding and related processes, special devices are used to create a plasma flow—plasmatrons (Fig. 1.14).

A ***plasmatron*** consists of a plasma generating nozzle 1 with a cooling water channel 2, a shielding nozzle 3 and an electrode 4. The electrode is usually a non-consumable cathode made of tungsten or copper with refractory inclusions. The plasma-forming gas 5 is supplied through the plasma-forming nozzle, and the shielding gas 6 is supplied through the shielding nozzle. Since argon is usually used as the plasma-forming gas and shielding gas, various design combinations of the plasma-forming and shielding nozzles are possible, including their combination. Depending on the current strength, the diameter of the plasma torch nozzle is selected so that it is smaller than the diameter of the freely expanding electric arc, and thus ensures the compression of the arc column and the formation of a plasma jet with an increased degree of ionization and temperature.

In addition to the geometric factor, an increase in the degree of ionization and plasma temperature occurs as a result of *heating and ionization of the plasma-forming gas*.

As a result of heating, the *volume of the plasma jet* with the plasma-forming gas increases by several orders of magnitude, which leads to a significant increase:

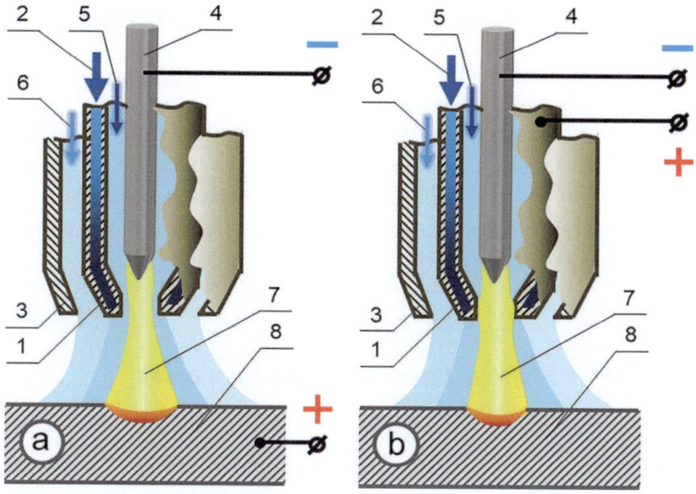

Fig. 1.14 Schematic diagram of a plasmatron (**a**—direct action plasma jet; **b**—plasma jet indirect action): 1—plasma-forming nozzle of the plasmatron; 2—cooling water; 3—protective nozzle; 4—electrode; 5—plasma-forming gas; 6—protective gas; 7—plasma jet; 8—product

- plasma outflow ***rate*** (can be 700–4000 m/s, depending on the flow rate and composition of the plasma-forming gas, current, and the design of the plasma torch nozzle);
- ***gas-dynamic pressure*** of the plasma jet on the electrodes (P_{eac} and P_{eaa} in Fig. 1.13).

In welding and related processes, a distinction is made between direct and indirect plasma jets (and, accordingly, electric arcs).

In a ***direct-acting*** plasma jet, one of the electrodes is the *product* (Fig. 1.14a), and the other is the electrode.

An ***indirect*** arc burns between *two electrodes* (non-consumable, consumable, or a combination of non-consumable and consumable) or between a non-consumable electrode and a current-carrying *nozzle* (Fig. 1.14b). In this case, plasma is blown by the plasma-forming gas onto the product. The product is excluded from the electrical circuit.

The efficiency of heating a product with plasma using argon as a plasma-forming and shielding gas is in the range of 55–80%. Replacing argon as a plasma-forming gas with helium leads to a decrease in the effective efficiency within 46–74% due to the higher thermal conductivity of helium.

1.4 Physical Phenomena of Hybrid Laser—Plasma Processes

This section presents material based on the work of the book's authors and other researchers [7–9] who study the interaction processes between laser radiation and plasma jets, as well as the possibility of combining them to perform technological operations. It also presents approaches to modeling the combined effect of hybrid laser-plasma processes.

1.4.1 Scheme of Hybrid Laser-Plasma Processes

Hybrid laser-plasma (L-P) processes are the interaction of laser beam photons and plasma particles (free electrons, positive and negative ions, and neutral gas atoms).

In the L-P plasmatron (Fig. 1.15), a plasma jet (Col) is generated in the flow of plasma-forming gas between the refractory (tungsten) tubular cathode 1 and the product 2 (anode). At the initial section, the plasma jet is stabilized by a copper plasma-forming nozzle 3 with a channel for cooling water 4. The plasma-forming gas is supplied to the plasma jet in two streams: through the channel of the tubular cathode 5 and through the plasma-forming nozzle 6. The flow of the plasma-forming gas through the plasma-forming nozzle also protects the plasma jet and the product heating zone from air. At the same time, the flow rate of the plasma-forming gas in the two streams is controlled independently.

A laser beam 7, focused by the optical system, is introduced into the plasma jet through the channel of the tubular cathode and propagates along the L-P axis of the plasmatron. Position correctors are used to ensure the alignment of the laser beam with the plasma flow.

In the cathode region, the processes of ionization and charge transfer are similar to those in an electric arc (Sect. 1.3).

In addition to the processes occurring in the electric arc, ***refraction*** (a change in the direction of laser radiation during the transition to plasma and a related change in the propagation velocity, change in wavelength and phase velocity) occurs ***in the plasma jet*** and ***absorption****of laser radiation by the plasma jet* is associated with the transfer of photon energy to plasma particles (atoms and gas ions), which leads to:

- *photoionization* of gas atoms 22 in Fig. 1.15;
- *increase in the speed* of plasma particles, hence an increase in temperature and thermal ionization 17.

In the anode region, in addition to the processes occurring in the electric arc:

- photons ***additionally heat*** the central zone of the surface 23 with the formation of a significant temperature gradient;

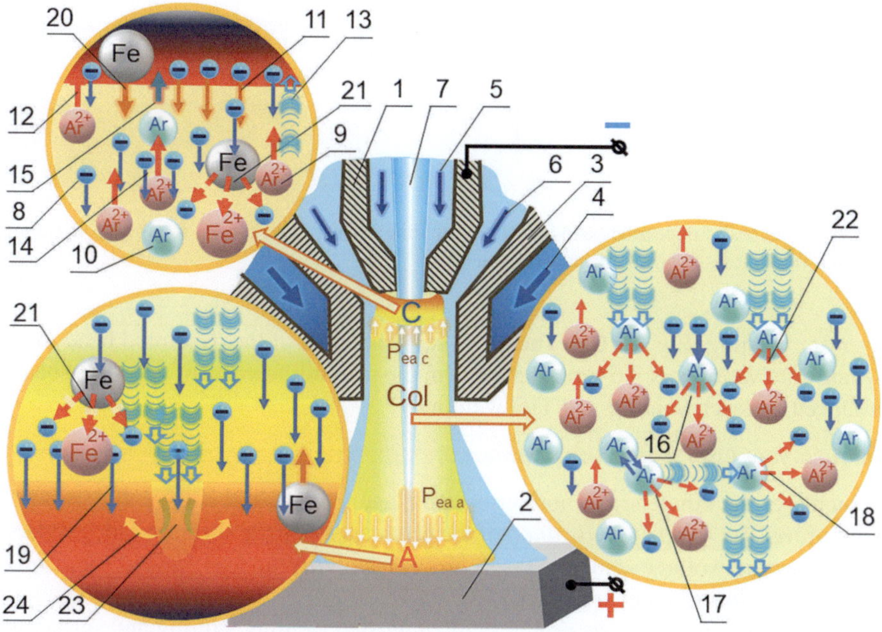

Fig. 1.15 Scheme of hybrid laser-plasma processes in a plasmatron (C—cathode region; Col—plasma jet; A—anode region; $P_{ea\,c}$ and $P_{ea\,a}$—gas-dynamic pressure of the plasma jet on the cathode and anode, respectively): 1—refractory tubular cathode; 2—product; 3—plasma-forming nozzle; 4—cooling water; 5—plasma-forming gas supply through the channel of the tubular cathode; 6—plasma-forming gas supply through the plasma-forming nozzle; 7—laser beam; 8—free electrons; 9—positive argon ions; 10—argon atom; 11—thermoelectron emission; 12—secondary emission by positive ions; 13—photoelectronic emission at the cathode; 14—neutralization of ions by free electrons; 15—inhibition of argon atom on the cathode surface; 16—ionization by electron; 17—thermal ionization; 18—photoionization from plasma's own light radiation; 19—inhibition and heating of anode by free electrons; 20—evaporation of metal atoms; 21—ionization of metal atoms; 22—photoionization from laser beam; 23—zone of additional heating by photons; 24—mixing in the zone of additional heating

- the ***gas-dynamic pressure*** of the plasma jet on the anode increases significantly ($P_{ea\,a}$ Fig. 1.15);
- the formation of an additional heating zone in combination with an increase in the gas-dynamic pressure of the plasma jet leads to a significant increase in the ***intensity of mixing processes*** 24 in the case of metal melting.

Thus, the interaction of the laser beam with the plasma changes the mechanisms of the processes and leads to a ***synergistic effect***—the useful energy input to the product from the hybrid L-P process *exceeds* (by 2 or more times) the *sum* of the energy inputs of the laser process and the plasma process separately. The synergistic effect is most important when low-power lasers are used in hybrid L-P processes.

The synergistic effect is clearly demonstrated by the shapes of the deposited beads produced using laser welding (Fig. 1.16a), TIG welding (Fig. 1.16b) and hybrid

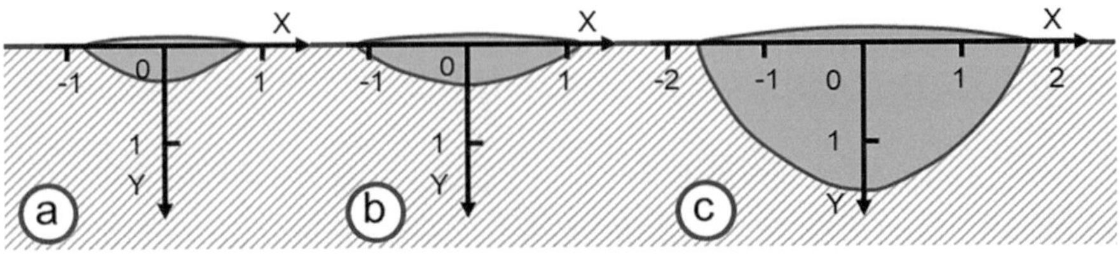

Fig. 1.16 Shapes of deposited beads: **a**—laser welding (power Q = 285 W; focus r_F = 0.7 mm); **b**—TIG welding; **c**—hybrid laser-arc welding [10]

laser-arc welding (Fig. 1.16c). The cross-sectional area of the deposited bead made by laser-arc welding is 3.6 times larger than the sum of the areas of deposited beads produced by laser welding and TIG welding.

Interaction of laser beam with plasma:

- increases the efficiency of both laser and plasma processes;
- improves the *spatial stabilization* of the plasma spot on the surface of the product;
- reduces arc voltage and increases *process stability* at *low currents* and *high* welding and surfacing *speeds*;
- significantly increases the *productivity of plasma* welding, cutting and surfacing;
- increases the energy density of the anode area, hence the *temperature and depth* of heating of the anode spot.

1.4.2 *Mathematical Modeling of Hybrid Laser-Plasma Processes (Coaxial Internal Arrangement of the Laser Beam Relative to Tubular Tungsten Cathode)*

The ***object*** of mathematical modeling is the interaction of a focused beam of a CO_2 -laser of continuous action with a plasma jet in an LP-plasmatron at a coaxial internal location of the laser beam relative to a tubular tungsten cathode (Fig. 1.17).

The ***purpose*** of the mathematical model is to obtain the desired ***output*** parameters: *temperature distributions* along the longitudinal axis and in the cross-sections of the hybrid laser-plasma channel; at the required ***input*** parameters: physical properties of the plasma jet and laser beam (specific heat capacity of the plasma, thermal conductivity, plasma density, electrical conductivity of the plasma, viscosity coefficient, laser absorption coefficient, universal magnetic constant, etc.), technological characteristics of the hybrid laser-plasma process (axial and radial components of the plasma speed; current density in the hybrid charge channel, radiation intensity in the laser beam, power density of plasma intrinsic radiation, pressure, magnetic field strength, etc.) and structural dimensions of the plasmatron (diameter of the hollow tubular cathode, diameter of the plasma-forming nozzle, etc.)

The mathematical model is based on [11].

Accepted *assumptions*:

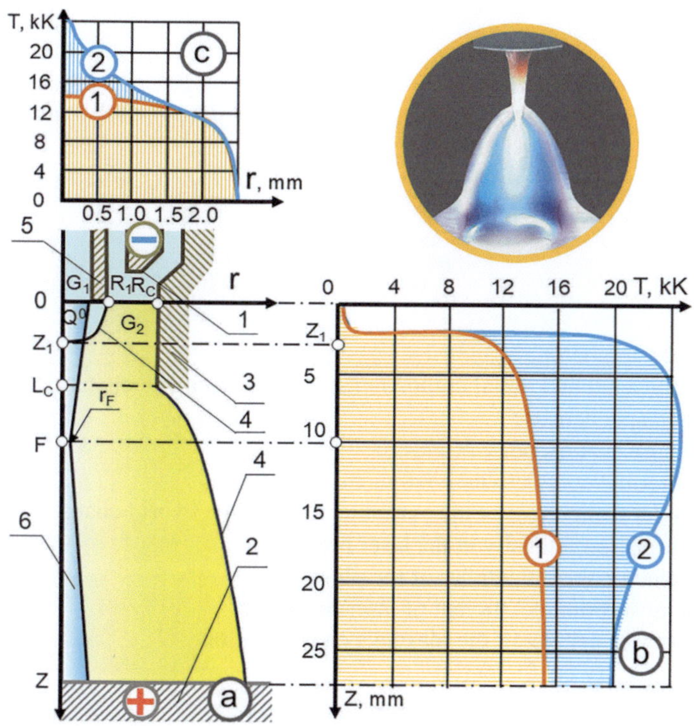

Fig. 1.17 Scheme of interaction of a focused laser beam with a plasma jet in an LP plasmatron with a coaxial internal arrangement of the laser beam relative to a tubular tungsten cathode: **a**—calculation scheme (1—tubular cathode; 2—anode-product; 3—plasma-forming nozzle; 4—plasma jet boundaries; 5—cathode nozzle for the laser beam; 6—laser beam boundaries); **b**—temperature distribution along the longitudinal axis (1—in the plasma jet at I = 100 A, $R_C = 2.5$ mm, $R_1 =$ 1 mm, G = 0.1 g/s; 2—in the hybrid LP channel at $Q^0 = 1.5$ kW, F = 10 mm, $r_F = 0.2$ mm); **c**—temperature distribution in a cross-section at a distance of z = 10 mm (1—in the plasma jet; 2—in the hybrid LP channel) [11]

- the plasma jet has cylindrical symmetry, there are no external magnetic fields, and the processes are stationary;
- the laser beam is Gaussian, and the axis of the plasma jet coincides with the axis of the plasmatron;
- the plasma is in a state of local thermodynamic equilibrium, and the plasma's own radiation is volumetric;
- the main mechanisms of plasma heating are joule heat generation and reverse braking absorption of laser radiation; energy transfer in the plasma jet occurs as a result of thermal conduction and convection;
- plasma flow is viscous, subsonic, and laminar.

A system of magnetogas-dynamic equations is used to calculate the temperature distribution along the axis of the hybrid charge channel and in the radial directions of the cross-sections:

$$\rho C_p\left(\mathrm{v}\frac{\partial T}{\partial r}+u\frac{\partial T}{\partial z}\right)=\frac{1}{r}\cdot\frac{\partial}{\partial r}\left(r\gamma\frac{\partial T}{\partial r}\right)+\frac{j^2}{\sigma}+k_\omega S-\psi \tag{1.7}$$

$$\rho\left(\mathrm{v}\frac{\partial u}{\partial r}+\mathrm{u}\frac{\partial u}{\partial z}\right)=\frac{1}{r}\cdot\frac{\partial}{\partial r}\left(r\eta\frac{\partial u}{\partial r}\right)-\frac{\partial}{\partial z}\left(p+\frac{\mu^0H^2}{2}\right) \tag{1.8}$$

$$\frac{1}{r}\cdot\frac{\partial}{\partial r}(r\gamma\mathrm{v})+\frac{\partial}{\partial z}(\rho u)=0 \tag{1.9}$$

where T is the temperature;

u, v are the axial and radial components of the plasma velocity;

ρ is the plasma density;

C_p—specific heat capacity of plasma;

χ is the thermal conductivity coefficient;

j is the axial component of the current density in the hybrid charge channel;

S is the radiation intensity in the laser beam;

σ is the electrical conductivity of the plasma;

k_ω is the absorption coefficient of laser radiation;

ψ is the power density of the plasma's own radiation;

η is the viscosity coefficient;

p—pressure;

μ^0 is the universal magnetic constant;

H is the magnetic field strength.

Components j^2/σ and $k_\omega S$ in Eq. (1.7) describe the energy release in the plasma as a result of electric current flow and laser radiation absorption, respectively.

The current density distribution in the hybrid charge channel is determined by Ohm's law. The axial component of the electric field strength E is determined based on the integral condition for storing the total current I.

The radiation intensity in the laser beam S is related to the complex amplitude A_ω of the laser beam electric field:

$$-2ik\frac{\partial A_\omega}{\partial z}=\frac{1}{r}\cdot\frac{\partial}{\partial r}\left(r\frac{\partial A_\omega}{\partial r}\right)+k^2\cdot(\varepsilon_\omega-1)A_\omega \tag{1.10}$$

$$k=2\pi/\lambda \tag{1.11}$$

where k is the wave vector;

λ is the length of a minute of laser radiation;

ε_ω is the complex dielectric constant of the plasma at the laser frequency.

The pressure distribution in the hybrid discharge is calculated based on the magnetic pressure, and the gas-static pressure gradient on the walls of the L-P channel is calculated based on the condition of maintaining the total flow rate of the plasma-forming gas $G = G_1 + G_2$ (Fig. 1.17).

Boundary conditions for solving the system of differential Eqs. (1.7–1.9):

- on the L-P axis of the channel—the system has axial symmetry;
- on the outer boundary of the computational domain at $z \leq L_C$—gas temperature equal to the wall temperature of the plasma-forming nozzle;
- on the outer boundary of the plasma jet 4 at $z > L_C$ is a smooth conjugation with the environment;
- the radius of the laser beam is much smaller than the radius of the design area 4.

The dependence A_ω (r, 0) is set under the assumption that a focused Gaussian laser beam of power Q^0 is introduced into the discharge, which has a minimum half-width r_F in the section z = F (Fig. 1.17).

The results of calculations of the temperature distribution in the plasma jet and in the hybrid laser-plasma channel are shown in Fig. 1.17b and c.

The effect of laser radiation on the plasma jet leads to an increase in the temperature of the central region (Fig. 1.17b), which is explained by the local heating of the plasma by the laser beam.

Starting from the section $z = Z_1$ (Fig. 1.17a), which represents the beginning of laser-plasma interaction, the temperature T_0 on the hybrid discharge axis increases rapidly and reaches values that significantly exceed the temperature of the plasma jet and laser beam (Fig. 1.17c). This is a manifestation of the ***synergistic effect***.

As the laser radiation power Q^0 increases, the maximum temperature increases. Further, the temperature T_0 decreases asymptotically, which is explained by the weakening of the plasma jet during propagation in the absorbing medium.

A significant *increase in the temperature* T_0 on the hybrid discharge axis leads to:

- increase in the electrical conductivity of the hybrid discharge;
- reducing the electric field strength required to maintain a given plasma current;
- redistribution of the current density in a hybrid discharge—the current density increases in the center of the L-P channel and slightly decreases at the edges;
- reduction of plasma density;
- an increase in the plasma speed near the axis associated with a sharp decrease in the viscosity coefficient of argon at a temperature of T > 10 kK.

1.4.3 Mathematical Modeling of Hybrid Laser-Plasma Processes (Lateral Location of Tungsten Cathode Relative to the Laser Beam)

The ***object*** of mathematical modeling is the interaction of a focused beam of a CO_2-laser of continuous action with a plasma jet with a lateral location of a non-consumable tungsten cathode (Fig. 1.18). The anode is a low-carbon steel product.

The ***purpose*** of the mathematical model is to obtain the desired ***output*** parameters: *temperature distributions* along the longitudinal axis and in the cross-sections of the hybrid laser-plasma channel; with the required ***input*** parameters: physical properties of the plasma jet and laser beam.

The mathematical model is based on [12].

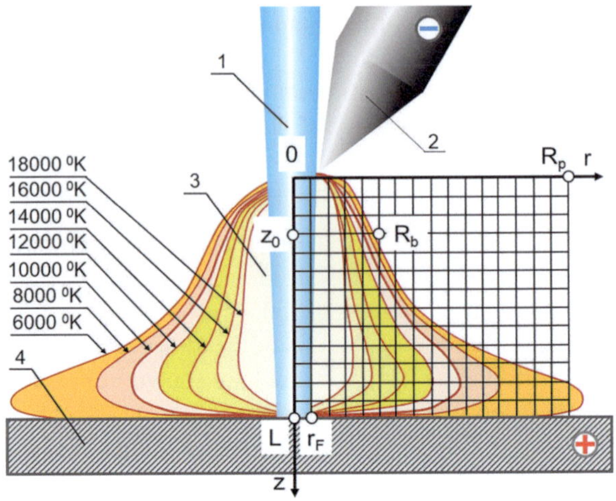

Fig. 1.18 Scheme of interaction of a focused laser beam with a plasma jet and isotherm at the lateral location of a non-consumable tungsten cathode (at $Q^0 = 1.5$ kW, I = 150 A): 1—laser beam; 2—non-consumable tungsten cathode; 3—plasma jet; 4—anode-product [12]

Accepted *assumptions*:

- the plasma jet is axially symmetrical starting from a certain distance from the anode ($z = z_0$);
- the characteristics of the plasma jet with a laterally inclined tungsten cathode are close to the corresponding characteristics of the plasma jet with the tungsten cathode perpendicular to the surface of the anode (product);
- the interaction of the laser beam with the plasma jet occurs only at $z > z_0$.

Assumed: cylindrical coordinate system (r, z); computational domains for the plasma jet $\Omega_P = \{0 \leq r \leq R_P$; computational domains for the laser beam $0 \leq z \leq L\}$; $\Omega_b = \{0 \leq r \leq R_b; z_0 \leq z \leq L\}$.

Basic equations

For mathematical description of the energy, mass and charge transfer processes in such a model system (axially symmetrical), we used a single-temperature model of the arc plasma [13], allowing for additional heating by focused laser radiation [14]. Respective stationary equations in the selected cylindrical coordinate system become:

$$\frac{1}{r} \cdot \frac{\partial}{\partial r}(r\rho v) + \frac{\partial}{\partial z}(\rho u) \tag{1.12}$$

where ρ—the mass plasma density;

v—the radial component of plasma velocity;

u—the axial component of plasma velocity.

Momentum Conservation:

$$p\left(v\frac{\partial v}{\partial r}+u\frac{\partial v}{\partial z}\right)=-\frac{\partial p}{\partial r}-j_z B_\varphi+2/r\frac{\partial}{\partial r}\left(r\eta\frac{\partial v}{\partial r}\right)$$
$$+\frac{\partial}{\partial z}\left[\eta\left(\frac{\partial u}{\partial r}+\frac{\partial v}{\partial z}\right)\right]-2\eta\frac{v}{r^2}$$
$$-2/3\frac{\partial}{\partial r}\left\{\eta\left[1/r\frac{\partial(rv)}{\partial r}+\frac{\partial u}{\partial z}\right]\right\} \quad (1.13)$$

$$p\left(v\frac{\partial v}{\partial r}+u\frac{\partial v}{\partial z}\right)=-\frac{\partial p}{\partial z}-j_z B_\varphi+2/r\frac{\partial}{\partial z}\left(r\eta\frac{\partial v}{\partial z}\right)$$
$$+1/r\frac{\partial}{\partial r}\left[r\eta\left(\frac{\partial u}{\partial r}+\frac{\partial v}{\partial z}\right)\right]$$
$$-2/3\frac{\partial}{\partial z}\left\{\eta\left[1/r\frac{\partial(rv)}{\partial r}+\frac{\partial u}{\partial z}\right]\right\} \quad (1.14)$$

where p is the pressure; j_z, j_r are the axial and radial components of electric current density in the arc plasma; B_φ—the azimuthal component of magnetic-field induction vector and η is the coefficient of plasma dynamic velocity.

Energy conservation:

$$\rho C_p\left(v\frac{\partial T_p}{\partial r}+u\frac{\partial T_p}{\partial r}\right)=\frac{1}{r}\frac{\partial}{\partial r}\left(r\chi\frac{\partial T_p}{\partial r}\right)+\frac{\partial}{\partial z}\left(\chi\frac{\partial T_p}{\partial z}\right)$$
$$\frac{k_B}{e}\left\{j_r\frac{\partial[(5/2-\delta)T_p]}{\partial r}+j_z\frac{\partial[(5/2-\delta)T_p]}{\partial z}\right\}$$
$$+\frac{j_r^2+j_z^2}{\sigma}+k_\omega S-\psi \quad (1.15)$$

where ρ—maximum plasma density;

C_p—specific heat capacity of the plasma taking into account the ionization energy;
v is the radial component of the plasma velocity;
u is the axial component of the plasma velocity;
T_p—plasma temperature;
χ is the thermal conductivity of the plasma;
k_B—Boltzmann's constant;
e—electron charge;
j_r is the radial component of the electric current density in plasma;
j_z is the axial component of the electric current density in plasma;
δ is the thermal diffusion constant of electrons;
σ is the specific electrical conductivity of the plasma;
ψ is the energy loss due to the plasma's own radiation;
$k_\omega S$—the energy release in the plasma jet due to the absorption of laser radiation;

k_ω is the coefficient of volumetric (back-braking) absorption of laser radiation by a plasma jet;

S is the spatial distribution of the radiation intensity in the laser beam interacting with the plasma.

Equation for arc electromagnetic field:

$$1/r\frac{\partial}{\partial r}\left(r\sigma\frac{\partial\varphi}{\partial r}\right)+\frac{\partial}{\partial z}\left(\sigma\frac{\partial\varphi}{\partial z}\right)=0 \tag{1.16}$$

$$B_\varphi(r,z)=\frac{\mu^0}{r}\int\limits_0^r j_z(\xi,z)\xi d\xi \tag{1.17}$$

$$j_r=-\sigma\frac{\partial\varphi}{\partial r};\, j_z=-\sigma\frac{\partial\varphi}{\partial z} \tag{1.18}$$

where φ—the scalar electric potential;

μ^0—the vacuum permeability.

In order to close the system of Eqs. (1.12–1.18), it is necessary to determine the distribution of radiation intensity S(r, z) in the laser beam, allowing for the absorption and refraction in the arc plasma. Assuming that relative change of plasma parameters at distances of the order of laser radiation wave length is small, and laser radiation reflected from the anode surface is absent (for example, due to formation of a keyhole absorbing practically all laser radiation, reaching the anode), quasi-optical approximation can be used to describe laser beam propagation in non-uniform plasma [14].

Equation for amplitude of the electric field of laser beam:

$$-2ik\frac{\partial A_\omega}{\partial z}=1/r\frac{\partial}{\partial r}\left(r\frac{\partial A_\omega}{\partial r}\right)+k^2(\varepsilon_\omega-1)A_\omega \tag{1.19}$$

where k ≡ kz = 2πλ—the wave vector of laser radiation in plasma absence; A_ω is the slowly changing in space complex amplitude of the laser beam electric field; $\varepsilon_\omega=\varepsilon_\omega'+\varepsilon_\omega''$ is the complex dielectric permittivity of the arc plasma at laser radiation frequency ω. The sought value S is the time-averaged value of the axial component of the beam electromagnetic energy flux density, and it is connected to complex amplitude of laser beam electric field by the following relationship [14]:

$$S=\frac{1}{2}\sqrt{\frac{\varepsilon^0}{\mu^0}}|A_\omega|^2 \tag{1.20}$$

where ε^0 is the vacuum dielectric permittivity.

To solve the system of Eqs. (1.12–1.20), it is necessary to determine thermodynamic characteristics ρ, C_p; transport coefficients η, χ, δ, σ; electromagnetic properties K_ω, ε_ω and energy losses for arc plasma radiation ψ, depending on plasma temperature and pressure. For isothermal atmospheric argon plasma, dependencies of thermodynamic characteristics, transport coefficients and energy losses for inherent plasma radiation are provided in [15]. Respective coefficient dependencies of inverse Bremsstrahlung absorption and complex dielectric permittivity of argon plasma at the laser radiation frequency can be calculated by the procedure described in [14].

Thus, we have a closed system of equations describing the influence of a focused laser beam on an electric arc column plasma, allowing for the reverse influence of the arc plasma on laser radiation beam propagating in it.

Boundary condition

The solution of the system of differential Eqs. (1.12–1.16, 1.19) requires assigning the respective boundary conditions. We will define boundary conditions for sought functions $\overline{V} = \{v, 0, u\}$, T_p, φ, A_ω as follows.

In the initial section of the calculation domain for the arc plasma column (plane $z = 0$), we will assign boundary conditions for values v, u, T_p, and φ similarly to what was done in [13] for an arc with a refractory cathode, located normal to the anode surface. In particular, for velocity vector components, we will take:

$$v|_{z=0} = 0;\ u|_{z=0} = u_0 \tag{1.21}$$

where value u_0 is determined by shielding gas flow rate and its feed nozzle diameter.

For temperature and electric potential in the near-cathode zone of the arc column, we will assign the following conditions:

$$T_p\big|_{z=0} = T_c(r);\ \sigma\frac{\partial\varphi}{\partial z}\bigg|_{z=0} = j_c(r) \tag{1.22}$$

where distributions of plasma temperature $T_c(r)$ and density of electric current near the cathode tip $j_c(r)$ are selected according to recommendations of [13]. In the shielding gas feeding zone, we will assume that

$$T_p\big|_{z=0} = T_0;\ \frac{\partial\varphi}{\partial z}\bigg|_{z=0} = 0 \tag{1.23}$$

where T_0 is the ambient temperature.

Moreover, considering the earlier assumption that the laser beam interaction with the arc plasma begins at $z = z_0$, in Eq. (1.16) at $z = z_0$ we will assign $S = 0$.

To assign an initial distribution of $A_\omega(r, z_0)$, we will assume that at $z = z_0$ a Gaussian beam of laser radiation focused by optical system, is introduced into the arc plasma, which in the absence of the plasma has a minimum radius r_F on the anode surface, i.e. in plane $z = L$ (see Fig. 1.18). The spatial distribution of complex amplitude of the electric field of such a beam at $z \leq z_0$ given by the following expression [14]:

$$A_\omega = A_{\omega F}\frac{r_F}{r_z}\exp\left[-\frac{r^2}{r_z^2} + i\left(k\frac{r^2}{2R_z} - \varphi_z\right)\right] \tag{1.24}$$

Here

$$r_z^2 = r_F^2\left[1 + \frac{(z-L)^2}{z_F^2}\right]$$

$$R_z = (z-L)\left[1 + \frac{z_F^2}{(z-L)^2}\right];\ \varphi_z = \operatorname{arctg}\left(\frac{z-L}{z_F}\right) \tag{1.25}$$

$$z_F = kr_F^2/2 \tag{1.26}$$

Constant $A_{\omega F}$ in (1.24) is found from integral relationship for total power of laser radiation:

$$Q = 2\pi_0\int_0^{R_b} Sr\,\mathrm{d}r \tag{1.27}$$

That, when allowing for (1.20), yields:

$$A_{\omega F} = \sqrt{\frac{4Q^0}{\pi r_F^2}\left(\frac{\mu^0}{\varepsilon^0}\right)^{1/2}} \tag{1.28}$$

where Q^0 is the power of the initial laser beam.

We will assume, that no-slip conditions are fulfilled on the anode surface (plane $z = L$) for plasma velocity $\overline{v}$:

$$\overline{v}|_{z=L} = 0;\ u|_{z=0} = u_0 \tag{1.29}$$

On the boundary of the arc column plasma with anode region taken to be infinitely thin [16], we can write the following condition of energy balance:

$$-\chi\frac{\partial T_p}{\partial z}\bigg|_{z=L} + j_a\frac{k}{e}\left(\frac{5}{2} - \delta\right)T_{pa} = \Delta\varphi_a j_a + q_a - j_a\overline{\varphi} \tag{1.30}$$

where $j_a = |j_z||_{z=L}$ is the absolute value of electric current density on the anode; $T_{pa} = T_{pz}|_{z=L}$ is the temperature of arc column plasma on the boundary with the anode layer; $\Delta\varphi_a$ is the difference between arc plasma potential on the boundary of the anode layer and anode surface potential; q_a is the heat flux contributed by the arc to the anode and $\overline{\varphi}$ is the anode metal work function. With good approximation, electric potential of the anode surface can be regarded as constant and can be assumed to be equal, for instance, to zero. Then, boundary condition for plasma potential on the boundary of the arc column with the anode layer can be written as [13]:

$$\varphi|_{z=L} = \Delta\varphi \tag{1.31}$$

Values $\Delta\varphi_a$ and q_a included into (1.29), (1.30), can be calculated using the arc anode region model, proposed in [16]. This model allows calculation of heat flux distribution q_a, contributed by the arc to the anode, and anode potential drop $U_a = -\Delta\varphi_a$ in the region of the arc anode binding, depending on distribution of anode current density j_a and plasma temperature near the anode T_a which can be determined proceeding from the model of the arc column (1.12–1.16) with self-consistent boundary conditions (1.29), (1.30). Boundary conditions for velocity, temperature and electric potential of the plasma, as well as the complex amplitude of electric field of the laser beam, assumed to be Gaussian, on the system axis of symmetry ($r = 0$) are assigned in the standard manner (see, for instance [13, 14]):

$$v|_{r=0} = 0; \quad \left.\frac{\partial u}{\partial r}\right|_{r=0} = 0; \quad \left.\frac{\partial T_p}{\partial r}\right|_{r=0} = 0; \quad \left.\frac{\partial \varphi}{\partial r}\right|_{r=0} = 0; \quad \left.\frac{\partial A_\omega}{\partial r}\right|_{r=0} = 0 \tag{1.32}$$

On the other boundary of calculation domain for the plasma ($r = R_p$), we can write boundary conditions for velocity and electric potential in the following form [13]:

$$\left.\frac{\partial(\rho v r)}{\partial r}\right|_{r=R_p} = 0; \; u|_{r=R_p} = 0; \; \left.\frac{\partial \varphi}{\partial r}\right|_{r=R_p} = 0 \tag{1.33}$$

We will determine boundary condition for temperature at $r = R_p$, depending on the direction of plasma flow:

$$T_p|_{r=R_p} = T_0 \textbf{if} \; v|_{r=R_p} \le 0; \; \left.\frac{\partial}{\partial r}(T_p)\right|_{r=R_p} = 0 \textbf{if} \; v|_{r=R_p} > 0 \tag{1.34}$$

Finally, assuming that laser beam transverse dimension is essentially smaller than the radius of the respective calculation domain R_b, we will write:

$$\mathrm{A}_\omega|\mathrm{r} = \mathrm{Rb} = 0 \tag{1.35}$$

The system of differential Eqs. (1.12–1.16), (1.19) with the above-given boundary conditions was solved numerically by the finite difference method. For numerical solution of gas-dynamic (1.12–1.14) and thermal (1.15) problems, the joint Lagrangian–Eulerian method was used, which was adapted to compressible medium conditions. When solving the parabolic equation for the complex amplitude of the laser beam electric field (1.19), a method was used which is described in [14].

Results of modeling

The laws of interaction of a focused laser beam with a plasma jet at the lateral location of a non-melting tungsten cathode are similar to the interaction at the coaxial internal location of the laser beam relative to of a tubular tungsten cathode.

The *main effect* is the appearance of a ***high-temperature plasma*** region (Fig. 1.19) of ***high conductivity*** (Fig. 1.20), which is rigidly connected to the laser beam axis. Conditions are created for the spatial stability of the plasma jet and increased welding and surfacing speeds. The maximum plasma temperature increases with increasing laser power. The radius of the high-temperature plasma region of high conductivity is about 0.5 mm.

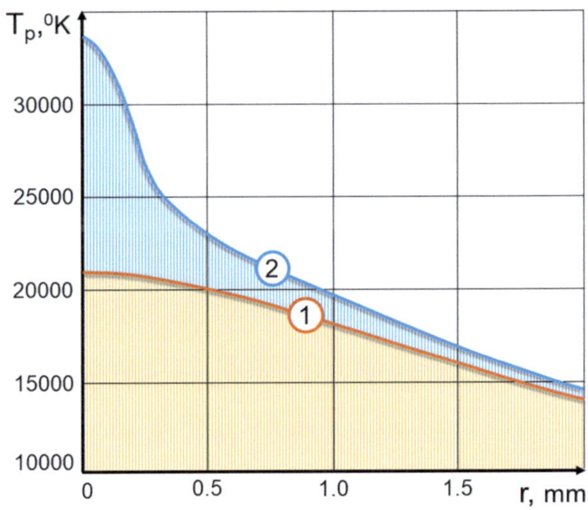

Fig. 1.19 Plasma temperature distribution in a cross-section at a distance of z = 2.5 mm: 1—in the plasma jet; 2—in the hybrid LP channel (I = 150 A; Q^0 = 1.5 kW; T_s = 500°K) [12]

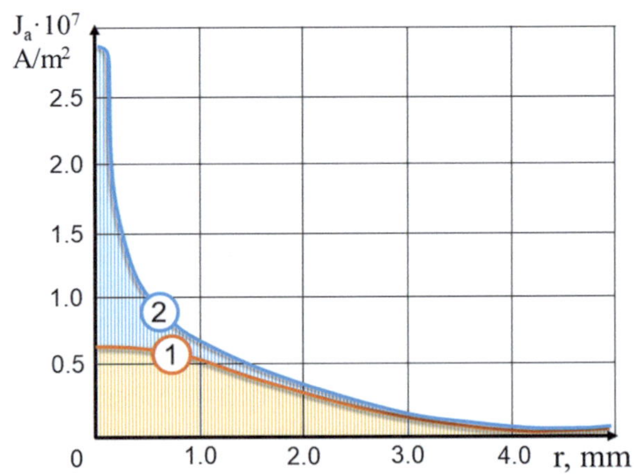

Fig. 1.20 Electric current density distribution on the anode surface (at z = L): 1—in a plasma jet; 2—in a hybrid LP channel (I = 150 A; Q^0 = 1.5 kW; T_s = 500° K) [12]

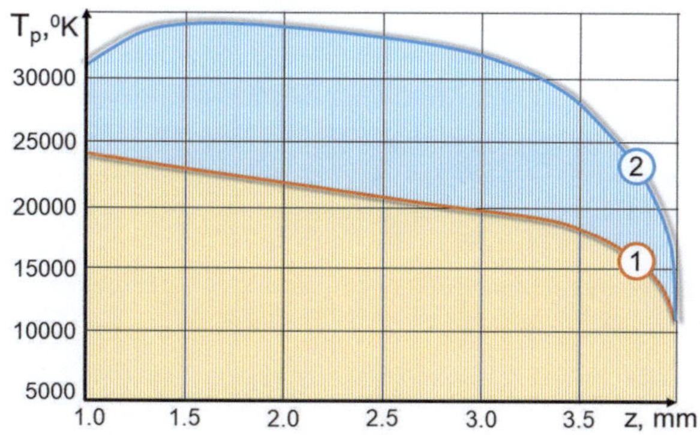

Fig. 1.21 Plasma temperature distribution along the plasma jet axis: 1—in the plasma jet; 2—in the hybrid LP channel (I = 150 A; $Q^0 = 1.5$ kW; $T_s = 500$ °K) [12]

A region of high-temperature plasma is formed along the entire length of the plasma jet (Fig. 1.21), which leads to a significant increase in the temperature of the near-anode zone on the product.

The *second regularity* of the interaction of a focused laser beam with a plasma jet is an increase (by a factor of 3 or more) in the ***dynamic impact*** of the plasma on the surface of the anode (molten metal pool). In the axisymmetric magnetic field of the plasma current, the Lorentz force arises. An increase in the electric current density leads to a significant increase in the Lorentz force. The vortex component of the Lorentz force has its maximum value on the surface of the molten metal pool and is directed toward the center of the plasma spot. The resulting pressure drop causes the molten metal to move from the surface deeper into the melting front and, consequently, to.

- increase the efficiency of absorption of thermal power of plasma and laser radiation;
- increase the intensity of mixing the molten metal;
- increase in the volume of molten metal.

The intensity of the *laser radiation* decreases along the beam length as it approaches the anode (Fig. 1.22) due to the absorption of laser energy by the plasma.

Refraction (change in the direction of propagation) of laser radiation in a heterogeneous plasma environment is negligible.

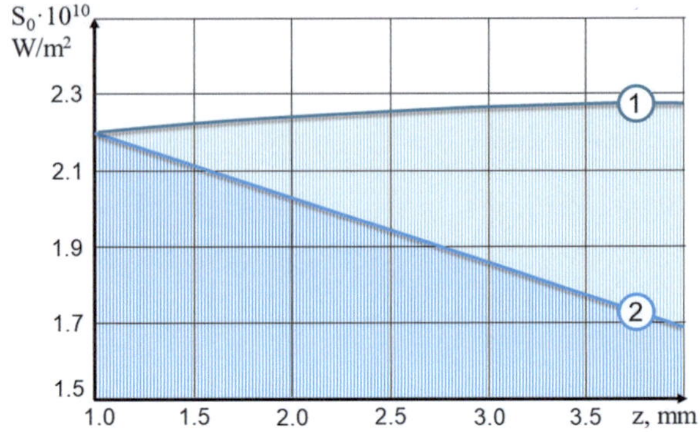

Fig. 1.22 Distribution of laser intensity along the laser beam axis: 1—in the original laser beam in air ($Q^0 =$ 1.5 kW); 2—in the hybrid LP channel ($Q^0 =$ 1.5 kW; I = 150 A) [12]

References

1. M. Eichhorn, Laser Phys. Springer International Publishing Switzerland (2014). https://doi.org/10.1007/978-3-319-05128-4
2. F. Träger, in *Springer Handbook of Lasers and Optics* (Springer Berlin, Heidelberg, 2012). https://doi.org/10.1007/978-3-642-19409-2
3. H.J. Eichler, J. Eichler, O.L. Lasers, in *Basics, Advances and Applications*, Springer Nature Switzerland AG (2018). https://doi.org/10.1007/978-3-319-99895-4
4. S. Yoshida, in *Fundamentals of Optical Waves and Lasers* (Springer Nature Switzerland AG, 2023). https://doi.org/10.1007/978-3-031-18188-7
5. V. Kovalenko, A. Verhoturov, L. Golovko, I. Podchernyaeva, in *Laser Surface Hardening and Electric-Sparc Surfacehardening of Materials*, (Plinum Publish. Corp., 1988), pp. 175
6. V. Kovalenko, L. Golovko, V. Chernenko, in *Hardening and Alloying of Machine Parts with a Laser Beam* (Kyiv: "Tehnika", 1990), pp. 192
7. P. Kumar, in *Quantum Plasma* (Springer Nature Singapore Pte Ltd., 2025). https://doi.org/10.1007/978-981-95-0481-7
8. A. Dinklage, T. Klinger, G. Marx, L. Schweikhard, in *Plasma Physics* (Springer-Verlag Berlin Heidelberg, 2005). https://doi.org/10.1007/b103882
9. A. Macchi, in *A Superintense Laser-Plasma Interaction Theory Primer* (Springer Nature Switzerland AG, 2025). https://doi.org/10.1007/978-3-031-90983-2
10. I. Krivtsun, Features of metal penetration during laser-arc welding using a YAG laser. Automat. Weld. **12**, 33–36 (2001)
11. V. Gvozdetsky, I. Krivtsun, M. Chizhenko, Laser-arc discharge: theory and application. in *Welding and Allied Processes*, vol. 1 (Harwood Acad. Publ., 1995), pp. 200
12. U. Reisgen, A. Zabirov, I. Krivtsun, V. Demchenko, I. Krikent, Interaction of CO_2-laser beam with argon plasma of gas tungsten arc. Weld. World **59**(5), 611–622 (2015)
13. I. Krikent, I. Krivtsun, V. Demchenko, Modelling of processes of heat-, mass- and electric transfer in column anode region of arc with refractory cathode. Paton Weld. J. **3**, 2–6 (2012)
14. P. Seyffarth, I. Krivtsun, Laser-Arc processes and their application in welding and material treatment, in *Welding and Allied Processes*, vol. 1 (Taylor and Francis Books, 2002), p. 200
15. M. Boulos, P. Fauchais, E. Pfender, in *Thermal Plasmas: Fundamentals and Application* (Plenum Press, New York, 1997), pp. 454
16. I. Krivtsun, V. Demchenko, I. Krikent, Model of the processes of heat, mass and charge transfer in the anode region and column of the welding arc with refractory cathode. Paton Weld. J. **6**, 2–9 (2010)

Chapter 2
Methods of Laser Surface Engineering

Abstract A general classification of methods of laser surface engineering and a brief overview of the basics of each method are provided: annealing, tempering, laser quenching, laser thermo-strain hardening, laser forming, liquid phase laser quenching, amorphization, changing the optical characteristics, laser shock hardening, surface roughness improvement, laser thermal cycling, laser thermochemical treatment, laser ion implantation, laser surface alloying, laser surfacing, laser sintering, laser thermal deformation sintering, laser-plasms surfacing, laser brazing, laser spraying. An algorithm is proposed for determining the technological parameters of laser irradiation depending on the required characteristics of the surface layer, taking into account the conditions of interaction and the characteristics of the part.

2.1 General Classification

In general, laser engineering is a broad field of technology [1] that allows for the modification of surface layer microstructures [2, 3], the alteration of chemical compositions through microalloying, and the application of various coatings [4]. In this section, the authors have systematized the methods of laser surface engineering based on their own work and the research of other scientists and have provided the basic definitions and characteristics of these processes.

Ultra-high *intensity* with a wide range of *adjustability* combined with an *ultra-small affected area* makes laser radiation a versatile surface engineering tool with no negative thermal and deformation effects on the material.

The ***versatility*** of laser radiation lies in its capabilities:

(a) to create *controlled thermal cycles* (thermal cycle is a curve of temperature change over time for a given point) in the surface zone, which, in turn, allows to obtain
- *the desired structure*,
- *the desired stress–strain state*;

(b) *to melt* the surface layer, which, in turn, allows:
- to *alloy* metal,

I. Krivtsun et al., *Laser and Hybrid Technologies for Surface Engineering and Welding*, Springer Series in Materials Science 358, https://doi.org/10.1007/978-3-032-15038-7_2

- *surfacing advanced materials* with predetermined properties, including composite materials;

(c) *to evaporate* the surface layer, which in turn allows:

- to obtain *the desired* surface *roughness*,
- *surfacing advanced materials* with predetermined properties;

(d) to form products of complex spatial shapes by *sintering them layer by layer* from powder materials;

(e) bend and stretch thin-sheet materials by locally *changing their stress state* and thus increasing the rigidity of thin-sheet structures.

There is no negative ***thermal and deformation effect*** of laser radiation on the material being processed:

(a) the practical absence of a heat-affected zone;

(b) ultra-small width of the elastic–plastic zone, which means ultra-small width of the tensile stress zone at the yield strength and ultra-small compressive stresses (in the case when the formation of the desired stress state is not the task of machining).

The versatility of laser radiation combined with the absence of negative thermal and deformation effects on the material results in a wide variety of laser surface treatment methods (Fig. 2.1), which can be classified into *three groups* according to their intended purpose [5]:

(1) modification of the surface layer structure;
(2) surface microalloying;
(3) coating application.

Within each group, laser surface treatment methods are divided into *three subgroups* depending on the surface heating temperature:

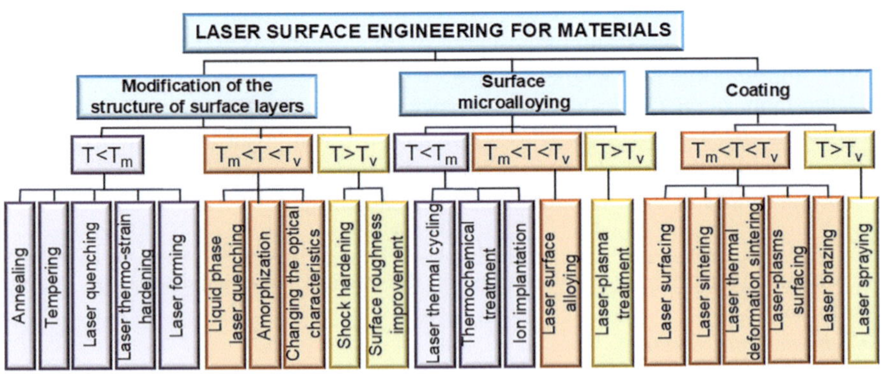

Fig. 2.1 General classification of laser surface engineering for materials (T—surface heating temperature during treatment; T_m—material melting temperature; T_v—material vaporization temperature)

(a) the process takes place in the solid state of the material at a temperature below the melting point $T < T_m$;
(b) the process takes place in the liquid state of the material in the temperature range $T_m < T < T_v$;
(c) the process takes place in the solid–liquid state of the material at an ultra-high temperature above the vaporization point $T > T_v$.

2.2 Modification of the Structure of Surface Layers

Modification of the structure of surface layers is aimed at targeted *structural and phase transformations* by heat treatment or deformation *without changing the chemical composition.*

Annealing is a type of heat treatment consisting in heating to a temperature (at full annealing) 30°–50° above the critical point A_3 (temperature of change in the type of crystal lattice, i.e. polymorphic transformation) with holding until pearlite turns into austenite and slow cooling (50°–100° /hour) to form an equilibrium structure.

Annealing *provides*:

- homogeneous structure;
- reduced hardness, increased ductility;
- increased impact toughness;
- elimination of stresses of the first (macro-level) and second (grain-level) kind.

Since laser heating is characterized by rapid cooling, a corresponding slow cooling mode must be enforced during laser annealing:

- by use of programmable heating and cooling from an additional source;
- by worsening the heat dissipation conditions due to a certain shape and size of the product;
- by cooling carried out in a thermal cabinet.

The *advantage* of laser annealing compared to traditional furnace annealing is the *controllability* of the process in terms of *temperature*, *time* and *depth.*

Areas of application of laser annealing:

- in the electronics industry for recrystallization of alloys at high temperatures and short holding times to eliminate structural defects formed during the cooling of metallic films;
- to improve the mechanical properties of certain areas of products to make them more plastic, for example, the ends of coiled and flat springs.

Tempering is a type of heat treatment of pre-hardened steel and involves heating to a temperature below the critical point A_1 (about 250 °C—low-temperature quenching; about 600 °C—high-temperature quenching) and subsequent slow cooling (in laser quenching, in air).

Tempering *provides*:

- decomposition of martensite, structural stability;
- high ductility with a *slight* decrease in tensile strength;
- reduction of internal stresses;
- reduction of the impact of stress concentrators.

Laser tempering is *used* to improve the physical and mechanical properties of products operating under cyclic and impact loads:

- gears;
- shafts;
- cutting tools.

Laser quenching is a type of heat treatment of martensitic steels which consists of:

(1) heating to a temperature 30°–50° above the critical point A_3 with the formation of austenite;
(2) holding for dissolving carbide phases existing at low temperatures;
(3) rapid cooling with the transformation of austenite into martensite.

In traditional hardening methods (furnace, induction, etc.), cooling is carried out in oil or water.

Features of laser quenching:

(a) *Ultra-fast heating.*
(b) *Intensive cooling* at speeds of 10^5–10^6 deg/s, which far exceed the critical ones. Rapid cooling occurs due to the localization of the volume of heated metal and efficient heat transfer to adjacent zones of cold metal with high thermal conductivity.
(c) The process is *superficial*, not volumetric.
(d) Possibility to harden only the *working* surfaces of parts.

Quenching *provides*:

- high hardness (while reducing plasticity and toughness);
- high wear resistance.

Areas of application of laser quenching:

- the working surface of the cutting tools;
- excavator teeth.

Laser thermo-strain hardening [6] consists in:

(1) heating to a temperature 30–50° above the critical point A_3 with the formation of austenite;
(2) rapid free cooling to 600–400 °C;
(3) mechanical plastic deformation at a temperature of 600–400 °C;
(4) subsequent free cooling to room temperature.

Laser thermo-strain hardening provides:

- formation of a fine-grained structure characterized by high hardness with increased plastic properties;
- increasing the density of dislocations and complicating their spatial configuration;
- formation in the surface layers of guaranteed compressive residual stresses of the first kind (acting at the macro level)
- high hardness;
- high wear resistance.

Laser thermo-strain hardening is *used* for surface layers of parts operating in friction under cyclic alternating loads.

Laser forming is the process of giving the required shape to a *thin sheet* product by controlling the creation of stresses on the surface [5, 6]:

- a certain *distribution pattern* caused by the trajectory of the laser beam;
- of a certain *value*, which is caused by changing the number of heating–cooling cycles and laser radiation modes.

Under the influence of controlled stresses, the workpiece plastically deforms and takes the required shape.

Laser forming provides:

- manufacturing of spatial thin-walled products in a wide range of materials and shapes (including complex and large-sized ones);
- ease of reorganization of the process to manufacture new products;
- elimination of expensive high-alloy dies and forming tools;
- and thus high productivity and efficiency of the process.

Modulation of the ***stress–strain state*** [5, 7] is a type of laser forming technology and consists in creating *compensating residual stresses* in certain parts of the structure that:

- in terms of magnitude — are equal to the maximum operational stresses that occur in these areas under the influence of working loads;
- are opposite to the operational stresses.

Modulation of the stress–strain state *allows*:

- to increase the permissible operating loads;
- to increase the rigidity of the structure.

Liquid phase laser quenching is a type of heat treatment of martensitic steels and *consists of*:

(1) Heating to the melting point. Impurities in the form of inclusions have time to dissolve in the overheated metal.
(2) Holding for dissolving carbide phases existing at low temperatures;
(3) Rapid cooling with transformation of austenite into martensite.

The *features* of liquid phase laser hardening are ***the same*** as those of solid phase laser hardening. In addition, since the heating time, molten layer thickness, and degree of heating of the base material are very small, cooling during the crystallization of the liquid layer occurs at a high rate—up to 10^6 degrees per second. Such cooling rates are about 10^3 times higher than the rates typical for conventional quenching. As a result, martensite is formed with a higher carbon content than in conventional quenching.

Liquid phase laser quenching *allows*:

- to evenly distribute the alloying elements, prevent the formation of impurity clusters (for example, graphite clusters in cast irons, or silicon clusters in silicones);
- to heal microcracks on the surfaces of parts.

Amorphization (metallic glass) is the process of obtaining *amorphous structures* on the surface layer of certain alloys:

- heating to the melting point;
- ultra-fast cooling (10^7 deg/s and more).

The ***amorphous structure*** of metals is the presence of a metallic bond between metal ions (a common negatively charged zone is formed by the release and sharing of valence electrons), in which the ions are in a disordered arrangement.

The ***model of an amorphous metal*** can be represented as follows. As a result of ultrafast cooling, metal ions do not have time to organize themselves into a crystal lattice. The process can be thought of as crystallization with an infinite number of crystallization centers. An ultra-fine-grained structure is formed, in which a "nano-grain" consists of several, usually five (the basis of the crystal cubic lattice) metal ions (Fig. 2.2)—the close order is preserved. The "nano-grains" are randomly arranged—no long-range order. Impurities are located between the "nano-grains", so they are evenly distributed. This structure determines the unique properties of amorphous metals.

The *unique* properties of amorphous metals are a combination of high strength and hardness with high fracture toughness and corrosion resistance, as well as phase homogeneity and isotropy of properties.

The widespread use of amorphous metals is still limited, despite their unique mechanical properties, due to the technological complexity of traditional production (shooting portions of metal onto a rapidly rotating heat sink) and the high cost.

Three groups of materials can be distinguished by their ability to amorphize:

(a) *easily amorphized*—binary and ternary alloys containing boron, nickel, etc. [6];
(b) *able to amorphize*—cast irons that contain sufficient amounts of carbon, silicon and phosphorus.
(c) *difficult to amorphize*—steels, as this requires cooling rates of more than 10^8 deg/s.

Changing the optical characteristics involves laser *melting* of the surface layer of an optical material (special types of glass, sitalls, quartz, ceramics) to obtain a surface with zero roughness and low optical anisotropy after hardening.

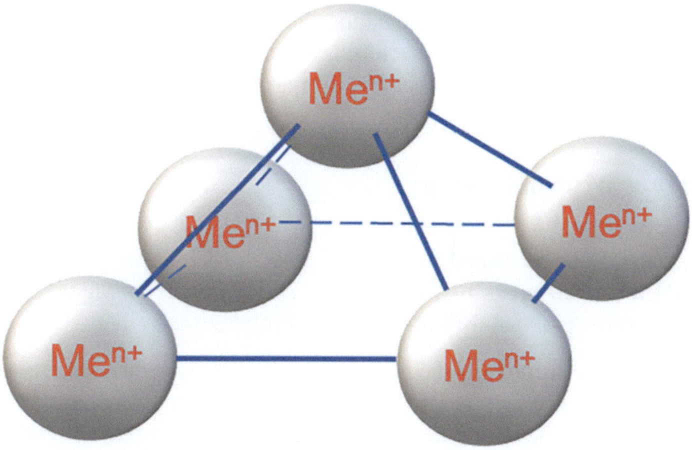

Fig. 2.2 "Nano-grain" of an amorphous metal

The laser beam is an effective tool for controlling the melting parameters because optical materials are usually dielectrics and absorb laser radiation at any wavelength well.

Changing the optical characteristics improves the refractive, reflective and transmittance coefficients of integrated optics elements.

Laser shock hardening involves treating the surface with *powerful* (10^9–10^{12} W/cm^2) *ultra-short* (in the range of 10^{-9}–10^{-15} s) radiation pulses that initiate the following *processes*:

(a) Instant *evaporation* (sublimation) of a thin layer of metal and *formation of plasma*.
(b) *Absorption* (full or partial) of laser radiation by plasma due to shielding and heating of the plasma.
(c) *Explosive increase* in plasma volume due to radiation absorption and heating.
(d) A *shock wave* propagates deep into the material. The pressure in the shock wave front can reach 10,000–20,000 MPa.
(e) Indirect *heating of metal* with plasma.
(f) *Plastic deformation* (increase in dislocation density).

Laser shock hardening *provides*:

- metal *strengthening*;
- Formation of *compressive residual stresses* in the surface layer; increase in *fatigue strength*.

Surface roughness improvement is based on the evaporation of sharp protrusions of micro-irregularities by *ultra-short* (10^{-15} s) *pulsed* laser radiation.

Improvements of the roughness of surfaces are:

- obtaining surface roughness at the nanoscale level;
- eliminating the dependence of machining results on the material structure;
- uniformity of physical and mechanical properties of the material surface.

2.3 Surface Microalloying

Surface microalloying is aimed at targeted *structural and phase transformations* of individual localized areas of materials by *changing the chemical composition.* Microalloying takes place at different heating temperatures:

- *in the solid phase* stimulates the occurrence of *impulsive* thermal elastic stresses in the surface layer, which leads to redistribution of alloying elements;
- *in the liquid phase*, new alloying elements are introduced into the melt.

Laser thermal cycling *involves* cyclically heating a section of the product's surface to temperatures no higher than the melting point and free cooling.

During laser thermal cycling, diffusion mass transfer occurs in the solid phase at rates much higher than diffusion under normal conditions.

The most effective treatment is laser thermal cycling of *coatings*, which *provides*:

- Significant *redistribution of alloying elements* and impurities, thus giving individual parts of the products the desired properties;
- Mutual diffusion of coating elements into the substrate, and of substrate elements into the coating, thus increasing the *adhesive strength* of coatings by 5–7 times;
- A significant reduction in the porosity of thermal gas coatings from 8–10 to 2–3%;
- Increasing the plasticity of gas-thermal coatings.

Laser thermochemical treatment (nitriding, boronizing, cementation, etc.) is a process that involves:

- preliminary *laser* hardening of the part surface;
- heating, holding and cooling in a suitable saturating medium according to the modes of conventional thermochemical treatment (for example, nitriding of carbon steels is carried out in ammonia at a temperature of 800–860 °C).

Laser thermochemical treatment *provides*:

- Increase in the depth of the nitrided layer in the heat-affected zone of laser radiation by 3–4 times (from 0.115 mm without laser treatment to 0.461 mm with laser treatment);
- Increasing the dispersion (grain size) of the structure;
- Increase in microhardness (in the nitrided carbon steel layer without preliminary laser treatment, the microhardness is 3.5 GPa, in the layer with laser treatment—8 GPa).

Laser ion implantation consists of:

(1) Laser vaporization of metal with the formation of ions in the plasma.

(2) Acceleration of ions in the electromagnetic field of an ion accelerator.
(3) Bombardment of the surface of the product with accelerated ions and, as a result:
- penetration of ions of any chemical element with alloying at a depth of 1–9 μm;
- deformation of the surface layer with a change in the dislocation structure at a depth of up to 100 μm.

Laser ion implantation *makes it possible* to create an alloy of metals in the surface layer that do not mix in the molten state, or to alloy one metal with another with a concentration that cannot be achieved by any other means.

Laser ion implantation *provides* increased corrosion resistance, wear resistance of surfaces, etc.

Laser-plasma implants have small overall dimensions, and they make it easy to switch from one type of ion to another and to produce multi-element ion beams with the desired stoichiometric composition [5].

Laser surface alloying *involves* melting the surface of a product and feeding alloying elements into the resulting melt. The alloying elements can be introduced into the melt:

- from a pre-sprayed metal layer;
- from the gas or liquid environment;
- powder material injection (injection of powder alloying elements into the melt with a transport gas).

The alloying elements in the molten pool are mechanically mixed with the base material by means of:

- hydrodynamic flows caused by uneven temperature distribution;
- diffusion processes.

After curing, solid solutions, chemical compounds or compositions of the required composition are formed in the surface layer. A prerequisite is that a significant amount of the base material is retained in the laser-treated areas of the surface.

Laser surface alloying *provides* high hardness, corrosion resistance, wear resistance and heat resistance.

Laser surface alloying is widely used in the manufacture of *semiconductors*. Donor and acceptor impurities (P, B, Sb, etc.) are introduced into semiconductors. The surface layer is melted by laser radiation and the alloying element diffuses deep into the part. The process of impurity diffusion can come from a surface solid pre-applied layer, from a liquid layer, or from a gas source.

2.4 Coating

Coatings are applied to fix a layer of another material on the surface of a metal product. Coatings are divided into two groups according to their purpose:

(a) *functional* coatings—possess the properties necessary to perform the operational functions of the product;
(b) *technological* coatings—possess properties that improve the manufacturing process of a product, for example, increase the efficiency of laser surface treatment.

Methods of coating include [8]:

- ***surfacing***, in which the surface layer is formed by *melting* the metal to be deposited and the base metal to *form a pool*;
- ***sintering***, in which the surface layer is formed by *diffusion* of metal in a *heated state*, or by melting a powder with subsequent *adhesion* without the formation of a pool;
- ***spraying***, in which the surface layer is formed by *high-speed deformation of particles* in the form of powders, or deposition of molten metal droplets on a heated substrate without melting the base metal;
- ***brazing***, in which the surface layer is formed by the *adhesion of a metal* in a solid heated state and a binding alloy—*solder* in a liquid state.

When applying coatings to iron-carbon and other alloys, laser heating and subsequent ultrafast cooling creates structures with unique properties. The main thing is to promote a significant increase (by 3–4 times) in the *adhesion of the deposited* metal.

Laser surfacing *involves* melting the surface layer of the metal and the powder mixture to be deposited to form a pool.

The laser beam used is either partially defocused or focused with scanning.

When surfacing by any method, one of the main tasks is to ***minimize the penetration depth and the proportion of the base metal in the deposited material*** [8]. Laser surfacing achieves the minimum possible penetration depth of 20–50 microns, which ensures metallurgical bonding and eliminates the mixing of the coating with the base metal.

The coating material is supplied by:

- pre-spraying of a metal layer;
- injection of powder material (blowing powder into the melt with a transport gas).

Laser sintering *involves* heating the metal surface layer and applying a composite powder material:

- heated—sintering in the solid phase;
- molten—sintering in the liquid phase.

Sintering in the liquid phase is of greater practical importance, especially for diamond-containing tool composite materials. There is a certain range of sintering *parameters* that ensure the preservation of the initial strength of diamonds—high-speed heating of the powder layer to temperatures of 1500–1600 °C for 0.2–0.4 s. At the same time, diamond particles do not lose their strength. Outside of this temperature, diamonds are subject to cracking, oxidation and other negative processes.

Laser thermal deformation sintering takes place in two stages:

- laser sintering in the liquid phase;
- plastic deformation of the surface at the cooling stage.

Plastic deformation allows the composite material to be compacted. During plastic deformation, the shape and dimensions of the product are simultaneously set in the mold.

The method makes it possible to produce working diamond-containing surfaces directly on tool bodies, which significantly increases production productivity.

The ***laser-plasma surfacing*** process *consists* of:

- melting of the thin surface layer of the product by laser penetration;
- feeding the powder material into the surfacing zone with simultaneous heating by an electric arc to a temperature of 0.9 T_{mel};
- final molding of the deposited pool by laser flushing.

The method can significantly increase sintering performance without significantly increasing the cost.

Laser brazing involves heating without melting the surface of the parts to be joined, applying an activating flux to the surface, and introducing molten solder.

The laser brazing process *is characterized by*:

(a) high *heating rate* of brazed joints—0.3–0.5 s;
(b) high *cooling rate* of brazed joints;
(c) minimum *time of heat exposure*;
(d) *localized* effect;
(e) *uniform* heating of the brazing zone.

The consequence of the features a, b, c, d of the brazing process is:

(f) formation of a *highly dispersed* solder *structure* and *increase* of its strength;
(g) practical absence of *residual stresses*;
(h) significant *reduction* in the risk of *intermetallic compounds and scale* formation.

Laser brazing *provides*:

- *preventing overheating* of adjacent heat-sensitive objects, which is critical for integrated circuits and electronic equipment;
- high *accuracy* of the deposition pool—up to 0.025 mm;
- greater *flexibility* of temperature and time parameters of the brazing mode;
- high *stability* of parameters during the brazing process;
- *no contact between* the tools and the brazing zone;
- the ability to use a *wide range of solders* with both low and high melting points.

The following are used for brazing:

- YAG:Nd solid-state lasers with a wavelength of 1.06 μm and an output power of 2 to 150 W (mostly);
- fiber lasers;

- lasers operating in the ultraviolet spectrum (this solves the problem of absorption of laser radiation by metals).

Brazing *modes*: pulsed (preferred); pulse-periodic; continuous.

The brazing *speed* is in the range of 1 to 3 mm/s.

The *scope* of laser brazing is extremely wide—from the manufacture of electronic and electron-vacuum equipment elements, metal-ceramic sealing gaskets in nuclear power plants to the connection of car body parts, stainless steel dentures in dentistry, jewelry, etc.

Laser spraying is carried out using two technologies:

(1) Laser vaporization of a "target" (metal to be sprayed) in a chamber with subsequent deposition on the surface of the part. The "target" can be made up of various materials that determine the desired stoichiometry of the resulting film. High energy input rates make it possible to preserve the stoichiometric composition of the target and obtain thin films of various compositions on the surfaces of parts. The use of frequency lasers makes it possible to build up the material in small portions, layer by layer, while maintaining the crystal structure.
(2) Feeding the powder to be sprayed into the laser-plasma jet of the LP plasma torch (Sect. 1.4.1) with subsequent melting, acceleration, and directing the particles to the part to be sprayed. The plasma formed by laser radiation has a power density of 10^8–10^{12} W/cm^2.

Areas of *application* of laser spraying:

(a) Preparation of complex semiconductor compounds.
(b) Creation of high-temperature multicomponent superconductor compounds.
(c) Growing single crystals by activating the electron shells of atoms during the crystallization process with concomitant recombination radiation. This stimulates the growth of crystals with more energy-intensive orientations. For example, thin diamond films can be produced in the process of carbon condensation during evaporation from a graphite target.

2.5 Technological System of Laser Surface Treatment

Laser surface treatment is based on *physical and chemical processes* (Fig. 2.3) in the surface layer when laser radiation interacts with the workpiece material. The physical and chemical processes are of a thermal nature, i.e., their entire variety is controlled by four *main parameters* [5, 6]:

(1) heating temperature (T_h);
(2) heating time (t_h);
(3) heating rate (V_h);
(4) cooling rate (V_c).

Physicochemical processes *provide* the surface layer the *required quality characteristics* (Fig. 2.3):

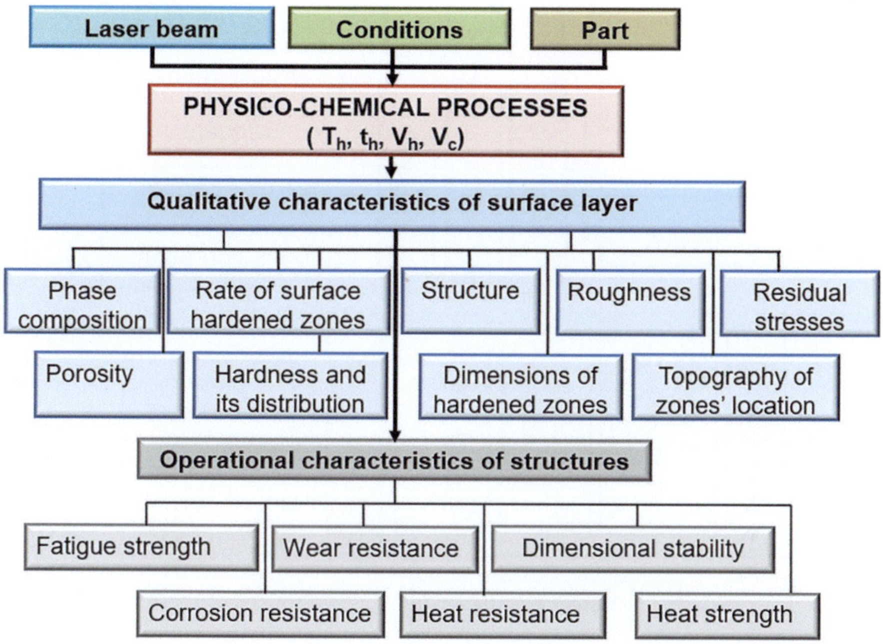

Fig. 2.3 Schematic of the technological system of laser surface treatment. General view

- phase composition and structure;
- hardness and its distribution;
- the proportion of hardened areas on the surface;
- the size of the reinforced zones;
- topography of the zones;
- porosity;
- roughness;
- residual stresses.

The qualitative characteristics of the surface layer *determine the performance characteristics of products* (Fig. 2.3):

- wear resistance;
- fatigue strength;
- corrosion resistance;
- heat resistance;
- strength at elevated temperatures;
- dimensional stability.

The physical and chemical processes of laser surface treatment are multifactorial and multicriteria. The factors that influence the *choice of the main parameters* of physical and chemical processes can be divided into three groups:

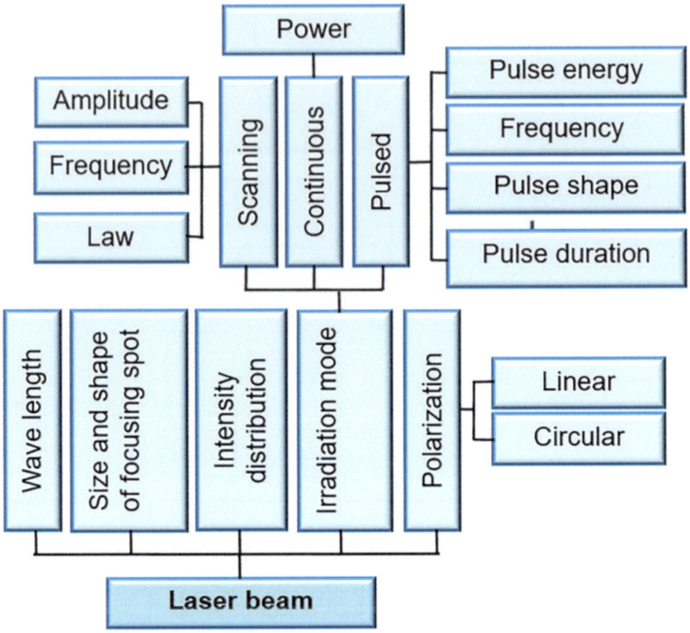

Fig. 2.4 Schematic of the laser surface treatment system. Characteristics of the laser beam

(1) characteristics of the laser beam that forms the heat source;
(2) characteristics of the workpiece;
(3) characteristics of the conditions of interaction between laser radiation and the product.

The most important *characteristics of a laser beam* during surface treatment are (Fig. 2.4):

(a) The wavelength of the radiation.
(b) The size and shape of the focus spot.
(c) Character of radiation intensity distribution in the focusing spot
(d) Depending on the irradiation mode:
- for continuous—power;
- for pulsed—energy, frequency, shape, and duration of the pulse;
- when scanning—amplitude, frequency, and scanning law.

(e) The polarization is either linear or circular.

The *characteristics* of the workpiece are as follows (Fig. 2.5):

(a) The geometric parameters of the part are shape and weight.
(b) Material properties, including chemical and phase composition, structure, and thermal properties (heat capacity, thermal conductivity, and thermal conductivity).

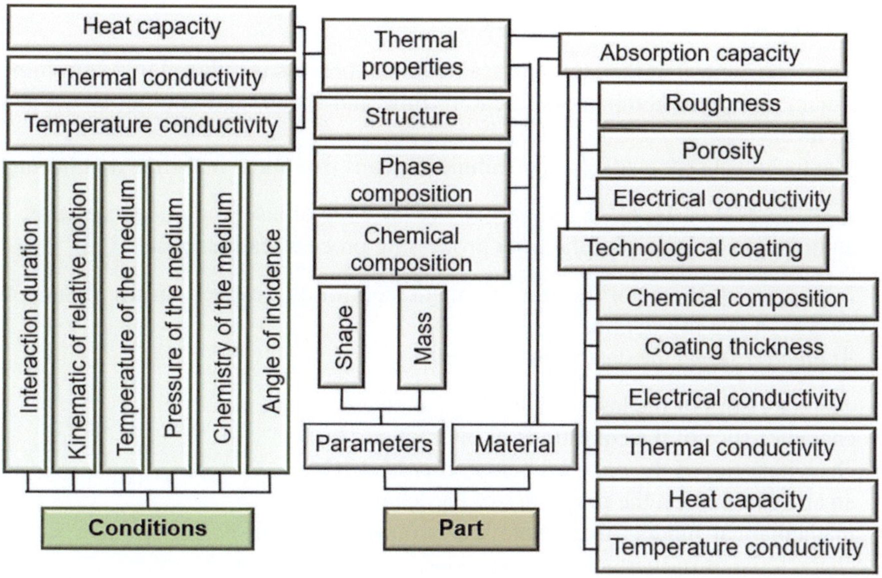

Fig. 2.5 Schematic of the laser surface treatment system. Characteristics of interaction conditions and characteristics of the part

(c) The absorption capacity of the surface, which is determined by surface roughness, porosity, electrical conductivity and the presence of a technological absorbing coating.
(d) In the presence of a technological coating, the process of absorbing and transmitting radiation energy is affected by the chemical composition of the coating, its thickness, electrical conductivity, and thermal properties.

The main *characteristics of the conditions of interaction* of laser radiation with the product (Fig. 2.5) are [9]:

(a) Interaction time.
(b) Kinematics of relative motion.
(c) Temperature, pressure, and chemical composition of the medium.
(d) The angle of incidence of the beam on the surface of the material.

All the factors shown in Figs. 2.4 and 2.5 affect the physicochemical processes that form the surface layer, so all of them should be taken into account when studying, mathematically modeling and developing laser surface treatment technology. However, the following characteristics are the most *effective for controlling* physical and chemical processes:

(1) radiation power or energy;
(2) area and shape of the focus spot;
(3) the nature of the radiation intensity distribution in the focusing spot;
(4) surface absorption capacity;

(5) initial temperature of the material.

Thus, the control of the laser surface treatment process is reduced to determining and maintaining the temperature, heating time and speed, and cooling speed at a given level.

The technical task is to develop technical means that allow real-time monitoring:

- measurement of T_h, t_h, V_h, V_c directly in the zone of laser radiation exposure;
- automated adjustment of the laser processing process via feedback.

A diagram of the algorithm for automatic control of the laser surface treatment process is shown in Fig. 2.6.

At the first stage, the databases that form:

- characteristics of the laser beam $\{A_n\}$;
- characteristics of the conditions of interaction $\{In_m\}$;
- characteristics of the material being processed $\{C_k\}$;
- an array of data on the properties of hardened materials;
- mathematical models that relate the parameters of the hardened layer to the characteristics of thermal processes, such as $Z = F_n$ (T_h, t_h, V_h, V_c);
- models that relate thermal process characteristics to technological characteristics, such as $T = F_n$ (A, B, C).

At the second stage, the processing goal is selected and the initial values of the quality characteristics of the hardened surface layer are entered:

(a) The calculation model T_{hc}, t_{hc}, V_{hc}, V_{cc} (index 'c' means 'calculated') is selected.
(b) They following are determined by limiting the process parameters:
  - [Z]—penetration depth;
  - [HV]—hardness;
  - [Rz]—surface roughness;
  - [σ_{res}]—surface residual stresses;
  - [σ_{-1}]—fatigue strength;
  - [ΔU]—wear resistance;
  - [T^*]—corrected heating temperature.
(c) If the calculated values do not correspond to the specified values, then T_{hc} and t_{hc} are adjusted by ΔT and Δt, respectively: $T_{hc1} = T_{hc} \pm \Delta T$ and $t_{hc1} = t_{hc} \pm \Delta t$.
(d) Discrepancies are being eliminated.
(e) Process parameters P, V, d_0 are calculated.
(f) A control signal is generated and the system is started.

At the third stage, to automatically control the process, one needs to measure the current values of the heating parameters T_h and t_h.

If the current values of the parameters T_h and t_h do not match the calculated values, a control error is detected and corrections are made:

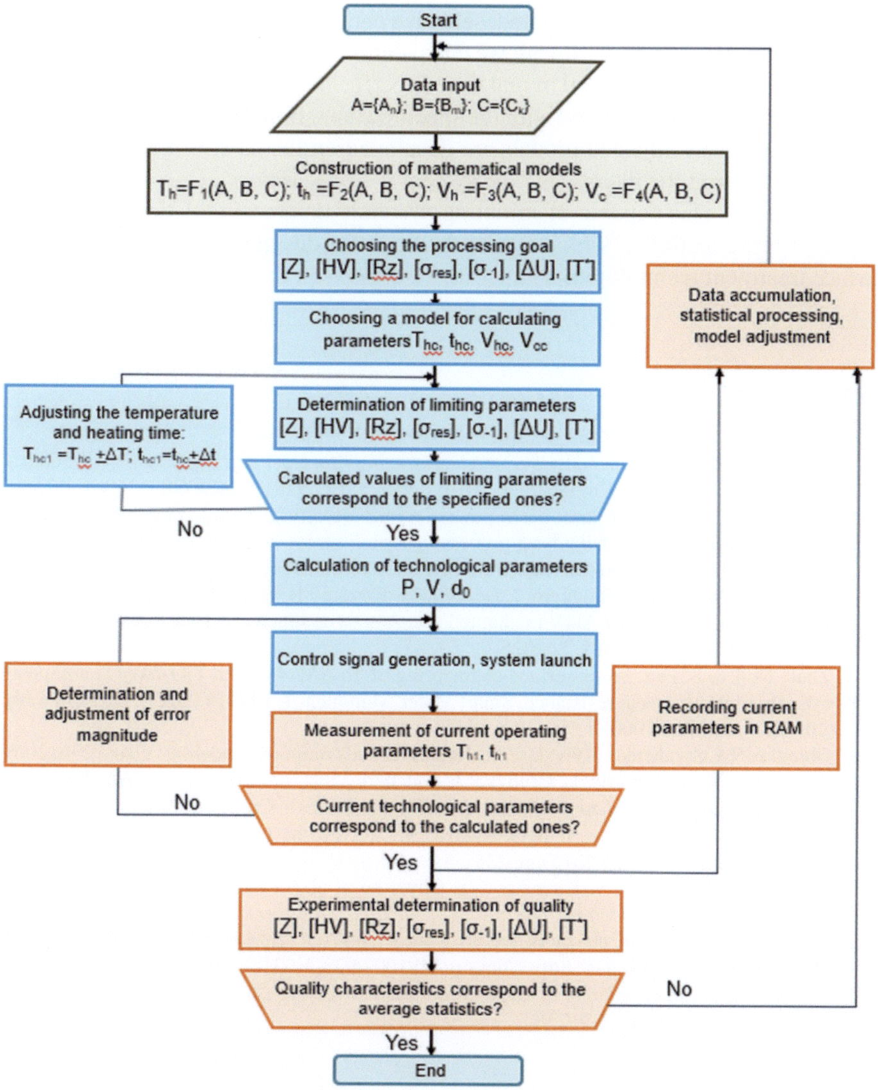

Fig. 2.6 Schematic of the algorithm for automatic control of the laser surface treatment process

(a) heating temperature T_h by:
- adjusting the diameter of the focused beam (first priority),
- control of radiation power (second priority),
- changes in the irradiation rate (third priority);

(b) heating time t_h by:
- adjusting the processing speed,
- the size of the focused beam.

After control errors are eliminated, selective control of the surface layer's qualitative characteristics (depth, hardness, wear resistance, fatigue strength, heat resistance, etc.) that cannot be determined in real time is performed. The obtained data is statistically processed and compared with the specified data. If the discrepancies approach critical values, new adjusted models are developed on the basis of the accumulated data to supplement the databases. The system is self-learning.

This algorithm is capable of initiating systemic data accumulation with a high degree of reproducibility, which can be the basis for designing new industrial laser surface treatment technologies.

References

1. J.R. Lawrence, D. Waugh, in *Laser Surface Engineering Processes and Applications* (Woodhead Publishing. Technology & Engineering, 2014) eBook ISBN 9781782420798
2. D. Bartkowski, Influence of laser beam power on microstructure and microhardness of Fe/ZrC coatings produced on steel using laser processing—preliminary study on the single laser tracks. MDPI Mater. **15**(3), 758 (2022). https://doi.org/10.3390/ma15030758
3. J.M. Vazquez-Martinez, J. Salguero, I. Del Sol, M. Ramirez-Peña, M. Batista, Endurecimiento superficial de la aleación Ti6Al4V mediante laser. Anales de Ingeniería Mecánica **1**(24) (2025). https://doi.org/10.63450/aim.1.72.2025
4. Y. Zhao, M. Lu, Z. Fan, Laser surface engineering of Ti–6Al–4V with TiO_2/Al_2O_3 composite powder for improved wear resistance. Smart Mater. Manufact. **1**, 100015 (2023). https://doi.org/10.1016/j.smmf.2023.100015
5. L. Golovko, S. Lukyanenko, Kyiv: laser technologies and computer modeling. Vistka 296 (2009)
6. V. Kovalenko, A. Verhoturov, L. Golovko, I. Podchernyaeva, in *Laser Surface Hardening and Electric-Sparc Surfacehardening of Materials* (Plinum Publish. Corp.", 1988), p. 175
7. L. Golovko, S. Lukyanenko, I. Mykhaylova, V. Tretyak, in *Computer Modeling in Laser Technologies* (Kyiv: "Text", 2015), p. 236
8. I. Ryabtsev, S. Fomichov, V. Kuznetsov, Y. Chvertko at al., in *Surfacing and Additive Technologies in Welded Fabrication: International Welding Engineers Textbook Series Under the Editorship of Borys Paton*, ed. by B. Paton (Switzerland AG: "Springer Nature", 2023), p. 229 P. ISBN 978–3–031–34389–6. https://doi.org/10.1007/978-3-031-34390-2
9. L. Golovko et al., in *Handbook of Laser Materials Processing* ("Laser Institute of America", USA, Orlando, 2001), pp. 249–258

Chapter 3
Thermal Processes in Materials During Laser Surface Treatment

Abstract Calculations of thermal fields during heating with pulsed and continuous laser radiation are presented. The obtained equations allow solving a wide range of technological problems. Calculations of laser treatment parameters that ensure heating to a given temperature at a given depth or ensure maximum hardening depth are illustrated. Based on calculations of thermal processes during heating with a laser beam scanning across the surface, the following has been determined: the minimum permissible scanning frequency to ensure a stable hardening depth along a hardening strip of a given width; the maximum heating temperature under the action of a heat source; the duration of the heat source with scanning, which ensures the cooling rate with hardening.

3.1 General Provisions

It is impossible to understand laser surface processing without understanding the thermal processes that occur in materials during rapid cycles of heating and cooling. Existing studies use mathematical methods to describe laser processing [1, 2], model the thermal influence of laser radiation on microstructure changes [3], and model combined thermomechanical processes [4]. Additionally, there are studies that use numerical and physical modeling to study thermal processes during laser processing [5]. However, given the importance of this issue, the authors of this book could not overlook it.

To determine the laser irradiation modes that provide the desired surface treatment result, it is necessary to quantitatively recognize the relationship between laser irradiation parameters and the characteristics of ***thermal processes***—heating, heat propagation and distribution (temperature field).

Knowledge of the characteristics of thermal processes allows:

1. to calculate laser surface treatment modes and to optimize them;
2. to determine the characteristics of technological equipment;
3. to predict the nature of structural and phase transformations in surface layers and the resulting material properties.

I. Krivtsun et al., *Laser and Hybrid Technologies for Surface Engineering and Welding*, Springer Series in Materials Science 358, https://doi.org/10.1007/978-3-032-15038-7_3

4. to assess the stress–strain state of surface areas and products as a whole (distribution of residual stresses and residual strains).

The characteristics of thermal processes in materials are affected by:

(a) Laser *parameters:*
- power P;
- power distribution P (x, y);
- is the energy E;
- is the duration of the action τ;
- size of the focus area d_0.

(b) *Characteristics of the material* being processed:
- absorption capacity A (T)—a nonlinearity of the second kind;
- thermal conductivity a (T);
- heat capacity c(T);
- thermal conductivity k(T)—a nonlinearity of the first kind;
- depth of perceived heating—*at*;
- size and shape of the product (semi-infinite array $h > 4\sqrt{at}$), thin plate $h \leq 4\sqrt{at}$, complex shaped products).

Thermal properties depend on temperature. And this dependence is non-linear. There are two ways to *calculate temperatures*:

(a) by solving a ***nonlinear*** heat conduction ***problem***, when the thermal properties are approximated by a certain *function*—polynomials (often cubic splines);
(b) by solving the ***linear*** heat conduction ***problem***, when the thermophysical properties are assumed to be averaged *constants* for the temperature interval at which the process is performed.

The ***hardening temperature*** T_h during high-speed heating (which occurs during laser surface treatment) increases by the correction value [6]:

$$T_h = T_{AC3} + \left(\frac{3k_1^2 a_0^2}{4D} V_h \right)^{1/3} \tag{3.1}$$

where T_{AC3} is the temperature of the critical point A_3 on the iron-cementite diagram at which the necessary structural and phase transformations occur;

V_h is the average heating rate;

k_1—coefficient (for steel $k_1 = 110$);

D is the diffusion coefficient of carbon in austenite;

a_0 is half the average distance between cementite grains in perlite.

In most calculations, the simplified *value of the correction for the heating rate* in Eq. (3.1) can be accepted with a sufficient degree of accuracy:

$$\left(\frac{3k_1^2 a_0^2}{4D} V_h \right)^{1/3} = 80 - 100^\circ \tag{3.2}$$

The *heat source* may be different depending on the *ratio of the diameter* of the focusing zone *to the heating depth*:

(1) point,
(2) distributed,
(3) voluminous.

A laser beam should be considered as a *heat source* during surface treatment of metal materials:

- *surface*, since the absorption coefficient for metals is very high—$\alpha = 10^5$–10^6 cm^{-1};
- *distributed*—since the radius of the focus area $r_0 \geq \sqrt{at}$), the length of the product $L > 4\sqrt{at}$.

This allows one to neglect heat propagation in the lateral directions (x and y) and solve the thermal problem as a *one-dimensional problem* (along the depth z).

3.2 Calculations of Thermal Processes During Heating by Pulsed Laser Radiation

Heat propagation along the z-axis is described by a one-dimensional differential equation:

$$\frac{\partial^2 T(z,t)}{\partial z^2} - \frac{1}{a} \cdot \frac{\partial T(z,t)}{\partial t} = -\frac{q}{k} \tag{3.3}$$

where T is the temperature, °K;
z is the coordinate in the direction of the heat source (depth), cm;
t—the current time value;
c—heat capacity, J/(g·K);
a—thermal conductivity coefficient, cm^2/s;
k—thermal conductivity coefficient W/(cm·K);
q is the intensity of the heat source W/cm^2:

$$q = AW_p\varphi(t) \tag{3.4}$$

φ (t) is a function that describes the change in the laser pulse over time (rectangular φ (t) = 1, triangular, bell-shaped);
A—absorption capacity;
W_p is the intensity distribution function of the heat source (Fig. 3.1), or power density [6, 7].

With a Gaussian distribution of radiation intensity (Fig. 3.1), the *effective radius of the heating spot* r_0 can be taken as the distance from the center of the zone at which the maximum radiation power P_{max} decreases by a factor of 'e' or 'e^2'. The

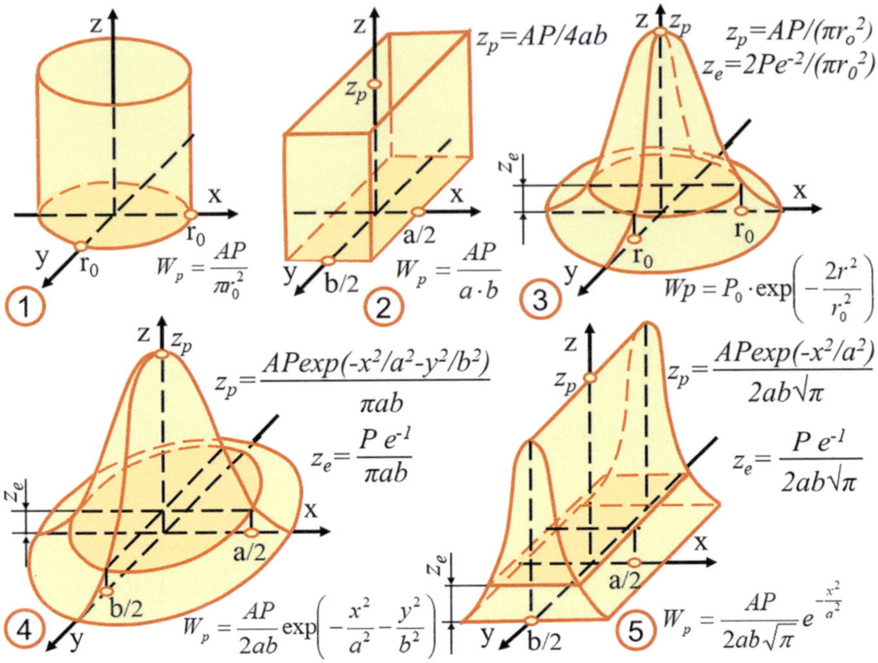

Fig. 3.1 Schemes of intensity distribution of the heat source W_p: 1—cylindrical; 2—rectangular; 3—Gaussian; 4—elliptical; 5—rectangular Gaussian (calculated heating spots are highlighted in dark color)

difference in the calculated areas of the heating spots determined for the condition $P = P_{max}/e$ and for the condition $P = P_{max}/e^2$ is 10–25%. Therefore, the calculated heating spot is determined in the plane in which $P = P_{max}/e$.

To solve the differential Eq. (3.3), it is necessary to set the *boundary conditions*—the absence of heat exchange with the environment:

$$\frac{\partial T(0, t)}{\partial z} = 0 \tag{3.5}$$

The *initial conditions* are:

(a) $T(\infty, t) = T_p$—the temperature at infinity does not change during the entire irradiation period (equal to the initial temperature T_p or zero);
(b) $T(z, 0) = T_p$—the temperature in the considered region at the initial time is equal to the initial temperature T_p or zero.

The ***general solution to the heat conduction equation*** over time $t \leq \tau$ is as follows:

$$T(z, t) = \frac{2AW_p\sqrt{at}}{k}\text{ierfc}\left(\frac{z}{2\sqrt{at}}\right) \tag{3.6}$$

$$\text{ierfc} = \int (1 - \text{erfy})\text{d}y \tag{3.7}$$

where: A is the absorption capacity;
τ is the pulse duration;
ierfc is the Bessel function;
erfy is an error function.
For the values of 'ierfc' and 'erfy', see the reference tables.

Temperature distribution on the surface:

$$T(0, t) = \frac{2AW_p}{k}\sqrt{\frac{a\tau}{\pi}} \tag{3.8}$$

To account for the shift of critical points of structural-phase transformations (austenitization), it becomes necessary to determine the heating rate:

$$\partial T/\partial t = \frac{AW_p\sqrt{a}}{k\sqrt{t}}\frac{1}{\sqrt{\pi}}e^{-\frac{z^2}{4at}} \tag{3.9}$$

In the above equations, we consider time comparable to the time of laser exposure. In real conditions, the process of heat conduction continues after the laser radiation ceases.

The isotherm of temperature T_i moves deeper into the material for a certain time t_0, reaches a maximum depth z_{max} and returns to the surface. The maximum depth z_{max} of heating to a temperature T_i is of practical interest because it determines the thickness of the layer in which the corresponding effects occur (e.g., quenching).

For time t > 0 and more than τ, the temperature distribution is described by the following dependence:

$$T(z, t) = \frac{2AW_p\sqrt{a}}{k}\left[\sqrt{t_0}\text{ierfc}\left(\frac{z}{2\sqrt{at}}\right) - \sqrt{t_0 - t}\text{ierfc}\left(\frac{z}{2\sqrt{a(t_0 - t)}}\right)\right] \tag{3.10}$$

where t_0 is the time for a given isotherm to reach depth z.

From dependence (3.10), we can determine z_{max}:

$$z_{max} = \sqrt{\frac{4at_0(t_0 - t)\ln\sqrt{\frac{t_0}{t_0 - t}}}{t}} \tag{3.11}$$

It is not possible to calculate z_{max} directly using (3.11) because t_0 is not known in advance. However, t_0 can be approximated by certain functions.

The relation of the isotherm T_i to the surface temperature is as follows:

$$b = T_i/T(0, t) \tag{3.12}$$

where b is an approximation factor that depends on the thermophysical properties of the material.

For $b < 0.3$ (low-carbon steels, eutectoid steels), the dependence (3.11) is approximated by the function:

$$z_{max}/\sqrt{(a\tau)} = 1/\sqrt{(2be)} \tag{3.13}$$

For $0.3 < b < 0.75$, dependence (3.11) is approximated by the function:

$$z_{\max} = 3.8\exp(-3.3 \cdot b) \tag{3.14}$$

For $0.75 \leq b \leq 1$ (tool steels), the dependence (3.11) is approximated by the function:

$$z_{\max} = 1.24(1 - b) \tag{3.15}$$

The ***cooling rate*** is the defining characteristic of structural and phase transformations during hardening and can be determined from the following relationship:

$$\partial T/\partial t = \frac{2AW_p\sqrt{a}}{k}\left[\frac{e^{-\frac{z^2}{4at_0}}}{2\sqrt{\pi t_0}} - \frac{e^{-\frac{z^2}{4a(t_0-t)}}}{2\sqrt{\pi(t_0-t)}}\right] \tag{3.16}$$

3.3 Calculations of Thermal Processes During Heating by Continuous Laser Radiation

The temperature dependence of continuous laser heating is more complex than that of pulsed laser heating.

Heat sources moving on the surface of a semi-infinite body along the X-axis can be of two types:

(a) *Slow-moving* heat source—when heat saturation occurs before the beam has traveled a distance equal to the radius of the focus area. The maximum heating temperature of the material is located in the center of the focusing zone. The problem of determining the temperature distribution can be solved as with a *fixed heat source*.

(b) *Fast moving* heat source—when the temperature maximum shifts to the edge of the focus zone in the direction opposite to that of the heat source movement. A moving temperature field is formed around the focus area. If the heat source moves at a constant speed, the temperature field will stabilize after a while. The temperature distribution will depend only on the ratio of the time of heat saturation of the sample and the time the focusing zone travels a distance equal to the radius of the focusing zone.

The ***mathematical model*** of the process of heating a ***semi-infinite body*** by a moving heat source [8] whose power density is distributed according to Gauss' law consists of differential equations:

$$\frac{\partial^2\Theta}{\partial\overline{x}^2}+\frac{\partial^2\Theta}{\partial\overline{y}^2}+\frac{\partial^2\Theta}{\partial\overline{z}^2}+2\mathrm{Pe}\frac{\partial\Theta}{\partial\overline{x}}-\frac{\partial\Theta}{\partial F_0}=0 \tag{3.17}$$

$$\left.\frac{\partial\Theta}{\partial\overline{z}}\right|_{\overline{z}=0}=-\overline{g}_0\exp(-(\overline{x}^2+\overline{y}^2)) \tag{3.18}$$

$$\Theta(\infty,\overline{y},\overline{z},F_0)=\Theta(\overline{x},\infty,\overline{z},F_0)=\Theta(\overline{x},\overline{y},\overline{z},0) \tag{3.19}$$

$$\Theta(\overline{x},\overline{y},\infty,F_0)=0 \tag{3.20}$$

where $\Theta=(T-T_p)/T_c$
$\overline{x}=x/r_0$;
$\overline{y}=y/r_0$;
$\overline{z}=z/r_0$;
$\overline{g}_0=q_0r_0/(kT_p)$;
$\mathrm{Pe}=\mathrm{v}r_0/(2a)$—Péclet number;
$F_0=at/r_o^2$- Fourier number;
T is the current body temperature;
T_p—initial body temperature;
$T_c=Q/(2kr_0)\cdot\sqrt{(1/\pi)}$—the limiting temperature of the center point of a stationary normally distributed source;
x, y, z—current coordinates of the source;
r_0—effective radius of the heating spot;
t is time;
v is the speed of the laser beam;
Q is the effective thermal power of laser irradiation;
q_0—heat flux density in the center of the heating zone;
k is the thermal conductivity of the material;
a is the thermal conductivity of the material.

The boundary condition for the process of heat propagation *in a semi-infinite body* corresponding to the quasi-stationary state from a moving normally distributed energy source is the temperature at infinity:

$$\theta_\infty=\frac{2}{\pi}\exp(-\mathrm{Pe}\overline{x})\int_0^\infty\frac{\exp\left[-\frac{\overline{z}^2}{4F_0}-\frac{\overline{r}^2}{1+4F_0}-\frac{\mathrm{Pe}^2}{4}(1+4F_0)\right]}{(1+4F_0)\sqrt{F_0}}dF_0 \tag{3.21}$$

where $\overline{r}=r/r_0$;
$r=\sqrt{x^2+y^2}$ is a flat radius vector.

Using dependence (3.21), it is possible to determine the temperature field in a semi-infinite body created *by continuous laser radiation* moving at *speed v.*

Determining the temperature field makes it possible to predict the nature of structural and phase transformations and material properties in the surface layer.

In practice, thick-walled parts where the presence of bounding surfaces does not affect heat distribution can be considered as semi-infinite ones.

Limit thickness h_b is the thickness of a part at which the presence of a bounding surface begins to affect the temperature distribution. The limit thickness depends on the energy parameters of the laser irradiation and the thermal and physical characteristics of the material. The beginning of the influence of the bounding surface can be considered the thickness at which the difference between the results of calculations with the semi-infinite body and flat layer scheme h is equal to the relative error (or permissible temperature deviation):

$$\Delta_{\Theta} = (1 - \Theta_{\infty}/\Theta_h) \cdot 100\% \tag{3.22}$$

where θ_{∞} is the temperature at a distance h in a semi-infinite body;

θ_h is the temperature on the bounding surface of a flat layer of thickness h.

The ***mathematical model*** of the process of heating a ***flat layer*** by a moving heat source [9], the power density of which is distributed according to Gauss's law, by analogy with a semi-infinite body consists of differential Eqs. (3.17–3.20).

The *boundary condition* for the process of heat propagation *in a flat layer* corresponding to the quasi-stationary state from a moving normally distributed energy source is the condition of adiabaticity on the lower surface of the unheated flat layer:

$$\left.\frac{\partial \Theta}{\partial \bar{z}}\right|_{\bar{z}=\bar{h}} = 0 \tag{3.23}$$

where $\bar{h} = h/r_0$

h—thickness of the flat layer.

The analysis of Eqs. (3.17–3.23) shows that the error in determining temperatures according to the semi-infinite body scheme and the flat layer scheme increases over time and reaches its maximum value at the quasi-stationary state. Therefore, the influence of the bounding surface on the heat transfer process during laser processing should be evaluated for conditions corresponding to the quasi-stationary state, i.e., the boundary state of the process.

To find the temperature distribution in a flat layer, the source method in combination with the principle of superposition and reflection can be used. The final determination of the *temperature of a flat layer* in the quasi-stationary state is as follows:

$$\theta_i = \exp(-\text{Pe}\bar{x}) \sum_{i=-\infty}^{+\infty} \int_0^{\infty} \frac{\exp\left[-\frac{(\bar{z}-2\bar{h}_i)^2}{4F_0} - \frac{\bar{r}^2}{1+4F_0} - \frac{\text{Pe}^2}{4}(1+4F_0)\right]}{(1+4F_0)\sqrt{F_0}} \text{d}F_0 \tag{3.24}$$

The differences between the temperature fields of a semi-infinite body and a flat layer decrease as the thickness h increases. It is important to note that at small values of h, there is an almost complete coincidence of temperature distributions in the compared regions. Thus, the determination of the conditions for identifying the temperature fields of a semi-infinite body and a flat layer can be reduced to finding the boundary thickness of the flat layer h_b. It is necessary to determine the minimum thickness at which the temperature field of a flat layer begins to differ with a given accuracy from that of a semi-infinite body.

When determining the limit thickness h_b, it must be borne in mind that the temperature difference depends significantly on the location of the point in question relative to the bottom plane of the unheated flat layer. The farther away the point is from the bottom plane, the less the influence of the limited size of the product is.

The limiting thickness h_b is based on the determination of the *temperature deviation* at a certain point C (x_c, y_c, $z_{c\,max}$) calculated for the semi-infinite body and flat layer schemes. The position of the point C determines the maximum depth $z_{c\,max}$ of the thermal influence zone in the semi-infinite body.

Taking into account the above, after substituting Eqs. (3.21) and (3.24) into expression (3.22), we obtain the dependence for calculating the boundary thickness of a flat layer as follows:

$$\Delta_\theta = \left\{ 1 - \frac{\int_0^\infty \frac{\exp\left[-\frac{\overline{z}^2}{4F_0} - \frac{\overline{r}^2}{1+4F_0} - \frac{\mathrm{Pe}^2}{4}(1+4F_0)\right]}{(1+4F_0)\sqrt{F_0}} \mathrm{d}F_0}{\sum_{i=-\infty}^{+\infty} \int_0^\infty \frac{\exp\left[-\frac{\left(\overline{z}-2\overline{h_b}(iT)\right)^2}{4F_0} - \frac{\overline{r}^2}{1+4F_0} - \frac{\mathrm{Pe}^2}{4}(1+4F_0)\right]}{(1+4F_0)\sqrt{F_0}} \mathrm{d}F_0} \right\} 100\% \tag{3.25}$$

Equation (3.25) is transcendental. The solution must be preceded by finding the coordinates of the point C (x_c, y_c, $z_{c\,max}$) as a result of the joint solution of the following system of equations:

$$\Theta_C = f(x_c, y_c, z_c, \mathrm{Pe}) \tag{3.26}$$

$$\frac{\partial \Theta_\infty}{\partial \overline{x}} = 0 \tag{3.27}$$

The relationship between θ_c and the coordinates x_c, y_c, z_c, which define the corresponding isothermal surface, is established by Eq. (3.21).

Figure 3.2 shows the results of the numerical solution of Eq. (3.25), taking into account (3.26), (3.27), which show that:

- the ultimate thickness h_b is significantly affected by the temperature θ_c;
- and the dependence of h_b on θ_c is more significant the smaller the number Pe is.

Figure 3.3 shows the results of calculating the depth of the thermal impact zone z_L of laser radiation for materials of different thicknesses.

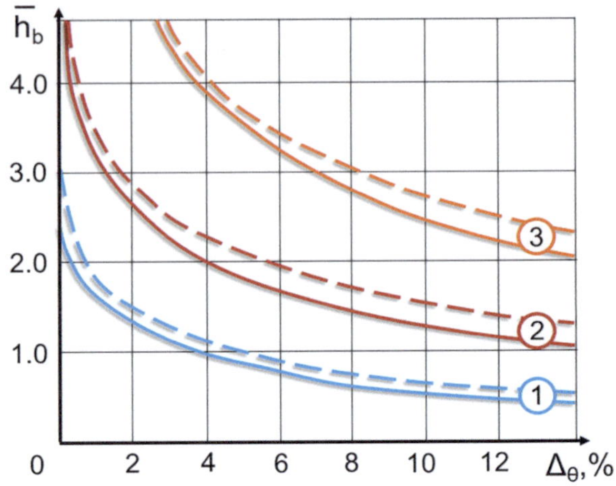

Fig. 3.2 Dependence of the boundary thickness $\bar{h}_b$ on the permissible temperature deviation Δ_θ: 1—at Pe = 1.555; 2—at Pe = 0.747; 3—at Pe = 0.404 (solid curves—at $\theta_C = 0.512$; dashed curves—at $\theta_C = 0.383$)

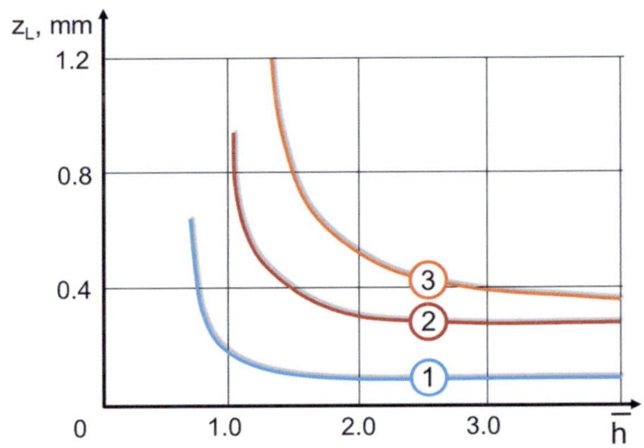

Fig. 3.3 Dependence of the depth of the thermal impact zone z_L on the thickness of the flat layer $\bar{h}$: 1—at Pe = 1.555, $\theta_c = 0.512$; 2—at Pe = 0.747, θc = 0.512; 3—at Pe = 0.404, θc = 0.512

Figure 3.4 illustrates the change in the ratio of the depth of the thermal influence zone z_L (at which microstructure changes occur) in a flat layer to the boundary thickness h_b depending on the permissible temperature deviation Δ_θ at different values of the Pe. The dependences show that:

- with increasing Δ_θ the ratio z_L/h_b increases with a dependence close to linear;
- the ratio z_L/h_b is not constant and depends on the Pe values;
- values of h_b can significantly exceed the limit values of the thickness of the flat layer, which correspond to the existing recommended condition $z_L/h_b = 0.2$.

Thus, the *limiting thickness h_b* defines the *minimum* thickness of parts at which temperature conditions can be calculated using simple dependencies for a *semi-infinite body*.

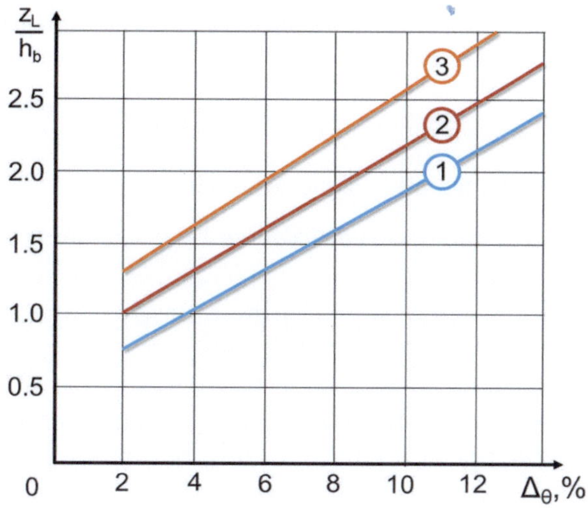

Fig. 3.4 Dependence of z_L/h_b on the permissible temperature deviation Δ_θ ($\theta_C = 0.512$): 1—at Pe = 1.555; 2—at Pe = 0.747; 3—at Pe = 0.404

3.4 Correlation of Laser Processing Parameters and Hardening Depth

Calculations of laser processing parameters that ensure a given hardening depth [6]

The ***purpose*** of the calculations is to obtain the parameters of the laser processing at the ***output***:

- intensity of the heat source W_{ph};
- radius of the focus spot r_{oh};
- speed of the laser beam v_h

with the required ***input*** parameter—the specified hardening depth z_h, by which the metal is heated to a temperature T_h, corresponding to the point A_3 of the iron-cementite diagram.

The ***initial equations*** for calculations:

(1) ***Intensity distribution*** W_p according to the selected scheme of a continuous heat source (Fig. 3.1). The simplest is to choose the scheme of a cylindrical heat source acting on a semi-infinite body:

$$W_p = AP/\left(\pi r_0^2\right) \tag{3.28}$$

where A is the absorption capacity of the material;

P is the power of the heat source;

r_0—effective radius of the heating spot.

(2) The *time* of the heat source acting t, which is equal to the time during which the beam passes its diameter (for a cylindrical heat source):

$$t = {}^{2r_0}/_{v} \tag{3.29}$$

where v is the speed of the heat source.

(3) The ***temperature distribution*** T(z,t) in a semi-infinite body along the z-axis with time t:

$$T(z,t) = \frac{W_p}{k} 2\sqrt{at}\left[\frac{1}{\pi} exp\left(-\frac{z^2}{2\sqrt{at}}\right) - \frac{z}{2\sqrt{at}}\Phi^*\left(\frac{z}{2\sqrt{at}}\right)\right] \tag{3.30}$$

where $\Phi = 1 - \Phi^*$ error function (probability integral).

If the heating depth $z \ll 2\sqrt{(at)}$, the expression (3.30) is significantly simplified:

$$T(z,t) = \frac{W_p}{k}\left[\sqrt{\frac{4at}{\pi}} - z\right] \tag{3.31}$$

The simplified dependence (3.31) allows one to calculate the temperature with an error of no more than 10% in relation to dependence (3.30).

It is convenient to ***calculate*** the ***parameters*** of the laser processing ***mode*** using the simplified dependence (3.31), for which the concept of *generalized speed* ξ is introduced:

$$\xi = {}^{vr_0}/(2a) \tag{3.32}$$

Further calculations are made in one of two ways:

(1) by introducing a correction factor γ;
(2) by introducing an auxiliary function F(ξ).

Method 1—calculating the processing parameters by introducing the *correction factor γ*, which is the ratio of dimensionless temperatures calculated according to the simplified (T_1) and full (T_2) dependence:

$$\gamma = \theta 1/\theta 2 \tag{3.33}$$

where $\Theta_1 = T_1 r_0 k/(AP_0)$; $\Theta_2 = T_2 r_0 k/(AP_0)$

The calculated dependence of the correction factor γ on the generalized speed ξ is shown in Fig. 3.5.

The required ***intensity of the heat source*** W_{ph}:

$$W_{ph} = {}^{k(T_m - T_h)}/_{z_h} \tag{3.34}$$

where T_m is the melting point;

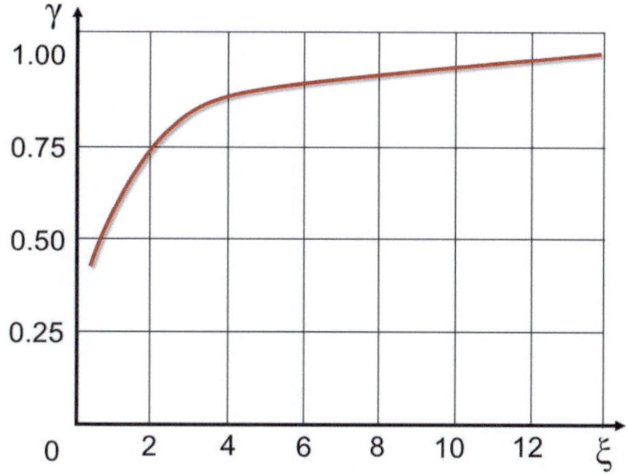

Fig. 3.5 Dependence of the correction factor γ on the generalized speed ξ

T_h is the temperature corresponding to point A_3 of the iron-cementite diagram; z_h—the specified hardening depth.

The ***radius of the focus spot*** r_{oh} for processing to a depth z_h is determined from expressions (3.28) and (3.31):

$$r_{0h} = \sqrt{\frac{\gamma A P_0 z_h}{\pi k (T_m - T_h)}} \tag{3.35}$$

The ***speed of the laser beam*** v_h is determined from expressions (3.29) (3.35) and (3.31):

$$v_h = \frac{2 r_{0h}}{t_h} = \frac{8a(T_m - T_h)^{3/2}}{\pi T_m^2 z_h^{3/2}} \sqrt{\frac{\gamma A P_0}{\pi k}} \tag{3.36}$$

Method 2—calculating the process parameters by introducing *an auxiliary function F(ξ)*:

$$F(\xi) = \frac{\xi}{\gamma(\xi)} = \frac{4 A P_0 (T_m - T_h)}{\pi^2 k z_h T_m^2} \tag{3.37}$$

A graphical representation of the auxiliary function F(ξ) is shown in Fig. 3.6.

In the calculations, the correction factor γ is introduced to determine the adjusted power P_a or the adjusted power density W_{Pa}:

$$P_a = \gamma \cdot A \cdot P_0;\ W_{P_a} = \gamma \cdot W_P \tag{3.38}$$

Taking into account Eqs. (3.42) and (3.35) for determining the ***temperature distribution*** on the beam axis will be as follows:

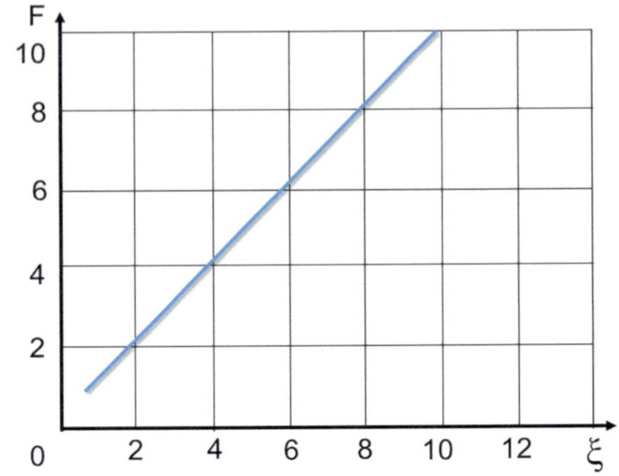

Fig. 3.6 Dependence of the auxiliary function F on the generalized speed ξ

$$T(z,t) = \frac{P_a}{\pi r_0^2 k}\left[\sqrt{\frac{4at}{\pi}} - z\right] \tag{3.39}$$

The maximum hardening depth will be reached when the material surface is heated to the melting point T_m. The specified power density that ensures the maximum hardening depth is determined by the expression:

$$W_{P_a} = T_m k\sqrt{\frac{\pi}{4at}} \tag{3.40}$$

The ***maximum depth of hardening*** on the beam axis when surface is heated without melting is determined from (3.39), (3.40) and has the form:

$$z_h = \sqrt{\frac{4at}{\pi}} - \frac{kT_h}{W_{P_a}} = \sqrt{\frac{4at}{\pi}} - \frac{T_m - T_h}{T_m} \tag{3.41}$$

The ***time of the heat source action*** is determined from Eq. (3.41):

$$t_h = \frac{\pi z_h^2}{4a} \cdot \frac{T_m^2}{(T_m - T_h)^2} \tag{3.42}$$

The ***width of the hardening zone*** b_h depends on the thermophysical properties of the material, irradiation time and intensity distribution; it can be larger or smaller than the focusing zone and is determined by the equation:

$$b_h = 2r_0\sqrt{[1 - (T_h/T_m)]} \tag{3.43}$$

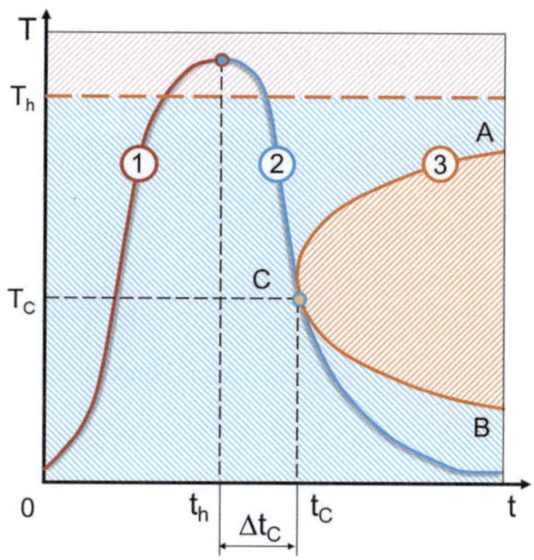

Fig. 3.7 Schematic of the conditions of steel hardening during laser surface treatment: 1—heating curve under laser irradiation; 2—cooling curve; 3—C-shaped austenite TTT curve

The width of the hardening zone is set in advance during design, so dependence (3.43) is usually used to calculate r_0.

Calculations of laser processing mode parameters that ensure maximum hardening depth

Two conditions must be met for the steel hardening process (Fig. 3.7):

(1) Heating (curve 1) to a *temperature above T_h* (critical point A_3) with the formation of austenite and holding to dissolve carbide phases existing at low temperatures.
(2) Cooling (curve 2) at a *rate* that *touches* or intersects the C-shaped time–temperature transformation (TTT) curve of austenite. The C-shaped curve delimits the region of the beginning (CA) and end (CB) of the austenite to martensite transformation.

The ***purpose*** of the calculations is to *determine* the parameters of the laser processing ***at the output***—the maximum laser irradiation time t_{hm}, which ensures cooling and hardening rates;

with the required ***input*** parameter—the intensity of the heat source W_p (energy capabilities of the equipment).

The temperature distribution in a semi-infinite body for time $t > t_h$ is determined by dependence (3.30).

The temperature change on the surface (at $z = 0$) can be written as follows:

$$T(\tau) = T_m\left(\sqrt{\tau + 1} - \sqrt{\tau}\right) \tag{3.44}$$

$$\tau = {\Delta t_C}/{t_h} \tag{3.45}$$

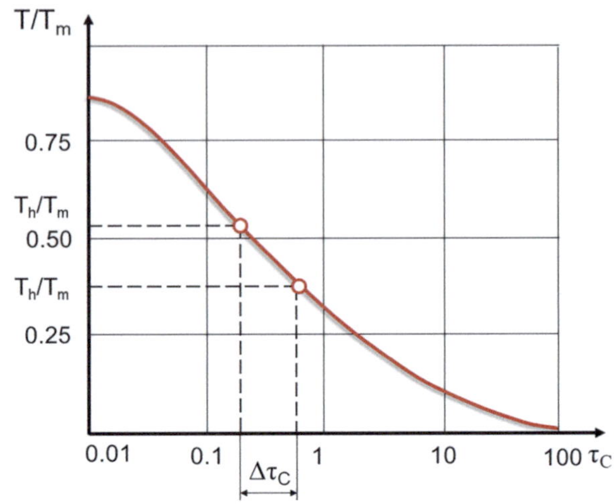

Fig. 3.8 Dependence of the normalized temperature (reduced to T_m) on the reduced cooling time

$$\Delta t_C = t_C - t_h \tag{3.46}$$

where τ is the reduced cooling time;

t_h is the heating time that ensures hardening during the cooling phase;

Δt_C is the cooling time to reaching a C-shaped austenite TTT curve (Fig. 3.7).

The critical value of $\Delta \tau_C$, at which hardening is ensured, is determined by the dependence of the normalized temperature (reduced to the melting point T_m) on the reduced cooling time (Fig. 3.8).

Let's determine the maximum laser irradiation time t_{hm}, which ensures cooling rates with hardening:

$$t_{hm} = {}^{\Delta t_C}/_{\Delta \tau_C} \tag{3.47}$$

Thus, the maximum laser irradiation time t_{hm}, which ensures cooling rates with quenching, is equal to the minimum time that ensures supercooling of austenite before the onset of martensitic transformation.

3.5 Calculations of Thermal Processes During Heating by a Laser Beam Scanning the Surface

A *homogeneous surface* heat source must be used to obtain a uniform hardened layer with no defects. A laser beam, whether single-mode or multimode, cannot provide uniform heating of the surface.

A homogeneous distribution of the heat source intensity without changing the beam parameters composition ensures *scanning* of the surface with laser beam.

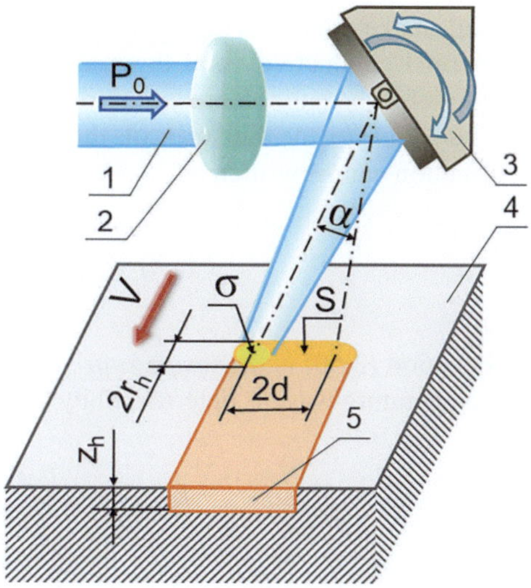

Fig. 3.9 Schematic diagram of heating by a laser beam scanning the surface of a product: 1—laser beam; 2—focusing system; 3—mirror scanner; 4—product; 5—cross section of the hardened zone

Scanning complicates the calculation of laser radiation modes. It is advisable to transit from the three-dimensional problem of heating by a rapidly moving heat source to a *one-dimensional* thermal problem [6]. The calculation scheme of heating by a laser beam scanning the product's surface is shown in Fig. 3.9.

A laser beam (Fig. 3.9) with radiation power P_0 and cross-sectional area $\sigma = \pi r_0^2$ is scanned with a frequency f at a deflection angle α over an area S. The product moves across the scan path at a speed V.

To calculate the temperatures, instead of the scheme of heating by continuous laser radiation, we will adopt a *simplified* scheme of heating *by pulsed-periodic* laser radiation of uniform intensity I(t) of the *zone S.*

The action of the heat source can be represented as a sequence of rectangular pulses, the squareness of which is determined by the ratio of the heated area S to the area of the focusing area σ:

$$G = {}^{S}/_{\sigma} \tag{3.48}$$

Such a heat source can be represented as a Fourier series:

$$\begin{aligned} I(t) &= I_0\left[1 + \sum_{k=1}^{a} \xi_k \cos(2f\pi kt)\right] \\ &= \frac{q_0}{G}\left\{1 + \sum_{k=1}^{\infty}\left[\frac{2G}{\pi k}\sin\left(\frac{\pi k}{G}\right)\right]\cos(2f\pi kt)\right\} \end{aligned} \tag{3.49}$$

Let's make the following *assumptions*: the absorption capacity A, the coefficient of thermal conductivity and thermal conductivity are *constants* and are taken as *averages* for the entire temperature range under consideration.

Then, due to the linearity of the heat conduction equation, the temperature distribution can be written as:

$$T(z,t) = T_0(z,t) + \sum_{k=1}^{\infty} T_k(z,t) \quad (3.50)$$

Equation (3.50) is the *temperature wave* of a pulse. The constant component of the temperature wave is determined by the dependence:

$$T_0(z,t) = \frac{I_0\gamma}{k}\left[\sqrt{\frac{4at}{\pi}} - z\right] \quad (3.51)$$

where γ is the dependence of the fraction of energy directed deep into the material on the dimensionless parameter $\xi = Vr/(2a)$ and the shape of the focusing zone.

In Fig. 3.10 shows the values of γ for heat sources with different intensity distribution shapes depending on ξ.

For a harmonic heat source, the established ***temperature distribution*** for time t > 1/f takes the form of a temperature wave:

$$T_k(z,t) = \frac{I_0\xi_k}{k}\sqrt{\frac{a}{2f\pi k}}\exp\left(-z\sqrt{\frac{f\pi k}{a}}\right)\cos\left(2f\pi kt - \frac{\pi}{4} - z\sqrt{\frac{f\pi k}{a}}\right) \quad (3.52)$$

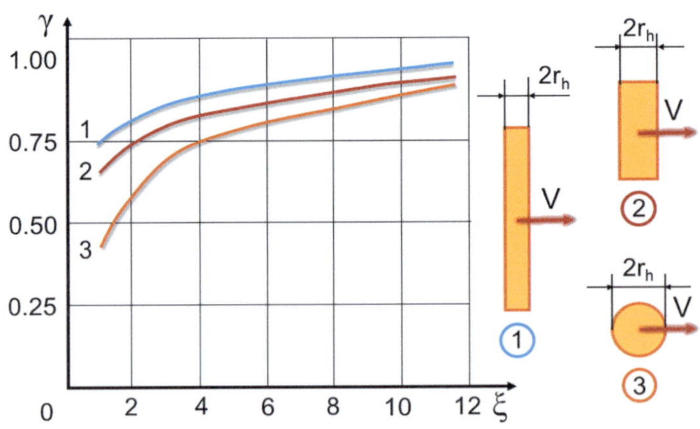

Fig. 3.10 The fraction of thermal energy γ that propagates in the z-axis direction deep into the material, depending on the parameter ξ: 1—a surface heat source in the form of a strip with a length/width ratio > 10; 2—a surface heat source in the form of a rectangle with a length/width ratio of 3–5; 3—a surface heat source in the form of a circle

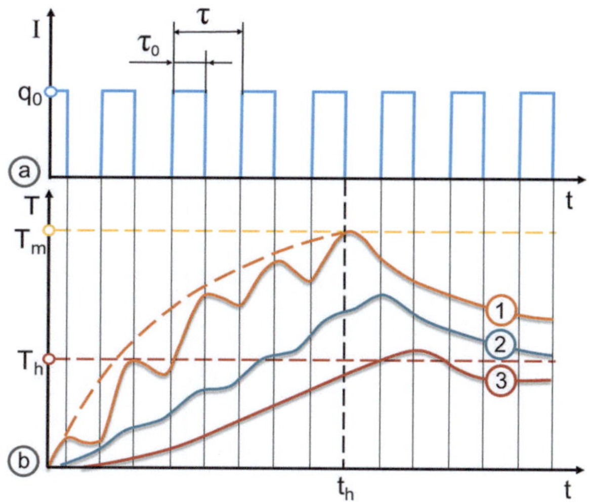

Fig. 3.11 Time variation t of the characteristics of an equivalent pulsed-periodic heat source: **a**—intensity of I pulses when scanning with a laser beam (pulse period $\tau = 1/f$; pulse duration $\tau_0 = \tau/G$; pulse amplitude $q_0 = A/\pi r_h^2$); **b**—temperature T cycles (temperature waves) during heating (1—on the surface of the material; 2—in the middle of the hardened zone at $z = 0.2$ mm; 3—at the boundary of the hardened zone at $z = z_h$)

Due to the high speed of the beam scanning the surface, the coefficient ξ is quite large, so $\gamma \to 1$ (Fig. 3.10).

The appearance of temperature waves at different depths of the hardened surface is shown in the graphs (Fig. 3.11).

Minimum allowable scanning frequency

To obtain a stable hardening depth along the hardening zone (highlighted in yellow in Fig. 3.9), it is necessary that the scanning frequency f is not lower than the *critical value* f_c, at which the temperature wave is completely attenuated at the boundary of the hardening zone (at $z = z_h$—curve 3 in Fig. 3.11).

The *criterion for the attenuation* of a temperature wave is that the ratio of the amplitude of the first harmonic of the series to the temperature formed by the constant component of the heat source is less than a given value ε:

$$\frac{\frac{2G}{\pi}\sin\left(\frac{\pi}{G}\right)\sqrt{\frac{a}{2f_c\pi}}\exp\left(-z_h\sqrt{\frac{f_c\pi}{a}}\right)}{\frac{4at}{\pi}-z} < \varepsilon \tag{3.53}$$

Examples of calculations using Eq. (3.53) of the dependence of the minimum allowable scanning frequency f_c on the hardening depth z_h of carbon steel (curve 1) and gray cast iron (curve 2) are shown in Fig. 3.12.

The ***maximum*** heating ***temperature*** when exposed to a scanning heat source is determined by the following relationship:

$$T_{max}(t) = \frac{\gamma I_0}{k}\sqrt{\frac{4at}{k}} + \frac{I_0}{k}\sum_{k=1}^{\infty}\left[\frac{2G}{\pi k}\sin\left(\frac{\pi k}{G}\right)\sqrt{\frac{a}{2f\pi k}}\right] \tag{3.54}$$

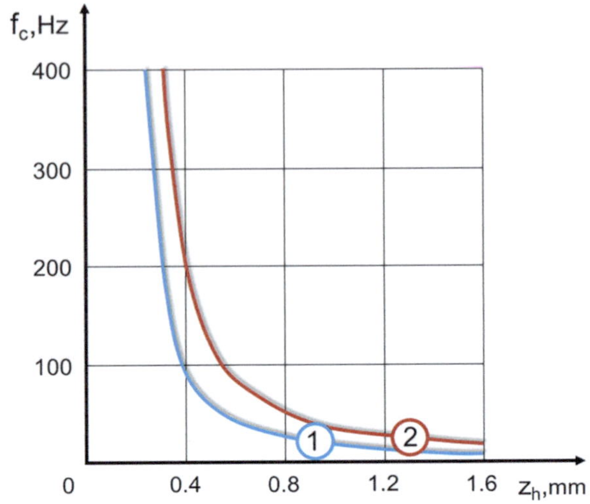

Fig. 3.12 Minimum permissible scanning frequency fc when hardening iron-carbon alloys to a given depth z_h ($G = 1$; $\varepsilon = 0.01$): 1—carbon steel; 2—gray cast iron

The scanning hardening mode should ensure the required depth of the hardened layer z_h without melting the surface, i.e. the following conditions must be met simultaneously:

$$T_{\max} = T_m(\text{at}z = 0);\, T = T_h(\text{at}z = z_h) \tag{3.55}$$

To use the dependence of temperature on the action of the constant component, the scanning frequency must be selected in accordance with the criterion ε (3.53), or the calculated dependence $f_c(z_h)$ in (Fig. 3.12) for a given quenching depth z_h. The quenching temperature T_h, which takes into account high heating rates and provides the required intensity of carbon diffusion in the lattice, is determined by dependence (3.1).

The ***operating time of a scanning heat source*** that provides cooling rates with quenching is determined by a relationship:

$$t_h = \frac{\pi}{4a}\left\{\frac{T_m}{T_m - T_h} z_h + \frac{T_h}{\gamma(T_m - T_h)}\sum_{k=1}^{\infty}\left[\frac{2G}{\pi k}\sin\left(\frac{\pi k}{G}\right)\sqrt{\frac{a}{2f\pi k}}\right]\right\}^2 \tag{3.56}$$

The ***intensity of the heat source*** is determined by the dependence:

$$I_0 = \frac{k(T_m - T_h)}{\gamma z_h + \sum_{k=1}^{\infty}\left[\frac{2G}{\pi k}\sin\left(\frac{\pi k}{G}\right)\sqrt{\frac{a}{2f\pi k}}\right]} \tag{3.57}$$

Dependencies (3.56) (3.57) with increasing scanning frequency ($f \to \infty$) and decreasing scanning amplitude ($G \to 1$) take the form for a continuous moving heat source.

The ***effective radius*** r_0 of a heating spot during scanning is given by the following relationship:

$$I_0 = {}^{AP_0}/_S = {}^{AP_0}/\left(G\pi r_0^2\right) \tag{3.58}$$

The ***speed*** V of the product across the scan path:

$$V = {}^{2r_0}/_{t_h} \tag{3.59}$$

Scan amplitude d:

$$d = r_0 \cdot (G - 1) \tag{3.60}$$

Productivity E is the criterion of the laser processing process and is determined as:

$$E = (2r_0 + 2d) \cdot V = {}^{4r_0^2 G}/_{t_h} \tag{3.61}$$

References

1. J.M. Dowden, The mathematics of thermal modeling. An introduction to the theory of laser material processing. Sub. Eng. Technol. Math. Stat. Physic. Sci. 286, eBook ISBN 9781032684758 (2024). https://doi.org/10.1201/9781032684758
2. G. Abilsiitov, V.S. Golubev, V.G. Gontar et al., in *Technological Lasers: Reference Book*, ed. by G.A. Abilsiitov, Calculation, Design, and Operation/[in Russian], vol. 1 (Moscow, Mashinostroenie, 1991), p. 432, ISBN 5–217–01269–2
3. C. Liu, Z. Zhang, S. Sun, Modeling of thermal behavior and microstructure evolution during laser cladding of AlSi10Mg alloys. Opt. Laser Technol. **123**, 105926 (2020). https://doi.org/10.1016/j.optlastec.2019.105926
4. L. Linjie, C. Quanwei, Z. Jianxing, S. Wenlei, L. Zhicheng, S. Haoran, Research on the temperature rise mechanism of ultrasonic field-assisted laser cladding. Int. J. Mater. Form **17**, 56 (2024). https://doi.org/10.1007/s12289-024-01859-3
5. A. Piccininni, G. Palumbo, Numerical modelling of the annealing determined by short-term laser treatment using a physical simulation-based approach. CIRP J. Manufact. Sci. Technol. **45**, 210–224 (2023). https://doi.org/10.1016/j.cirpj.2023.07.006
6. V. Kovalenko, L. Golovko, V. Chernenko, Hardening and alloying of machine parts with a laser beam. Kyiv: "Technica" 192 (1990)
7. V. Kovalenko, A. Verhoturov, L. Golovko, I. Podchernyaeva, Laser surface hardening and electric-sparc surface hardening of materials. Plenum Publish. Corp., p. 175 (1988)
8. V. Kovalenko, J. Shvets, L. Golovko, N. Fialko, A. Krasavin, G. Sherenkovskaya, N. Meranova, On the influence of the size factor on the process of hardening parts by continuous laser radiation. Electron. Mater. Process. **1**, 23–25 (1988)
9. V. Kovalenko, L. Golovko, N. Fialko, in *Laser Hardening of Components and Tools with Complicated Working Elements Shaped for Heat Transfer* (ICALEO'93 Orlando, Florida, 1993)

Chapter 4
Structural-Phase Transformations in Surface Layers During Laser Treatment

Abstract The features of the formation of the microstructure of characteristic surface layers of remelted metal and the heat-affected zone during laser treatment of the main groups of metals are analyzed: technical iron, carbon unalloyed steels, alloyed steels, cast irons, titanium alloys, aluminum alloys, copper alloys. For each group of metals, graphs are provided and the distribution of microhardness by depth in the initial state, after initial normalization or hardening and subsequent laser treatment at different modes, is analyzed. The mechanisms for increasing the microhardness of metals after surface laser treatment are revealed. The directions for controlling the microstructure, microhardness, and their distribution in the surface layer by adjusting the energy density and the speed of laser radiation movement are shown.

4.1 Industrial Iron and Carbon Unalloyed Steels

One of the key factors that determines the results of surface laser treatment is the structural and phase transformations that occur in the upper layers of the material when it is heated by a laser beam and cooled by heat transferring to the lower, unheated layers [1]. The resulting material structure depends on treatment process parameters, such as laser beam power, focusing area size, beam movement speed over the workpiece surface, and related conditions [2]. While the study of structural and phase changes in modeling these processes using software packages is helpful [3], the experimental study of this issue is also important. The results of this study are presented in this section. It is also worth mentioning the possibility of combining laser radiation with other influences, which enhances the effect of laser processing and achieves the desired surface layer structure [4].

Industrial iron under laser irradiation with both pulsed and continuous radiation with melting is characterized by the formation of two layers with an interface along the fusion line:

(1) The ***remelted layer*** has a structure formed as a result of directional crystallization of the melt and contains columnar crystals—*dendrites*. The main axes of the dendrites are directed perpendicular to the interface—the fusion line. The

I. Krivtsun et al., *Laser and Hybrid Technologies for Surface Engineering and Welding*, Springer Series in Materials Science 358, https://doi.org/10.1007/978-3-032-15038-7_4

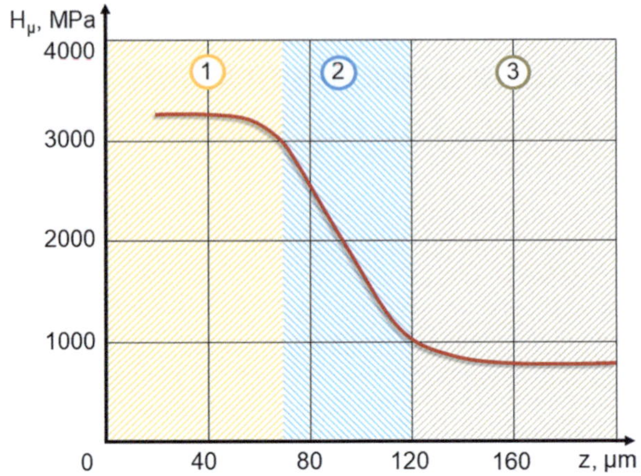

Fig. 4.1 Distribution of microhardness H_μ along the depth z of the heat affected zone during laser irradiation of industrial iron: 1—melted layer; 2—heat affected layer; 3—base metal

microhardness is 2800–3200 MPa, which is significantly higher than that of the original ferrite structure (800–1000 MPa) (Fig. 4.1).

The significant increase in hardness is due to two factors:

- *auto hardening* with the formation of low-carbon rail martensite due to ultra-fast cooling;
- austenite phase *embrittlement* caused by the volume effect of martensitic transformation.
- The ***thermal impact layer*** is formed where the heating temperature has exceeded the upper *critical point* A_3, but has not exceeded the *melting point*.

The second layer is where the following takes place:

- Complete phase *recrystallization* with grain size reduction to 3–20 microns (compared to 80–100 microns in the original structure).
- *Auto-hardening* with increased hardness. The hardness depends on the laser heating temperature: the higher the temperature, the more complete the phase transformation and the higher the hardness. For this reason, the hardness decreases monotonically with the depth of the heat-affected layer (zone 2 in Fig. 4.1).

Carbon hypo-eutectoid steels (with a carbon content of $0.02\% < C < 0.8\%$ and a primary ferritic-pearlite structure), which have been pre-annealed or normalized, form layers similar to industrial iron when laser heated.

Features of the structure and properties of layers in hypo-eutectoid steels:

(1) The ***melted layer is*** characterized by *increased hardness*—up to 6500 MPa for C22 steel (C content = 0.17–0.24%) and 7500–8500 MPa for C45 steel (C content = 0.42–0.50%). The increase in hardness is due to the carbon content: the more carbon in the steel, the higher the hardness of martensite and the greater

the degree of its tetragonality. Liquid-phase quenching under laser irradiation fixes all the carbon in the α-phase crystal lattice. The resulting structure is characterized by fine grain and uniform properties, despite the small amount of carbon.

(2) In the ***heat-affected layer***, the *pearlite component* of the structure (Fig. 4.2a) turns into austenite, and with further auto hardening it turns into martensite. The *ferrite component* is either completely preserved (at temperatures not much higher than the A_1 point) or partially transformed into austenite (without further homogenization due to the short duration of the laser pulse). Austenite formed from ferrite has a lower carbon content compared to austenite formed from pearlite.

In order to form a *homogeneous fine-grained structure of high hardness* during laser surface treatment of hypo-eutectoid steels, it is necessary to:

- carry out laser heating to temperatures close to, but *not exceeding* the melting point;
- increase the *holding time* at the heating temperature.

Technological recommendations for laser processing to ensure a homogeneous structure in the hardened layer:

- increased duration of pulses;
- increase in intensity in the first third of the pulse;
- use of a laser beam that scans the surface.

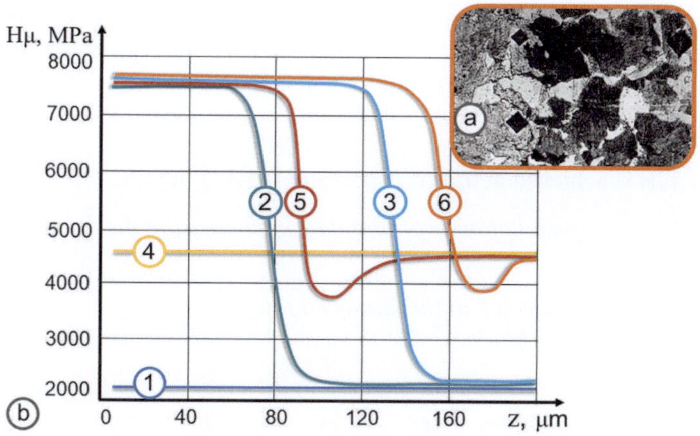

Fig. 4.2 Change of properties in the surface layer of steel type C45 (C content = 0.42–0.50%) under laser irradiation: **a**—microstructure of the laser heating zone (×500); **b**—distribution of microhardness H_μ along the depth z (1—initial state after normalization; 2—after initial normalization and subsequent laser treatment at $W_E = 2$ J/mm^2; 3—after initial normalization and subsequent laser treatment at $W_E = 5$ J/mm^2; 4—initial state after quenching; 5—after initial quenching and subsequent laser treatment at $W_E = 2$ J/mm^2; 6—after initial quenching and subsequent laser treatment at $W_E = 5$ J/mm^2)

Localization and clear boundaries are important advantages of laser processing. The phase transformation boundary can run along the middle of a pearlite grain. In this case, martensite is formed in the part of the grain where the temperature exceeded the A_1 point, and pearlite remains unchanged in the part of the grain where the temperature was lower. Therefore, the laser heating zone in annealed (normalized) hypo-eutectoid steels has a clear lower boundary.

Heat treatment of carbon hypo-eutectoid steels (bulk quenching, quenching and tempering) before laser irradiation promotes *high-speed tempering*. Tempering occurs where the temperature of laser irradiation exceeds the temperature of pre-tempering but is less than the critical point A_1 [5]. Due to the short duration of laser heating, the decomposition product of martensite in the laser tempering layer is a highly dispersed mixture of ferrite and cementite—tempering *troostite*. The hardness of the high-speed laser tempering layer is 4000–4900 MPa.

Laser treatment of *pre-hardened* steels of type C45 (C content = 0.42–0.50%) in comparison with previously normalized steels gives:

- a *homogeneous structure* of the heating zone in the temperature range between the melting point and the lower critical point A_1;
- *increased depth of hardening* by 20–30% (Fig. 4.2b).

The ***irradiation time*** can be easily adjusted in a wide range and allows one to control changes in the properties of the surface layer. *Increasing* the irradiation time (by reducing the speed of the laser beam):

- *Increases the depth* of the hardened layer. For example, at a radiation power of 1000 W, a laser beam movement speed of 0.8 m/min, a focused beam diameter of 4 mm, transformations occur in the solid phase and the depth of the hardened layer reaches 1 mm (compared to 0.15 mm when treated with pulsed radiation in a similar mode with a pulse duration of 4–6 ms).
- Ensures *uniformity of microhardness* distribution due to *complete homogenization of austenite* from primary ferrite. For example, in C22 steels (C content = 0.17–0.24%), this is achieved at traverse speeds of 0.5–0.8 m/min.

NOTE: Reducing theirradiation exposure time by increasing the scanning speed to 2–3 m/min results in:

- *reduction of the total depth of the hardened layer*
- *reduction of structure hardness due to incomplete homogenization of austenite.*

Figures 4.3 and 4.4 illustrate an increase in the *depth* of the hardened layer (up to 0.5 mm and above) and an *increase* in the zone of *homogeneous microhardness* with an increase in the laser irradiation time (by reducing the laser beam speed from 0.07 m/s—curves 1 to 0.01 m/s—curves 4). Laser irradiation provides:

- Hardness at the level of 6000–7000 MPa for steel type C22 (C content = 0.17–0.24%)—Fig. 4.3. During conventional quenching (heating in an oven and cooling in oil), this steel does not harden.

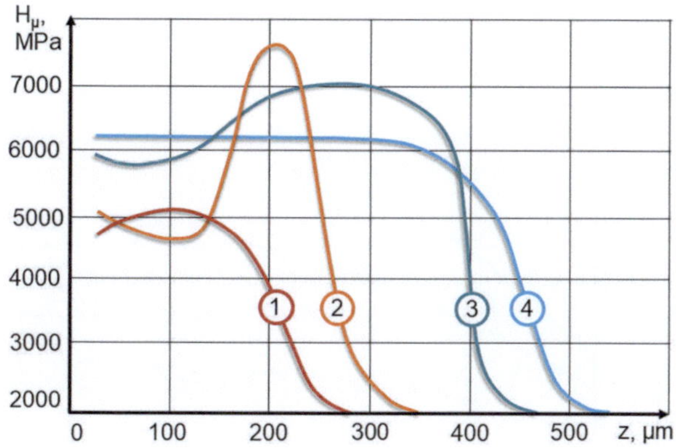

Fig. 4.3 Distribution of microhardness H_{μ} along the depth z of the heat-affected zone of C22 steel (C content = 0.17–0.24%) at different laser beam speed values v: 1—at v = 0.07 m/s; 2—at v = 0.05 m/s; 3—at v = 0.03 m/s; 4—at v = 0.01 m/s

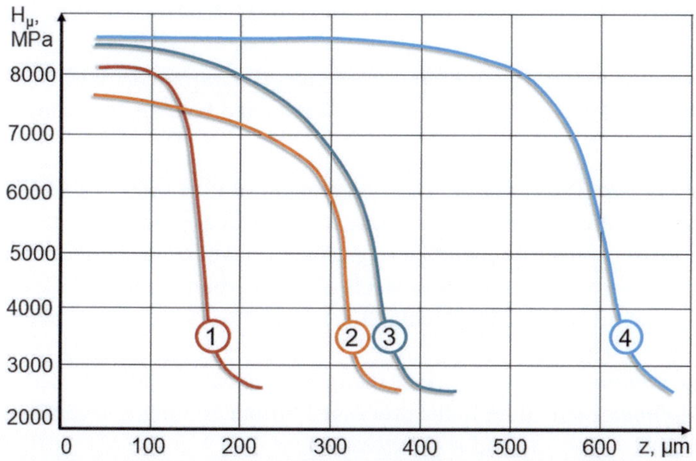

Fig. 4.4 Distribution of microhardness H_{μ} over the depth z of the heat-affected zone of C45 steel (C content = 0.42–0.50%) at different laser beam speed values v: 1—at v = 0.07 m/s; 2—at v = 0.05 m/s; 3—at v = 0.03 m/s; 4—at v = 0.01 m/s

- Hardness at the level of 8000–9000 MPa of steel type C45 (C content = 0.42–0.50%)—Fig. 4.4. Conventional quenching (heating in an oven and cooling in oil) does not produce such a high hardness.

The *mechanism of exceeding the hardness* of steels after laser quenching compared to conventional quenching:

- significant *grinding of ferrite and austenite grains*, as well as *martensite* rails, batches and plates;

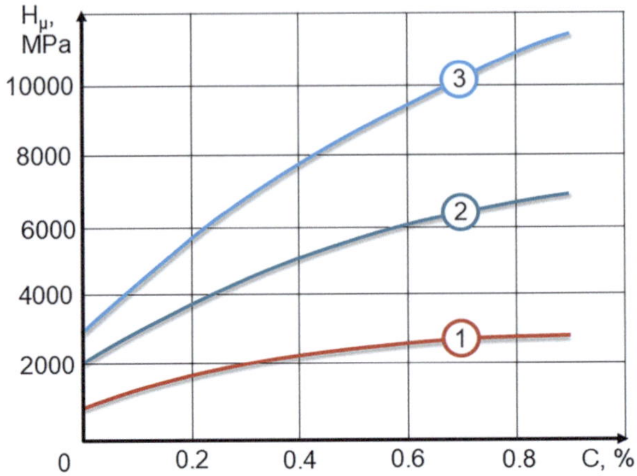

Fig. 4.5 Variation of microhardness H_μ of hypo-eutectoid steels depending on the carbon content C: 1—without heat treatment; 2—conventional quenching (heating in an oven and cooling in oil); 3—laser quenching

- increase in the proportion of *high-carbon lamellar martensite* with high hardness (Fig. 4.5) instead of low-carbon phases (ferrite and batch martensite) with lower hardness.

Carbon eutectoid steels (with a carbon content of C = 0.8%) of C80W type are characterized by the presence of only one lower critical point A_1. The equilibrium structure consists of pearlite (lamellar or granular) and contains neither excess ferrite (which is typical for hypo-eutectoid steels with a carbon content of C < 0.8%) nor secondary cementite (which is typical for hypereutectoid steels with a carbon content of 0.8% < C < 2.0%).

The ***laser heating*** zone with surface melting in annealed C80W steel has a primary *dendritic-columnar* structure. In the process of further cooling, a secondary structure is formed from *martensite* and *residual austenite*.

After laser hardening, C80W steels contain a significant amount of residual austenite—from 5 to 70% [6]. At the same time, the *microhardness* of the structure is *high* and reaches 11,000–12,000 MPa. Compensation for the low hardness of residual austenite is explained by:

- austenite phase carburization as a result of the bulk effect of martensitic transformation;
- saturation of residual austenite with lattice defects;
- high density and complex configuration of dislocations.

Carbon hypereutectoid steels (with a carbon content of 0.8% < C < 2.0%, with a structure of pearlite and secondary cementite) of C105W1 (C content = 0.96–1.00%) or 1.1654 (C content = 1.1–1.3%) types retain their structure in the annealed state after laser quenching, i.e., pearlite grains and a fine network of secondary cementite.

During auto-hardening from the liquid state, these steels form a primary *dendritic columnar* structure in the process of high-speed crystallization, followed by a secondary structure—*lamellar high-carbon martensite* and *residual austenite* in a significant amount of 30–48%.

The *microhardness* of the structure is 11000–13,000 MPa, which is much better than that of bulk hardening. The density of dislocations in the heating zone is 5–6 times higher than in the original steel structure after bulk quenching and tempering.

In the heating zone, in the temperature range between the *melting point* and the critical *point* A_m in the surface layer, the following occurs:

- transformation of pearlite into *austenite*,
- dissolution of *secondary cementite* in *austenite*.

The austenite grains that were formed at the site of pearlite and the austenite grains in which cementite dissolved contain different amounts of carbon. This is reflected in the amount of martensite and the uneven distribution of hardness in the structure (Fig. 4.6).

In the heating zone, in the temperature range between the critical *point* A_{Cm} and the critical *point* A_1, the secondary cementite dissolves only partially. As it moves away from the surface, the following is observed:

- Reduction in the thickness of light carbon-enriched areas containing an increased amount of residual austenite;

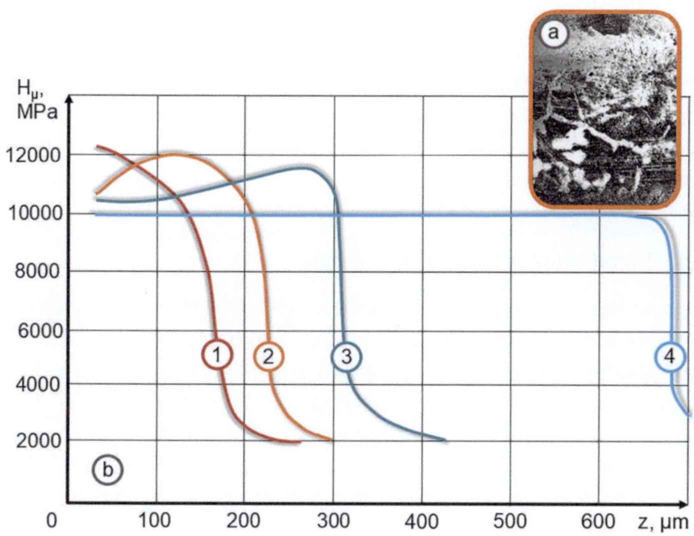

Fig. 4.6 Changes in properties in the surface layer of steel type 1.1654 (C content = 1.1–1.3%) under laser irradiation: **a**—microstructure of the laser heating zone (×500); **b**—distribution of microhardness H_μ along the depth z (1—at v = 0.07 m/s; 2—at v = 0.05 m/s; 3—at v = 0.03 m/s; 4—at v = 0.01 m/s)

- Gradual transition of residual austenite into a network of secondary cementite that did not dissolve during laser heating;
- Reduction of hardness.

A *high-speed laser tempering* layer is formed in the *heating zone* in the temperature range between the critical *point A_1* and the pre-tempering temperature. The layer has a highly dispersed ferrite-cement structure of the troostite type. The microhardness is 4000–4500 MPa [7].

To increase the impact toughness and fatigue strength, tempering is performed after quenching.

Tempering is a type of heat treatment that involves heating hardened steels to temperatures not exceeding the austenite formation temperature (A_1), holding them at these temperatures to transform quenching martensite into equilibrium structures of tempering sorbite and tempering troostite (which contain cementite in a granular form), and then cooling them.

Tempering of carbon steels after laser hardening:

(a) at a temperature of 100 °C (low tempering)—does not cause changes in the heating zones;
(b) at a temperature of 200 °C (low tempering)—leads to a noticeable decomposition of martensite and residual austenite, accompanied by a decrease in hardness;
(c) at a *temperature of 300–400 °C*—leads to *complete decomposition of martensite* and residual austenite;
(d) at temperatures of 500–600 °C (high tempering):
   - in hypo-eutectoid steels of C22 (C content = 0.17–0.24%) and C45 (C content = 0.42–0.50%) types—grain enlargement in the ferrite matrix, structure approaches equilibrium, and dislocation density decreases;
   - in hypereutectoid steels of the C105W1 type (C content = 0.96–1.00%), structural changes extend to the carbide phase, i.e., cementite coagulation occurs.

Features of tempering after laser quenching:

- at any tempering temperature in the range of 100–600 °C, the laser hardened zones have a higher hardness than that of conventional hardening;
- At any tempering temperature in C105W1 steel (C content = 0.96–1.00%), after initial hardening and subsequent laser treatment, the hardness of the laser heating zone is higher than that of the original hardened structure;
- initial quenching followed by laser treatment of C105W1 steel (C content = 0.96–1.00%) provides better characteristics after tempering than initial annealing.

4.2 Alloy Steels

Chromium-silicon steels of the 90CrSi type (composition: C = 0.9%; Cr = 0.95–1.25%; Si = 1.2–1.6%) are deep-hardening tool steels:

- in the normalized state have a structure of granular *pearlite* and *secondary cementite*;
- in the hardened state have a *martensitic* structure with more residual austenite than carbon steels.

Due to its increased heat resistance, 90CrSi steel has better tool cutting properties than carbon, chromium and chrome-manganese-tungsten steels.

Critical energy density range—the lower and upper critical energy density levels at which a hardened layer is formed and the surface is not melted.

The critical range of energy density for 90CrSi steel is shifted towards higher values ($1.8 < W_E < 3.7$ J/mm^2) compared to carbon steels.

The ***heat-affected zone*** during laser irradiation of 90CrSi steel (Fig. 4.7) has a depth of 80–160 μm and consists of two layers:

(a) The *upper surface* layer is a hardened white layer with a martensitic-austenitic structure and a microhardness of 11,000–12,000 MPa. At energy densities above 3.7 J/mm^2 (upper critical level), the surface is heated above the melting point. A melt layer is formed on the surface, which has a characteristic dendritic structure of high microhardness (curve 3)

(b) The *lower transitional* layer is a high-speed tempering layer, as indicated by the "dips" in curves 2 and 3.

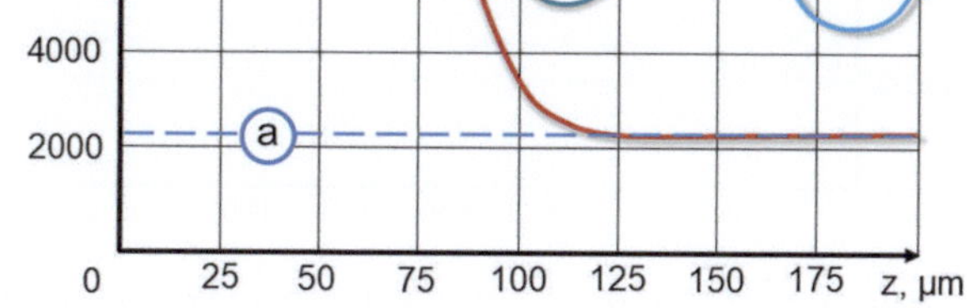

Fig. 4.7 Distribution of microhardness H_μ along the depth z of the heat affected zone of 90CrSi steel at different densities energy W_E of pulsed laser radiation: 1—after initial normalization and subsequent laser treatment at $W_E = 2$ J/mm^2; 2—after initial quenching and subsequent laser treatment at $W_E = 2$ J/mm^2; 3—after initial quenching and subsequent laser treatment at $W_E = 4$ J/mm^2 (**a**—level of microhardness in the initial state after normalization; **b**—level of microhardness in the initial state after quenching)

Chromium-manganese-tungsten steels of the 107WCr5 type (composition: C = 1.25%; Cr = 1.5%; Mn = 0.95%; W = 1.4%) are tool steels with high hardenability. When annealed, they have a *pearlite-cementite* structure.

Due to the presence of tungsten in the steel, the lower and upper critical energy density levels for this steel are shifted towards lower values and are equal to: 0.75 < W_E < 3 J/mm^2—for pre-normalized steel and 1.7 < W_E < 3.7—for pre-hardened steel.

The *energy density* is decisive in the formation of the structure of the heat-affected zone under both *pulsed* and *continuous* laser radiation. However, the nature of the formation of structures in pulsed and continuous modes has its own characteristics.

During ***pulsed*** laser irradiation of 107WCr5 steels, an increase in *energy density* changes the *structure* of the heat-affected zone.

The ***heat-affected zone*** of 107WCr5 steel under *pulsed* laser irradiation with energy density ***W_E < 3 J/mm²*** (curves 1 and 3 in Fig. 4.8) consists of *two layers*:

(a) The *upper surface* "white" hardened layer has a martensitic-austenitic fine structure with evenly distributed carbides. The structures of the white hardened layers in pre-normalized and pre-tempered steels do not differ. The microhardness is evenly distributed and reaches 10,000 MPa. The depth of the eat-affected zone of the pre-hardened steel is greater than that of the pre-normalized steel under the same laser irradiation conditions.
(b) The *lower transition* layer, the microhardness of which decreases to its original values. In pre-hardened steel, the transition layer undergoes rapid tempering (curve 3).

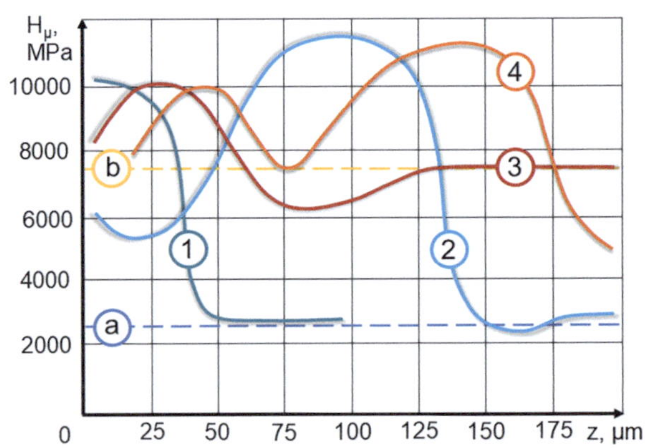

Fig. 4.8 Distribution of microhardness H_μ along the depth z of the heat affected zone of 107WCr5 steel at different energy densities W_E of pulsed laser radiation: 1—after initial normalization and subsequent laser treatment at W_E = 2.5 J/mm^2; 2—after initial normalization and subsequent laser treatment at W_E = 4 J/mm^2; 3—after initial hardening and subsequent laser treatment at W_E = 2.5 J/mm^2; 4—after initial quenching and subsequent laser treatment at W_E = 4 J/mm^2 (**a**—microhardness level in the initial state after normalization; **b**—microhardness level in the initial state after quenching)

The ***heat affected zone*** of 107WCr5 steel under laser *pulse* irradiation with energy density $\boldsymbol{W_E > 3.7\ J/mm^2}$ (curves 2 and 4 in Fig. 4.8) consists of *three layers*:

(a) The *upper surface* layer is formed from molten metal and has a dendritic martensite-austenite structure. Carbides are practically absent, since the heating temperature exceeded the A_m point. The microhardness of the structure of this layer is about 6000 MPa (curve 2 in Fig. 4.8).
(b) The *intermediate* layer contains martensite, residual austenite, and carbides. The carbides did not dissolve in the austenite because the temperature did not exceed the temperature of the A_m point. The number and size of carbides increased to the limit of those of the original structure. The microhardness increased to 11,500–12,000 MPa.
(c) The *lower transitional* layer is the layer whose microhardness decreases to its original values. In pre-hardened steel, the transition layer undergoes rapid tempering, as indicated by the "dips" in curves 3 and 4.

During ***continuous*** laser irradiation of 107WCr5 steels, an increase in *energy density* significantly *increases the depth* of the heat-affected zone. Figure 4.9 shows that an increase in energy density from $W_E = 5\ J/mm^2$ (curve 1) to $W_E = 40\ J/mm^2$ (curve 4) increases the depth of the heat affected zone by 10 times—up to 500 μm.

When the energy density increases the following is *not observed*:

- changes in the maximum microhardness level;
- reduction of microhardness in the upper surface layer (in contrast to pulsed radiation—curves 2 and 4 in Fig. 4.8).

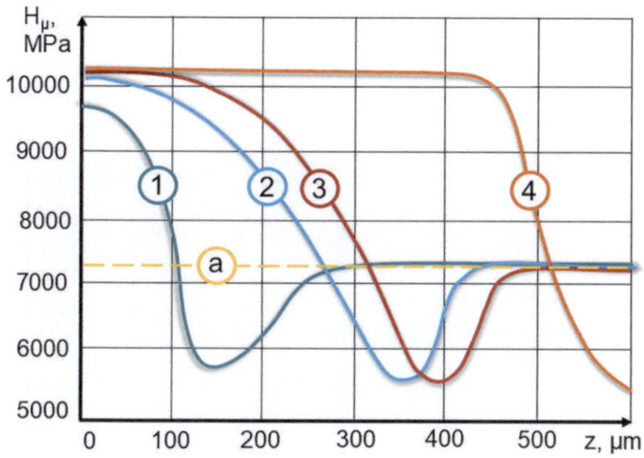

Fig. 4.9 Distribution of microhardness H_μ along the depth z of the heat affected zone of 107WCr5 steel at different energy densities W_E of continuous laser radiation: 1—at $W_E = 5\ J/mm^2$; 2—at $W_E = 16\ J/mm^2$; 3—at $W_E = 20\ J/mm^2$; 4—at $W_E = 40\ J/mm^2$ (a—microhardness level in the initial state after quenching)

The composition of the ***heat affected zone*** of 107WCr5 steel under laser continuous irradiation does not significantly depend on the energy density and consists of *two layers*:

(a) The *upper surface* layer, where the laser heating temperature exceeded the melting point, has clear boundaries. The structure of the layer is formed from the liquid state and consists of dendritic crystals oriented normal to the fusion line. Below, one can see a martensitic-carbide-austenitic structure formed during laser hardening. This explains the increase in the depth of the heat affected zone. The microhardness is evenly distributed and reaches 9600–10,200 MPa.

(b) The *lower transitional layer* is the layer whose microhardness decreases to its original values. In pre-tempered steel, high-speed tempering occurs in the transition layer (curves 1, 2, 3 in Fig. 4.9).

The effect of the laser beam travel speed on the depth of the heat affected zone and on the microhardness for steels of the 107WCr5 type (Fig. 4.10) is the same as that of carbon steels.

Chromium tool steels of the 100Cr6 type (composition: C = 0.95–1.05%; Si = 0.17–0.37%; Mn = 0.20–0.40%; Cr = 1.30–1.65%) of increased hardenability in the initial annealed state have a structure of granular pearlite and evenly distributed carbides.

The ***patterns*** of formation of heat-affected zones of 100Cr6 steels under pulsed and continuous laser irradiation are *similar* to those of 107WCr5 steels:

(a) Under *pulsed* radiation with an energy density *below* the critical value $W_E < 3.7$ J/mm^2 (curve 1 in Fig. 4.11), the upper layer *does not melt*. A *two-layer* heat-affected zone is formed. The *upper white* layer has a *martensitic-austenitic*

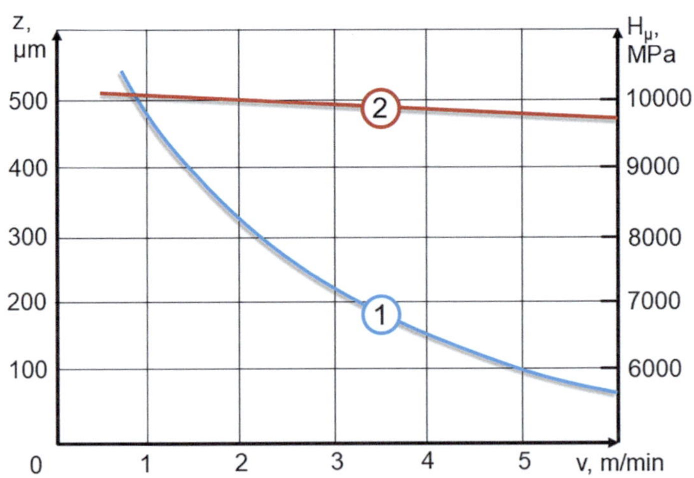

Fig. 4.10 Effect of the laser beam travel speed v on the processing of 107WCr5 steels: 1—on the depth of the heat affected zone z; 2—on the maximum value of microhardness H_μ

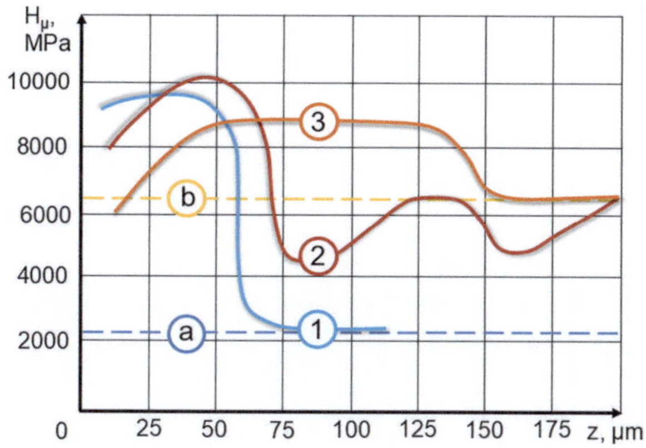

Fig. 4.11 Distribution of microhardness H_μ along the depth z of the heat affected zone of 100Cr6 steel at different energy densities W_E of pulsed laser radiation: 1—after initial normalization and subsequent laser treatment at $W_E = 2.5$ J/mm^2; 2—after initial quenching and subsequent laser treatment at $W_E = 2.5$ J/mm^2; 3—after initial quenching and subsequent laser treatment at $W_E = 4$ J/mm^2 (**a**—level of microhardness in the initial state after normalization; **b**—level of microhardness in the initial state after quenching)

fine structure with evenly distributed *carbides*. During laser treatment of *pre-hardened* steel (curves 2, 3 *in* Fig. 4.10), a *high-speed tempering zone* ("dip" in the microhardness distribution curves) with a tempered martensite structure will form at the site of the transition zone. The microhardness of the hardened layer is 11000 MPa. The depth of the hardened layer depends on the intensity of the laser radiation and, above all, on the pulse duration.

(b) Under *pulsed* radiation with an energy density *above* the critical value $W_E >$ 3.7 J/mm^2 (curve 3 in Fig. 4.11), the upper surface layer *is melted*. A *three-layer* heat-affected zone is formed with a layer of dendritic structure formed from the molten metal.

(c) With *continuous* energy density radiation, an increase in the *energy density* W_E significantly *increases the depth* of the heat-affected zone (Fig. 4.12).

The formation of heat affected zones of 100Cr6 steels has the following ***features***:

- Reduction of the maximum level of microhardness under pulsed laser irradiation with $W_E >$ 3.7 J/mm^2 (curve 3 in Fig. 4.11).
- Formation of *areas* with an abnormally high microhardness of 15,000–16,000 MPa in the layer bordering the tempering zone. The areas are clusters of *highly dispersed carbides*, have the shape of rays and are located radially relative to the upper white layer. Superhard carbide inclusions are located in the soft base and are parallel to the surface in the surface zone. The composition and location of the inclusions change the wear mechanism from brittle to plastic and significantly increases wear resistance.

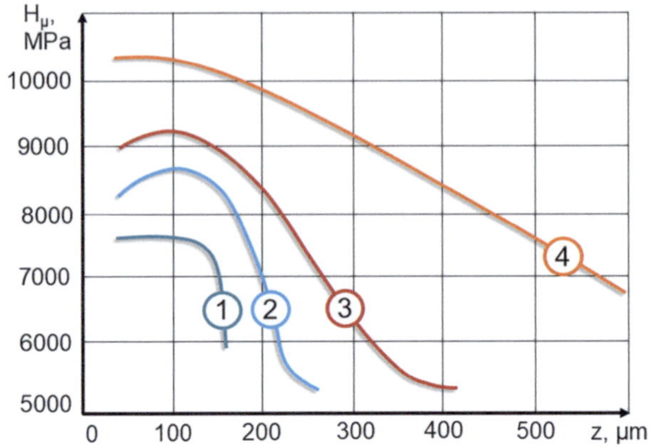

Fig. 4.12 Distribution of microhardness H_μ along the depth z of the heat-affected zone of 100Cr6 steel at different energy densities W_E of continuous laser radiation: 1—at $W_E = 8$ J/mm^2; 2—at $W_E = 13$ J/mm^2; 3—at $W_E = 20$ J/mm^2; 4—at $W_E = 70$ J/mm^2

High chromium tool steels of the ledeburite class of X165CrMoV12 type (composition: C = 1.45–1.65%; Si = 0.15–0.35%; Mn = 0.15–0.40%; Cr = 11.0–12.5%; Mo = 0.4–0.6%; V = 0.15–0.3%) in the initial annealed state have a structure of pearlite and excess carbides.

During ***pulsed*** laser irradiation of *pre-normalized* steel (curve 1 in Fig. 4.13), the ***heat affected zone*** consists of *two layers* (at an energy density of $1 < W_E < 2.6$ J/mm^2—no melting occurs):

(a) The *upper surface* "white" hardened layer consists of *fine needle martensite*, residual *austenite*, and fine *carbides*. The structure is the same for all modes of the specified energy concentration range. The microhardness reaches 9000 MPa.
(b) The lower, *narrow transition* layer, the microhardness of which decreases to its original values.

During ***pulsed*** laser irradiation of *pre-hardened* steel (curves 2, 3, and 4 in Fig. 4.13), the features of the formation of the ***heat affected zone*** depend on the energy density:

- At an *energy density of* $0.9 < W_E < 3.6$ J/mm^2, no melting occurs. A two-layer heat-affected zone is formed. The upper white layer is characterized by *carbide grinding*—the number of carbide particles increases and the size of carbide particles decreases. The microhardness of the upper layer in hardened steel increases slightly to 11,000–11,500 MPa (curve 2 in Fig. 4.13). The lower layer is a high-speed tempering zone (microhardness 4000 MPa) and has a troostite structure.
- At an *energy density* $W_E > 3.6$ J/mm^2, another layer of austenite dendrites forms on the surface as a result of crystallization of the molten metal. The ***microhardness of the melted layer is either lower or does not differ from that of the original***

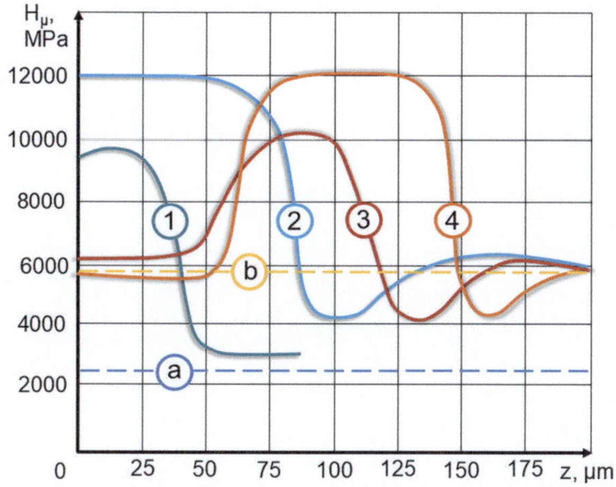

Fig. 4.13 Distribution of microhardness H_μ along the depth z of the heat affected zone of X165CrMoV12 steel at different energy densities W_E of pulsed laser radiation: 1—after initial normalization and subsequent laser treatment at $W_E = 2.5$ J/mm^2; 2—after initial quenching and subsequent laser treatment at $W_E = 2.5$ J/mm^2; 3—after initial quenching and subsequent laser treatment at $W_E = 3.1$ J/mm^2; 4—after initial quenching and subsequent laser treatment at $W_E =$ 4 J/mm^2 (**a**—microhardness level in the initial state after normalization; **b**—microhardness level in the initial state after quenching)

hardened material (curves 3 and 4 in Fig. 4.13). Below there are two layers: one has characteristics of the surface layer when radiated in the critical energy density range, and the other is a high-speed tempering zone.

At ***continuous*** laser irradiation of *pre-hardened* steel of the X165CrMoV12 type (Fig. 4.14) in a wide range of energy density $10 < W_E < 100$ J/mm^2 the ***heat affected zone***:

(a) Is *significantly larger* in size (depth and width) compared to that of pulsed irradiation. The dimensions increase with the increase of energy density (Fig. 4.15).
(b) It has a more pronounced *layered structure*. The boundaries of the layers are clearly located along the lines of isotherms.
(c) It consists of *two*, *three*, or even *four layers*, depending on the energy density of the radiation:

- At low energy densities $W_E < 5$ *J/mm*^{2}, a *two-layer* HAZ is formed: an upper white hardened layer of *needle martensite*, residual *austenite*, and fine *carbides* and a lower *transition* layer whose microhardness decreases to its original values.
- At an energy density of 5 *J/mm*$^2 < W_E < 13$ *J/mm*^{2}, a secondary hardening zone (microhardness $H_\mu = 11{,}500$ MPa—curve 1 in Fig. 4.14) is formed on the surface consisting of martensite, finely dispersed carbides, and a small amount

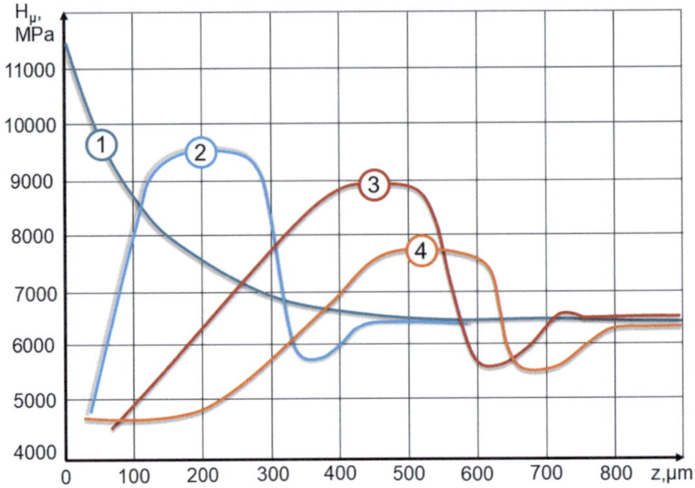

Fig. 4.14 Distribution of microhardness H_μ along the depth z of the heat-affected zone of X165CrMoV12 steel at different energy densities W_E of continuous laser radiation: 1—at $W_E = 13$ J/mm^2; 2—at $W_E = 20$ J/mm^2; 3—at $W_E = 40$ J/mm^2; 4—at $W_E = 80$ J/mm^2

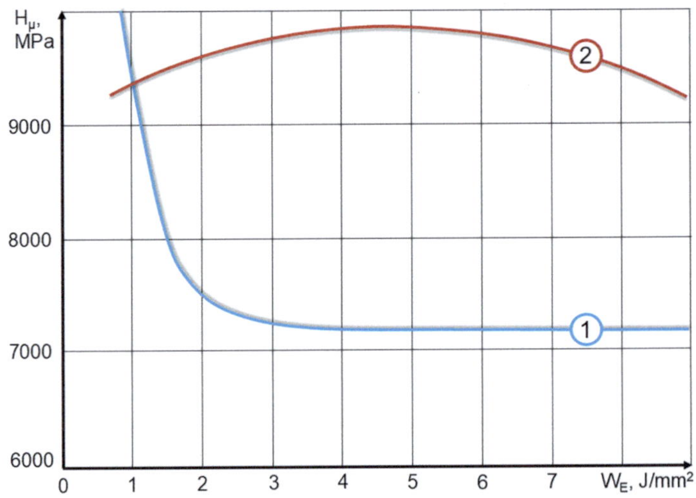

Fig. 4.15 Effect of energy density W_E of continuous laser radiation on the depth z of the heat affected zone of X165CrMoV12 steel

of residual austenite. A layer with a rapid tempering structure is formed between the secondary hardening zone and the base material.

- When the energy density increases to $W_E > 13$ J/mm^2, another intermediate narrow layer begins to appear on the surface, in which the amount of *residual austenite increases* sharply. A further increase in the energy density (curves 2,

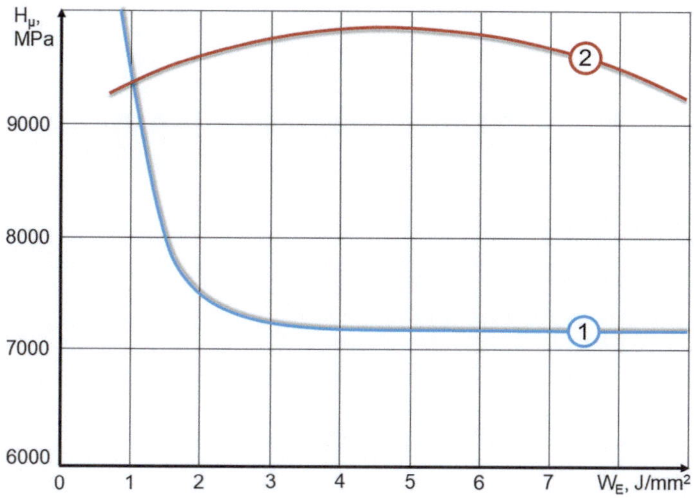

Fig. 4.16 Effect of energy density W_E of continuous laser radiation on the microhardness of X165CrMoV12 steel: 1—surface layer at a depth of z = 50 μm; 2—intermediate layer at a depth of z > 100 μm

3, and 4 in Fig. 4.14) leads, firstly, to an *increase in the depth of all layers*, including the layer with a high content of residual austenite, and secondly, to a *decrease in the microhardness* of the hardening zone.

When the *energy density* of continuous laser radiation *increases*:

- the microhardness of the surface layer (at a depth of z < 50 μm) decreases (curve 1 in Fig. 4.16);
- the microhardness of the intermediate layer (at a depth of z > 100 μm) slowly increases to a maximum at W_E = 4.5 J/mm^2 with a subsequent slow decrease (curve 2 in Fig. 4.16).

High-speed tool steels of X82WMoCrV6-5-4 type (composition: C = 0.82–0.90%; Si = 0.20–0.50%; Mn = 0.20–0.50%; Cr = 3.8–4.4%; Mo = 4.8–5.3%; V = 1.7–2.1%; Co = 0.5%; Ni = 0.4%; W = 5.5–6.5%) in the initial normalized state have a structure of sorbitol-like pearlite and excess carbides. The initial structure of the hardened steel P6M5 is needle martensite, residual austenite and carbides. The steels are characterized by high heat resistance. The critical energy density range is 0.8 < W_E < 3.0 J/mm^2.

When ***pulsed*** laser irradiates *pre-normalized* steel in the critical energy density range (without melting the surface—curve 1 in Fig. 4.17), the ***heat affected zone*** consists of *two layers*:

(a) The *upper surface* "white" hardened layer consists of *fine-needle martensite*, residual *austenite*, and evenly distributed *carbides*. The microhardness reaches 10,500 MPa.

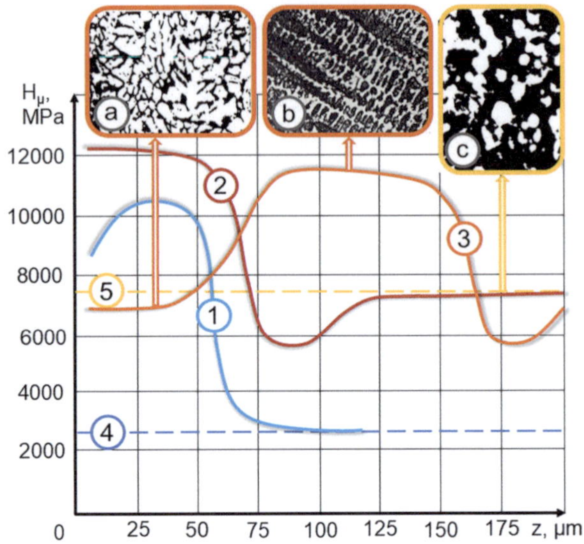

Fig. 4.17 Distribution of microhardness H_μ along the depth z of the heat-affected zone of X165CrMoV12 steel at different energy densities W_E of pulsed laser radiation: 1—after initial normalization and subsequent laser treatment at $W_E = 2.5$ J/mm^2; 2—after initial hardening and subsequent laser treatment at $W_E = 3.0$ J/mm^2; 3—after initial hardening and subsequent laser treatment at $W_E = 4.0$ J/mm^2; 4—microhardness level in the initial state after normalization; 5—microhardness level in the initial state after quenching (**a**—microstructure of the upper surface layer, × 6000; **b**—microstructure of the intermediate white layer, × 6000; c—microstructure of steel after initial quenching, × 6000)

(b) The lower, *narrow transition* layer, the microhardness of which decreases to its original values.

During ***pulsed*** laser irradiation of *pre-hardened* steel (curves 2 and 3 in Fig. 4.17), the features of the formation of the ***heat affected zone*** depend on the energy density:

(a) At an energy density in the critical range of $0.8 < W_E < 3.0$ J/mm^2 (curve 2 in Fig. 4.17), melting does not occur and a *two-layer HAZ* is formed:
- The *top layer* has no structural difference from the top layer of pre-normalized steel under the same irradiation conditions. However, it has a higher microhardness (up to 12,000 MPa) and a greater depth.
- The *lower layer* is a high-speed tempering zone ("dip" on curve 2 of Fig. 4.17) of ferrite and carbides formed as a result of dispersion hardening and not dissolved during heating.

(b) With an increase in the energy density $W_E > 3.0$ J/mm^2 (curve 3 in Fig. 4.17), the surface melts and a *three-layer HAZ* is formed:
- The *upper surface* layer is formed from molten metal and consists of two sublayers.

In a narrow *surface sublayer*, there is a *light band* with a maximum *carbon* concentration. This is due to the upward mass transfer of carbon from the lower layers to the surface where the melting occurred. The depth of the light band decreases with increasing processing speed.

In the sublayer below, the carbon content decreases. The fine-grained structure transforms into elongated *martensitic-austenitic dendrites* (Fig. 4.17a). The microhardness (6500–7000 MPa) is practically the same as the original one, but is distributed much more evenly.

- The *intermediate white* layer consists of highly dispersed martensite, residual austenite, and finely dispersed carbides such as Me_2C, Me_6C, $Me_{23}C_6$. Carbides are located both along the boundaries and inside the grains. The heating temperature of this zone was close to the melting point, resulting in significant dissolution of carbides and enrichment of the solid solution with alloying elements. The layer contains the smallest amount of carbon evenly distributed over the volume. It has a uniformly distributed microhardness in the range of 10,500–11,500 MPa. The main alloying elements (W, Cr, Mo, V, Co, Ni) are distributed very evenly, which is associated with a large dispersion of the structure (fine grain—Fig. 4.17b). When treated in *argon*, the area of the white layer increases compared to that of air treatment.
- The bottom *layer of the rapid tempering*, where the number of carbide particles is much higher than the base metal. The microhardness gradually decreases to 5000–5500 MPa, and then levels off with the microhardness of the base metal (7500 MPa). A significant unevenness in the distribution of alloying elements is observed in the high-speed tempering layer and in the base metal, which is due to the uneven distribution of carbides in the volume and less dispersion of the structure (coarse grain—Fig. 4.17c). The base metal has a fine-needle martensite structure with separated primary and small secondary carbides and a small amount of residual austenite.

During multi-pass laser surface treatment, a strip of white layer with high microhardness is formed at the boundaries of the overlapping zones. The configuration of the layers of the heat-affected zone fully reproduces the nature of the isotherms.

Thus, for strengthening steels of the X165CrMoV12 type, it is advisable to use pulsed laser radiation in the *critical* ***energy density*** *range* of $0.8 < W_E < 3.0$ J/mm^2.

The size, structure, and microhardness of the heat affected zones strongly depend on the ***irradiation time*** (Fig. 4.18), which is regulated by the processing ***speed***.

At ***low*** processing *speeds* (long irradiation times), the heat affected zone has the *largest dimensions* and the *highest microhardness* (curve 4 in Fig. 4.18), because of the following:

- Intensive melting of the metal is observed. Melting results in complete dissolution of the carbides of the original structure and an increase in carbon concentration. The high degree of dissolution of carbide phases at low processing speeds is illustrated in (Fig. 4.19). In the process of further crystallization, *excess phases of*

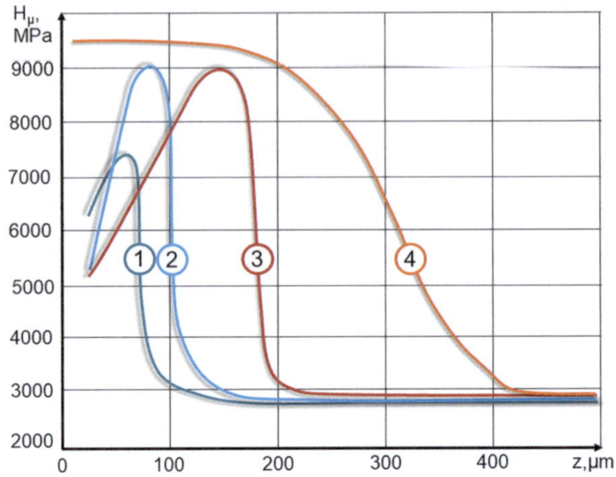

Fig. 4.18 Effect of laser treatment speed (radiation exposure time t) on the distribution of microhardness H_μ along the depth z of the heat-affected zone of X165CrMoV12 steel: 1—at $t = 2 \times 10^{-4}$ s; 2—at $t = 6 \times 10^{-4}$ s; 3—at $t = 45 \times 10^{-2}$ s; 4—at $t = 10^{-2}$ s

complex carbides such as Me_2C, Me_6C, $Me_{23}C_6$ are released, which have a clear *spherical shape.*

- A significant amount of alloying elements is transferred into solid solution and leads to the *alloying of austenite.*
- The highest content of residual austenite (up to 47%) is observed in the melting zone. With an increase in the processing speed, the amount of residual austenite decreases to 15–20% (Fig. 4.20).
- Complex carbides are also observed in the layer between the melt and the original structure. A layer with a martensitic-carbide-austenitic structure of high microhardness (9500–10,000 MPa) can be obtained on the surface by controlling the energy density and time of laser radiation without melting the surface. To do this,

Fig. 4.19 Influence of laser processing speed v on the content of carbide phases $I_c/I_{\alpha\text{-}Fe}$ of X165CrMoV12 steel

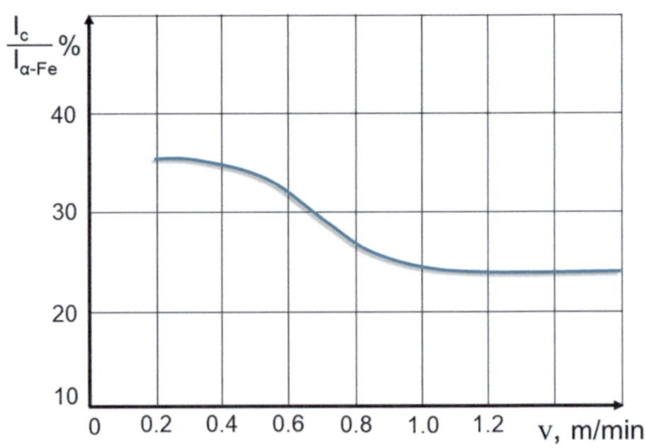

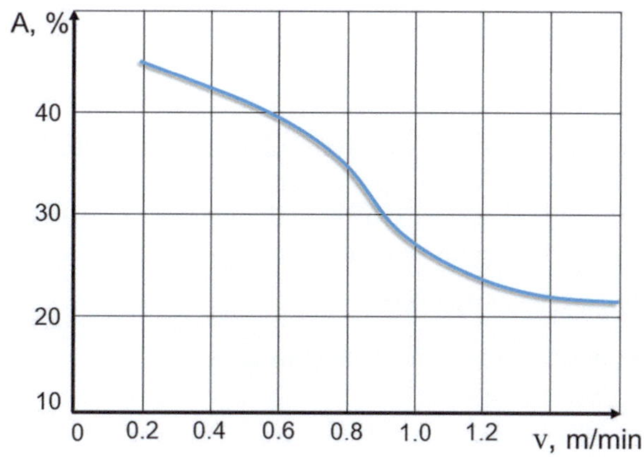

Fig. 4.20 Effect of laser processing speed v on the content of residual austenite A in X165CrMoV12 steel

it is necessary to reduce the power density to $W_P = 50$ W/mm^2 and increase the exposure time to 0.5 s.

Increasing the processing speed (reducing the time of exposure to radiation) causes a decrease in microhardness. The heat-affected zone has a finely dispersed structure. Carbides are present only in the transition layer.

Thus, high-speed tool steels of the X165CrMoV12 type are very sensitive to the time of exposure to laser radiation. The *maximum microhardness* of the surface melted hardened layer (about 10,000 MPa—Fig. 4.21) is provided at: *V = 0.8 m/min P = 1000 W; A = 0.75; d = 5 mm*. The structure of the hardened layer contains hidden needle martensite, alloyed austenite and carbides.

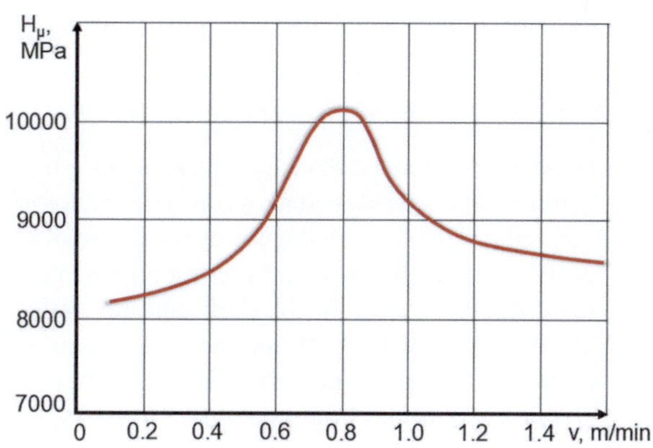

Fig. 4.21 Effect of laser processing speed v on the microhardness H_μ of the surface layer of X165CrMoV12 steel

4.3 Cast Irons

Cast irons are a continuation of the range of iron-carbon alloys ('technical iron' → 'carbon steels' → 'cast irons') and are characterized by a *carbon content* of ***2.14*** to ***6.67%*** (2.14% is the point of maximum carbon solubility in austenite on the iron-cementite diagram; 6.67% is the maximum carbon concentration for iron-carbon alloys, the rightmost point of the iron-cementite diagram).

Properties of cast iron:

- high hardness and wear resistance;
- high strength;
- high corrosion resistance;
- high metallurgical properties, including fluidity (primarily gray cast iron) and tempering resistance;
- are well processed by cutting;
- insensitivity to surface defects;
- high brittleness—low impact strength and fracture toughness.

The properties of cast irons determine their widespread use in industry and the multifactorial *response to laser irradiation*, which significantly increases the hardness, wear resistance, strength, and corrosion resistance of surface layers.

Cast irons are divided into two groups based on their carbon state:

(1) *White* cast irons, in which all carbon is chemically bound in the form of a cementite eutectic (iron carbide Fe_3C). The cementite provides the cast iron its white shimmer.
(2) *Gray* cast irons in which carbon is present predominantly in the free state in the form of graphite inclusions (C). The graphite provides the gray color to the fracture surface.

White cast irons (Fig. 4.22) are divided into:

(1) ***Hypo-eutectic*** cast irons contain from 2.14% to 4.3% C and have three structural components: pearlite—dark grains of a eutectoid mixture of ferrite and cementite; ledeburite—mottled areas of a mixture of small rounded or elongated dark pearlite grains evenly distributed in a white cementite base; secondary cementite—stands out within the dark pearlite grains (due to a decrease in carbon solubility) in the form of light needles or a grid.
(2) ***Eutectic*** cast irons—contain 4.3% C and have a ledeburite structure.
(3) ***Hyper-eutectic*** cast irons contain 4.3 to 6.67% C and have two structural components: ledeburite and primary cementite, which is released from the liquid solution in a needle-like form.

White cast irons are practically not amenable to cutting and are of less practical interest for industrial use and, consequently, for laser surface treatment compared to gray cast irons.

Laser irradiation of white cast iron *with melting* forms two layers on the surface:

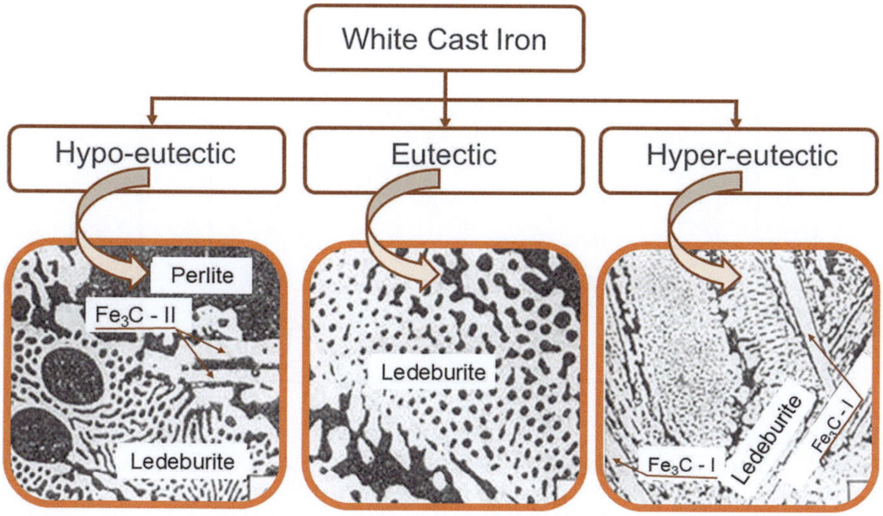

Fig. 4.22 Classification and microstructure of white cast iron

(a) *The upper surface* layer is heated to a temperature above the liquidus point and, upon high-speed cooling, has a predominantly highly dispersed *ledeburite* structure with a certain amount of dendritic (columnar) residual *austenite.* The formation of ledeburite during laser irradiation of white cast iron with the participation of not only cementite but also austenite is a new structural component of iron-carbon alloys—'bukovite'. The diameter of the austenite components in ledeburite is less than 0.5 microns, while in the original structure it reaches 2 microns. The austenite dendrites are also small, indicating that they crystallize under severe supercooling.

(b) *The lower transition* layer is heated to a temperature above the A_1 point but below the solidus temperature. During laser heating, pearlite is completely transformed into austenite. Upon subsequent rapid cooling, a certain part of the austenite turns into *martensite.* The size of the martensite needles decreases with distance from the surface (in accordance with the maximum heating temperature). The second component of the structure is *secondary cementite,* which crystallizes in the form of needles inside the austenite grains during accelerated cooling. This *creates a wimmenstätte structure.* In the vicinity of the melting zone, a modified cementite is found, which is well etched and has an increased microhardness of 12,700–14,800 MPa (cementite outside the heating zone has a hardness of 8900–9600 MPa.

A layer-by-layer X-ray diffraction analysis of the pulsed laser heating zone in hypo-eutectic white cast iron with a carbon content of about 3% showed [6] that only interference maxima of austenite are recorded on the surface X-ray diffraction patterns, and the lines of α-phase and cementite are absent. This means that all the carbon of the alloy is contained in the austenite. Starting from 0.5%, each 0.1%

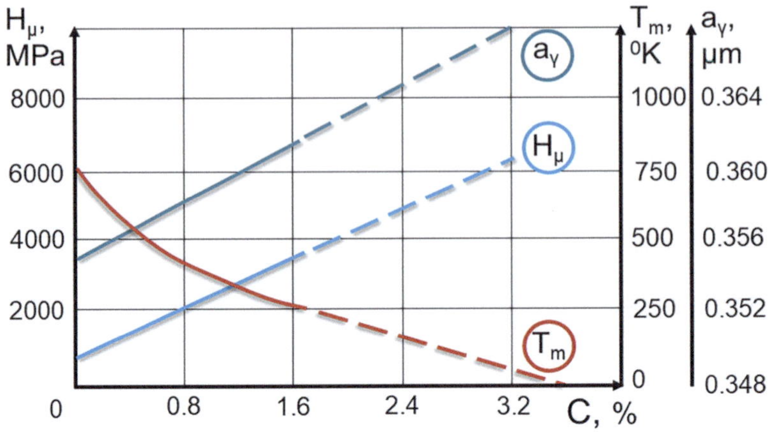

Fig. 4.23 Characteristics of austenite depending on the carbon content in the solid solution: H_μ—microhardness, a_γ—lattice period, T_m—point of onset of martensitic transformation

of austenite content reduces the point of onset of martensitic transformation T_m by 19.6 °C. Extrapolation of this dependence (Fig. 4.23) to higher carbon concentrations shows that for such an austenite, the point of onset of martensitic transformation should lie in the region of cryogenic temperatures. Indeed, the appearance of martensite in the laser heating zone is not recorded either by cooling the irradiated cast iron to room temperature (20 °C) or by supercooling it in liquid nitrogen (–196 °C). The calculations showed that there should be 3.1% C in austenite. Similar results (2.6–3.1% C) were obtained by precise changes in the austenite lattice period a_γ. Finally, the microhardness H_μ of this austenite (6370 MPa) also corresponds to the content of 3.1% carbon in the solid solution. Thus, the *pulsed laser irradiation of white cast irons* produces a very *stable austenite* with an *exorbitant carbon concentration*, which has never been seen before.

Grey cast irons depending on the metal base, are divided into (Fig. 4.24):

(1) *ferrite*—having the lowest strength and hardness with the highest ductility;
(2) *ferrite-pearlite*;
(3) *pearlite*—having the highest strength and hardness with the lowest ductility.

The metal base contains graphite inclusions, the shape of which also affects the mechanical properties, primarily strength and ductility:

(a) *lamellar* graphite inclusions weaken the strength of cast iron, especially under tensile loads;
(b) *flake* graphite inclusions characterize forging cast irons;
(c) *spherical* graphite inclusions define high-strength cast irons.

Laser irradiation is widely used for *surface hardening* of gray cast iron. Both *pulsed* and *continuous* laser radiation is used. In most cases, laser radiation is used to *melt* the surface layer.

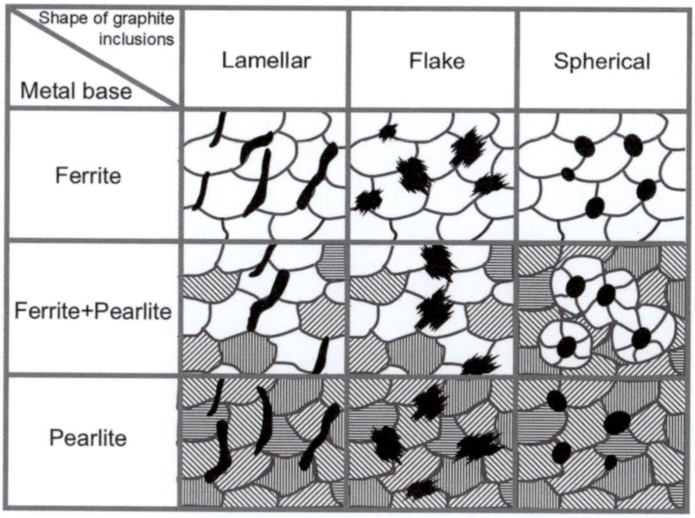

Fig. 4.24 Schematic of the macrostructure of gray cast iron

During laser irradiation ***with melting*** of gray cast irons with a *pearlite* or *pearlite-ferrite* metal base (the most common one), two layers are formed in the heating zone (Fig. 4.25)—the upper surface layer and the lower transition layer.

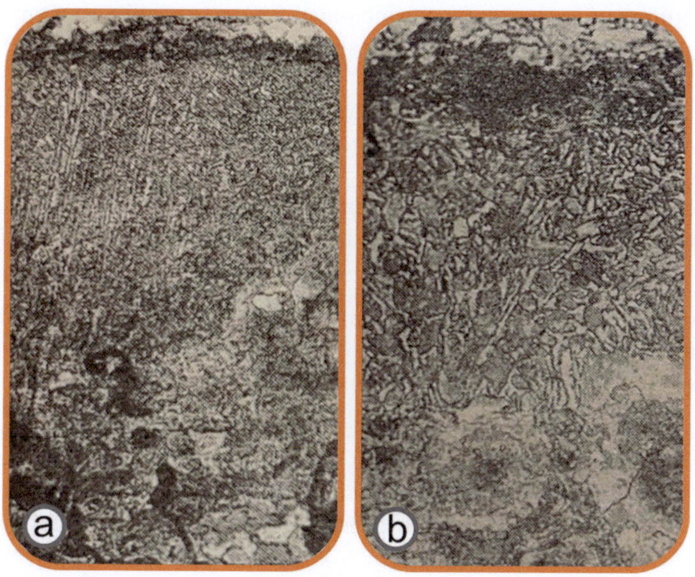

Fig. 4.25 Microstructure of the laser heating zone of ferritic-pearlite cast iron with spherical graphite inclusions: **a**– × 200; **b**– × 500

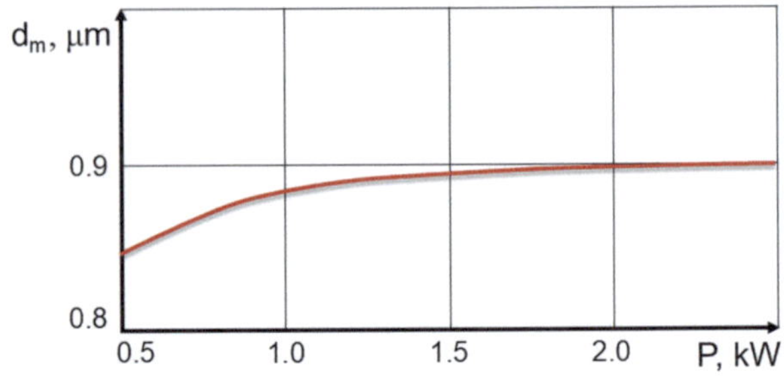

Fig. 4.26 Average length d_m of austenite dendrite axes as a function of laser radiation power P

The ***upper surface*** layer is in a liquid state at the time of laser irradiation. During the *directed* crystallization with a *high* cooling *rate* (5×10^3–10^5 K/s), a microstructure is formed in the form of *crushed dendrites of residual austenite* (up to 60%). The lengths of the dendrite axes are weakly dependent on the power of continuous laser radiation (Fig. 4.26). The angles of inclination of the dendrite axes to the metal surface slightly differ from 90° due to the fact that the movement of the laser beam creates an additional temperature gradient along the direction of movement.

Laser irradiation leads to:

- reduction of the distance between dendritic branches by 4 times;
- reducing the average length of the axles by at least 60 times.

The crystallization of the upper surface layer occurs according to the metastable diagram of the iron-cementite system with the formation of a hypo-eutectic white cast iron structure. Consequently, the interdendritic spaces in the structure are filled with *finely dispersed ledeburite*, which has increased wear resistance (like hardening martensite). Thus, the *hardening* of gray cast iron under *laser irradiation*, unlike steel, does not require hardening of the surface layer. High strength, hardness, and wear resistance are achieved through *bleaching* during accelerated cooling from the liquid state.

Despite the short duration of laser irradiation (both continuous and pulsed), *graphite inclusions* of the upper surface layer have time to completely *dissolve* in the liquid alloy, which leads to:

- increasing the carbon content of austenite (and hence the hardness) to the highest possible values;
- increase in corrosion resistance by an order of magnitude.

In the *lower* part of the upper surface layer near the melting point, large *graphite inclusions* are partially *insoluble*.

Along the undissolved graphite inclusions, the upper surface layer extends deep into the lower transition layer in wedge-shaped areas. This creates an *uneven boundary* between the two layers, which is the difference between laser heating zones of gray cast iron and laser heating zones of steels.

The microhardness of the upper surface layer is 6800–14,000 MPa. The presence of a significant amount of residual austenite does not lead to a significant decrease in hardness, which is explained by the high degree of dispersion of structural components, phase adhesion of austenite during high-speed cooling and a high density of crystal structure defects.

Tempering of the upper surface layer at a temperature of 150–200 °C for 2–4 h does not cause the decomposition of residual austenite. At a tempering temperature of 300 °C, the amount of austenite decreases sharply, which leads to an increase in the microhardness of the upper surface layer.

Pulsed laser irradiation produces an amorphous state in cast iron with a bleached structure containing 3.20% C; 2.67% Si; 0.64% Mn; 0.014% S; and 0.06% P. In the upper surface molten layer near the melting boundary, cementite-free areas are formed in the ledeburite. These areas contain about 6% C with a microhardness of 12,000 MPa and are amorphous. As a result of tempering at 400 °C for 2 h, the hardness increases by 500 MPa. Secondary crystallization of the amorphous phase during tempering causes the formation of a new structural component of iron-carbon alloys—bukovite (see above). Bukovite is a finely dispersed mixture of cementite and austenite.

The ***lower*** laser-irradiated ***transition*** layer of gray cast irons is thin compared to steels for two reasons:

(a) The upper boundary of the transition layer runs along the fusion line and is determined by heating to the solidus point T_S. The solidus point of gray cast irons is the eutectic temperature $T_S = 1153$ °C.

(b) The lower boundary of the transition layer is determined by heating to the critical point A_1 (similarly to steels). In gray cast irons, under the influence of silicon, which is part of their composition, the critical point A_1 shifts upward to higher temperatures.

In the lower transition layer, the laser heating causes complete phase *recrystallization* of the matrix:

- Pearlite turns into *austenite*.
- During subsequent autocarburization, a larger proportion of austenite turns into *fine-needle martensite*.
- *Graphite inclusions* are partially dissolved in austenite and therefore become noticeably *smaller*. In addition, *spheroidization* of graphite inclusions occurs.

The microhardness of the lower transition layer varies in the wide range of 7000–14,000 MPa.

In the lower transition layer, melted areas around large graphite inclusions are observed as a result of contact melting (Fig. 4.25b).

During laser irradiation ***without melting*** of gray cast irons with a *pearlite* or *pearlite-ferrite* metal base, a single hardened layer of small thickness is formed in the heating zone, since the temperature range for austenite formation is narrow and does not exceed 150 K (limited by the alloy solidus temperature at the top and the critical point A_1 at the bottom). Hardening is achieved by martensitic transformation in the matrix [5]. The small thickness of the hardening layer limits the use of laser hardening of gray cast irons without melting the surface.

Alloy cast irons of the HXMD type (2.8–3.2% C; 1.6–2.0% Si; 0.8–1.2% Mn; 0.7–1.6% Ni; 0.2–0.7% Cr; 0.2–0.7% Mo; 0.2–0.5% Cu) are hardened by the mechanism of gray cast irons under laser irradiation *with melting*. Alloy cast iron is less prone to melting, cracks and gas pores. The lower thermal conductivity of alloy cast iron makes it possible to use significantly *lower* laser processing *speeds* (compared to conventional gray cast iron) to guarantee the formation of a hardened layer.

When laser irradiated *without melting* the surface layer, the structure of alloy cast irons consists of *fine-needle bainite*, *carbides* and *lamellar pearlite*.

High-strength cast iron of the GGG60 type (3.0–3.6% C; 2.4–2.8% Si; 0.4–0.7% Mn; 0–0.40% Ni; 0–0.15% Cr; 0–0.30% Cu) has a basic structure: a metal matrix of sorbitol-like and thin-plate *pearlite* with *spherical graphite* inclusions.

In laser *continuous* irradiation, the peculiarities of the formation of the ***upper surface*** layer are:

- a *high degree of dispersion* of the metal base, which leads, on the one hand, to a reduction in carbon diffusion pathways during the transformation of pearlite into austenite and, on the other hand, to an increase in the length of ferrite-cementite boundaries on which austenite crystals are formed;
- *preservation* of some *graphite inclusions* due to the low dissolution rate of spherical graphite inclusions during the melting of the metal matrix (compared to lamellar graphite inclusions in conventional gray cast irons);
- the possibility of forming a high-strength ledeburite structure (Fig. 4.28).

The *effectiveness* of laser hardening of cast irons depends on the type of metal matrix and the degree of its dispersion more than on the shape of graphite inclusions. Therefore, the upper surface layer has a high hardness of 9000–10,000 MPa (compared to 7500–8500 MPa of ordinary gray cast irons).

The ***lower transitional*** layer of laser-irradiated high-strength cast iron consists of *martensite*, residual *austenite*, and spheroidal *graphite* inclusions, similar to gray cast iron. In the lower part of the layer, closer to the base metal, traces of the original pearlite (the so-called "pearlite spirit") are found, since the high laser heating rate and short duration of laser exposure leave the diffusion processes of austenite formation and its homogenization incomplete.

After *normalization* at a temperature of 890 °C, the metal base of cast iron consists of *pearlite* (85–95%) and *ferrite* (5–15%). The presence of "pearlite spirit" in the lower part of the layer leads to a decrease in the amount of residual austenite. In the upper part of the layer, regardless of the laser irradiation parameters, about 30% of martensite is formed and bainite structures are also observed. The microhardness of the lower transition layer varies within a wide range of 6000–9000 MPa.

During *tempering*, the microhardness significantly depends on the temperature (Fig. 4.27). At a quenching temperature of up to 500 °C, the residual austenite decomposes. This is accompanied by an increase in microhardness to a maximum at 350–400 °C. At tempering temperatures above 500 °C, the microhardness is significantly reduced due to cementite coagulation. In high-strength cast iron, an additional decrease in microhardness occurs due to graphitization. As a result, high-strength cast iron of the GGG60 type acquires a lower hardness than conventional gray cast iron of the SCh 24–44 type (2.9–3.2% C; 1.2–1.6% Si; 0.8–1.2% Mn).

High-speed graphitization of the laser heating zone of high-strength cast iron during tempering is due to:

- High content of a strong *graphitizing* agent, which is silicon (3.2%).
- The formation of a *heterogeneous* solid solution with a carbon and silicon content during rapid cooling after laser radiation, which accelerates the diffusion process during subsequent tempering.
- A high degree of *dispersion of ledeburite* in the upper surface layer, which increases the length of austenite-cementite interfaces and becomes the site of graphitization centers.
- During laser heating of high-strength cast irons of the GGG60 type with a structure of sorbitol-like and thin-plate pearlite, the dissolution rate of *spherical* graphite inclusions is *lower* than that of lamellar ones (Fig. 4.28a). Therefore, a part of graphite inclusions in the upper surface layer remains undissolved.

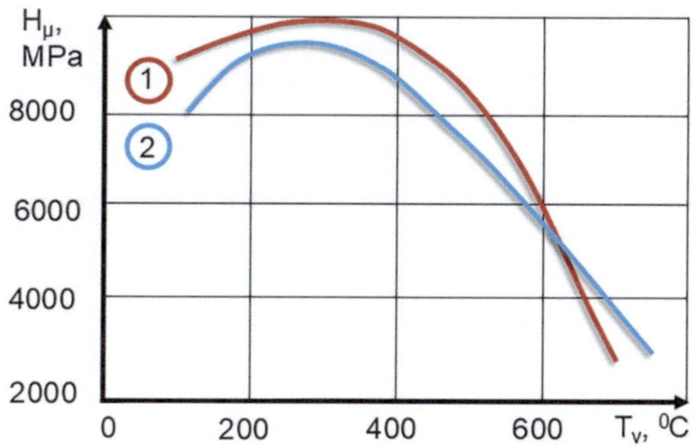

Fig. 4.27 Microhardness H_μ of cast irons after tempering at different temperatures Tv: 1—GGG60 cast iron; 2—SCh 24–44 cast iron

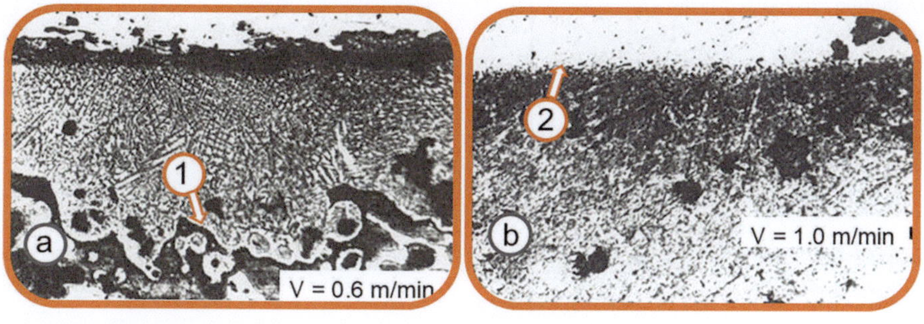

Fig. 4.28 Microstructure of high-strength cast iron GGG60 with spheroidal graphite after laser treatment at hardening parameters: **a**—with surface melting; **b**—without surface melting (1—liquid phase hardening zone; 2—solid phase hardening zone; V—laser beam speed)

Ferritic-based **ductile iron** with flake graphite inclusions of the GTS35-10 type (2.5–2.8% C; 1.1–1.3% Si; 0.3–0.6% Mn; 0.06% Cr) under continuous and pulsed laser irradiation has the following features of the ***upper surface*** layer formation:

- It crystallizes according to the metastable diagram of the iron-cementite system;
- The upper part of the layer is dominated by highly dispersed *ledeburite* with a hardness of about 9600 MPa;
- In the lower part of the layer, large *dendrites of austenite* containing about 2% C grow perpendicular to the melt boundary. Since the cooling rate after pulsed laser heating is much higher than after continuous irradiation, the hardness of austenite differs significantly and is 6700 MPa and 4700 MPa, respectively. The difference in hardness is due to the different number of crystal structure defects and different levels of fine structure distortion and residual stresses.

Austenite in the upper surface layer under continuous laser heating has greater resistance to tempering compared to pulsed laser irradiation (Fig. 4.29).

With an increase in the tempering temperature, austenite decomposes into a ferrite-graphite mixture. This is accompanied by diffusion of carbon to graphite inclusions and an increase in hardness. The maximum increase in hardness occurs at a tempering temperature of 200 °C under pulsed laser irradiation and 300 °C under continuous irradiation (Fig. 4.29). With a further increase in the tempering temperature, a rapid decrease in hardness is observed.

In the ***lower transition*** layer, ferrite grains that have undergone phase recrystallization are crushed and deformed during high-speed cooling. As a result, their hardness increases from 1500 MPa in the initial state to 2500 MPa under pulsed and 2200 MPa under continuous laser irradiation.

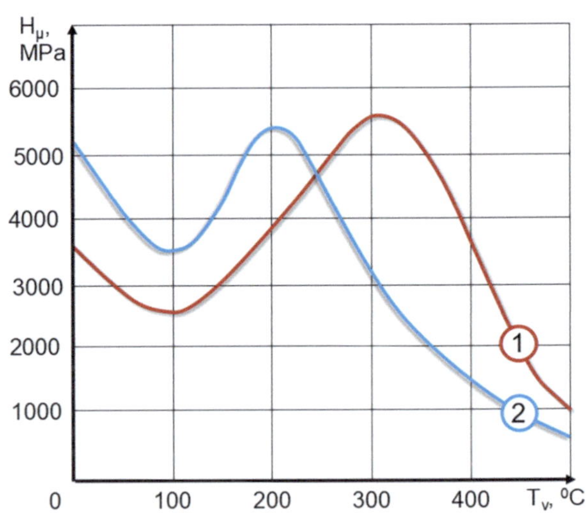

Fig. 4.29 Microhardness H_μ of ductile iron after tempering at different temperatures T_V: 1—continuous laser irradiation; 2—pulsed laser irradiation

4.4 Titanium Alloys

Titanium alloys and titanium-based composite materials have high *strength, corrosion resistance* and *heat resistance,* which is why they are widely used in aerospace, chemical, oil and shipbuilding.

Titanium alloys have a density of about 4.5 g/cm^3, similar to medium-strength steel, but they are only 60% of its weight. This leads to an almost twofold increase in specific strength, which helps reduce the weight of aircraft and missile structures. Titanium alloys effectively replace aluminum alloys and stainless steel, which are not suitable for high temperatures applications. Compared to aluminum and structural steel, titanium alloys have better corrosion resistance. Titanium alloys are superior to copper alloys in terms of corrosion resistance in seawater and other aggressive environments.

The structure and properties of titanium alloys (Table 4.1) are formed as a result of phase transformation at a temperature of 882 °C, depending on the alloying with α- and β-stabilizing elements: α-alloys (Fig. 4.30a) are ductile but have lower strength; β-alloys (Fig. 4.30b) are the strongest but have lower ductility. Two-phase (α + β) alloys have the best set of mechanical and technological properties.

The disadvantages of titanium alloys are poor machinability, high price, low tribotechnical properties, low elastic modulus, high chemical activity at high temperatures (above 500 °C) and a tendency to hydrogen embrittlement.

Alloying elements can be divided into the following according to their influence on the structure of alloys (Table 4.2): α-stabilizers (Al, Sn, Ga, In, O, N, C), which extend the region of existence of the α-phase and increase the temperature of polymorphic α → β transformation; β-stabilizers (Mo, V, Mn, Cr, W, Nb, Ta, Fe, Si, etc.), which extend the region of existence of the β-phase.

α-alloys (the main alloying element is Al) have low strength at 20 °C, but higher strength at low and elevated temperatures (400–500 °C). α-alloys cannot be strengthened by heat treatment. The ductility of α-alloys is low, so pressure treatment is performed only in the hot state. They are designed to work in a wide temperature range—from cryogenic to 450 °C (BT5) and 500 °C (BT5-1).

Pseudo-α alloys have a structure consisting of an α-phase and a small amount (1–4%) of a β-phase due to alloying with β-stabilizers. The presence of the β-phase

Table 4.1 Microstructures of titanium alloys

Alloy microstructure	Alloy grade
α-alloys	VT1-00, VT1-0, VT5, VT5-1, PT-7 M
α-pseudo alloys ($K_\beta < 0.25$)	OT4-0, OT4-1, OT4, VT20, AT3
(α + β)-alloys martensitic ($K_\beta = 0.3–0.9$)	VT6C, VT6, VT14, VT8, VT9, PT-3 V, VT3-1, AT3
(α + β)-alloys transitional ($K_\beta = 1.0–1.4$)	VT22, VT30
β-pseudo alloys ($K_\beta = 1.5–2.4$)	VT35, VT32, VT15
β- alloys ($K_\beta = 2.5–3.0$)	4201

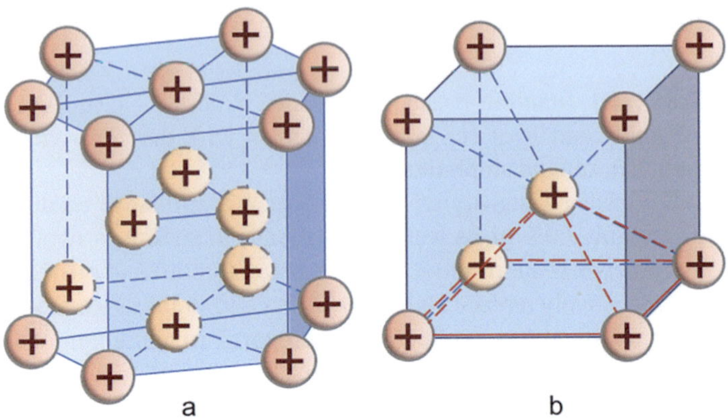

Fig. 4.30 Types of crystal lattices of titanium alloys: **a**—hexagonal close packed Ti_{α}; **b**—body-centered cubic Ti_{β}

Table 4.2 Chemical composition, structure and mechanical properties of titanium alloys

Alloy	Chemical composition, %				Mechanical properties			
	Al	V	Mo	Others	σ, MPa	σ, MPa	δ,%	KCU, MJ/m^2
α—alloys								
VT5	4.5–6.2	1.2	0.9	0.3 Zr	700–650	650–850	12	0.4
VT5-1	4.3–6.0	1.0	-	2–3 Sn	700–950	650–850	12	0.45
Pseudo-α alloys								
OT4-1	1.5–2.5	–	–	0.7–2 Mn	600–750	470–650	30	0.5
OT4	3.5–5.0	–	–	0,8–2 Mo	700–900	550–650	18	0.5
VT20	5.5–7.0	0.8–2.5	0.5–2.0	1.5–2.5 Zr	950–1150	950–1150	8	0.45
(α + β)—alloys								
VT6	5.3–6.8	3.5–5.3	–	–	1100–1150	1000–1050	15	0.3
VT14	3.5–6.3	0.9–1.9	2.5–3.8	0.3 Zr	1150–1140	1080–1300	8	0.5
VT16	1.6–3.8	4.0–5.0	4.5–5.5	0.8–1.2 Cr	1250–1450	1100–1200	5	0.5
VT22	4.4–5.7	4.0–5.5	4.0–5.5	0.8–1.2 Cr	1100–1250	-	9	0.4

Note Properties of VT6, VT14, VT22 alloys are given after quenching and aging; others—after annealing

increases the technological ductility. Alloys with a low aluminum content (OT4, OT4-1) are pressure treated in the cold state. With a higher aluminum content, hot deformation is required. Alloys alloyed with Zr, Si, Mo, Nb, V (BT20) have the highest heat resistance among titanium alloys.

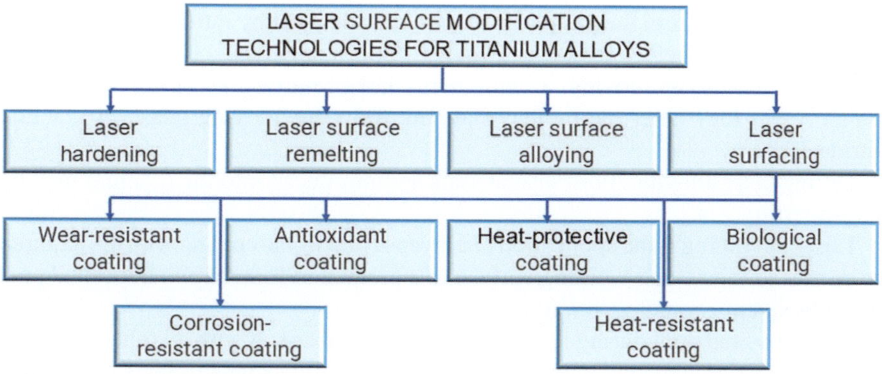

Fig. 4.31 Technologies for laser modification of titanium alloys

Two-phase (α + β) alloys contain an α-stabilizer (Al) and β-stabilizers (Cr, Mo, Fe, etc.) that reduce the temperature of polymorphic transformation. Alloys are strengthened by heat treatment (quenching + aging) and have high mechanical properties (see Table 4.2).

β-alloys have the structure of a solid solution of alloying components in β-titanium and contain β-stabilizers (W, Cr, Fe, Mo, etc.) that reduce the temperature of polymorphic transformation of titanium. The alloys are not used in industry due to their high cost and poor mechanical properties.

Technologies for laser surface modification of titanium alloys are shown in Fig. 4.31.

Laser surface **hardening** is a *structural-phase transformation* due to *ultrafast laser heating* and rapid self-cooling by a high-energy laser beam. The temperature of the local heat affected zone *does not exceed the melting point.* Laser hardening from the solid state:

- provides the fine-grained structure of the alloy;
- makes the chemical composition of the surface more homogeneous;
- reduces the segregation coefficient (from the initial 1.28 to 1.04);
- reduces the coefficient of friction;
- increases resistance to fretting (up to 100 times).

Laser surface **remelting** is a rapid *melting* followed by surface *crystallization* by irradiating the material in a protective argon atmosphere.

Laser remelting of the surface of two-phase (α + β) alloys such as UNS R56400 (3.5–4.5% V; 5.50–6.75% Al; 0.35 Fe; 0.08% C) reduces the crystals' size and makes it difficult for dislocations to move, which leads to a significant increase in surface hardness and wear resistance.

However, laser remelting of the surface does not improve the characteristics of the rest of the titanium alloys.

Laser alloying is a *change in the chemical* and phase composition by melting the surface of a material with a high-energy laser beam with the simultaneous supply of alloying elements from the gas phase (mainly nitriding) or by introducing powdered materials into the melt.

During *gas-phase* alloying, a high-energy laser beam *activates* nitrogen atoms, which, when they enter the molten titanium, form the alloying phase of titanium nitride TiN.

During laser alloying, *powder* materials are introduced into the titanium melt by a transporting gas.

Laser surfacing is the application of a layer of functional coating with the required properties to the *molten* working surface of a product to form a *joint* deposited pool and subsequent crystallization to form a metal bond.

The deposited pool during laser surfacing, in comparison with all other surfacing methods [8], is characterized by a short duration and small dimensions, which ensures:

- preserving the properties of the material being deposited;
- structural homogeneity;
- fine grain size;
- good adhesion of the deposited layer to the substrate (high adhesive strength).

According to the composition and properties of the deposited layer, functional coatings can be divided into wear-resistant coatings, antioxidant coatings (resistant to high-temperature oxidation), heat-protective coatings (heat-insulating ceramic materials), biological coatings (coating of artificial implants), corrosion-resistant coatings, and heat-resistant coatings (Fig. 4.31).

The **microstructure** of the laser modification of the surface of titanium alloys depends on the type of radiation (continuous or pulsed), the medium (air, nitrogen or shielding gases) and the processing parameters (primarily power density). In general, two laser impact zones are formed:

(a) Melting zone (upper surface layer);
(b) Heat-affected zone (lower transition layer).

The **melting zone** during laser surface modification by ***continuous*** irradiation has the following features (Fig. 4.32a):

- The *microstructure* during laser treatment in the medium of air, nitrogen, or shielding gases (helium, argon) has a dendritic structure with rod-like inclusions (Fig. 4.32b). In addition to dendrites, grains of martensitic α-phase and β -phase are formed. During laser treatment in helium, colonies of α-plate are located within the grains of the β-phase, stretched in the direction of the heat sink (Fig. 4.32c).
- During laser treatment in nitrogen, the liquid alloy is saturated with nitrogen, forming titanium *nitride* TiN.
- *Microhardness*, on the one hand, increases due to the formation of titanium nitride and the martensitic α-phase. On the other hand, with the growth of alloying elements in titanium alloys, the amount of the β phase increases. The microhardness of the β-phase is significantly lower than that of the α-phase. Consequently,

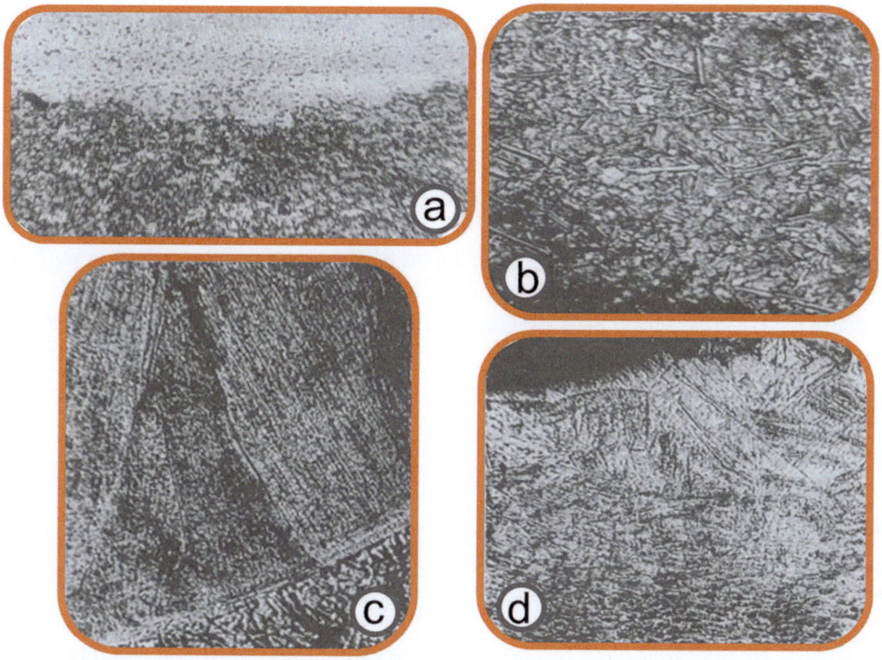

Fig. 4.32 Microstructure during laser modification of the surface of titanium alloys: **a**—melting zone during air treatment (OT4 alloy, × 100); **b**—melting zone during air treatment (OT4 alloy, × 400); **c**—melting zone during helium treatment (BT16 alloy, × 500); **d**—upper part of the heat-affected zone during air treatment (OT4 alloy, × 400)

the microhardness in the melting zone varies in a wide range. After laser treatment in helium, the microhardness is significantly lower compared to air treatment.

The **melting zone** during laser surface modification by ***pulsed*** irradiation consists of ***three*** layers:

(a) As a result of ultra-high heat transfer rates, a ***layer of large columnar dendrites*** of small thickness (20–26 μm) with high microhardness (7500–8200 MPa) is formed, oriented normal to the surface in the direction of heat transfer.
(b) When laser irradiation is performed in the medium of *nitrogen* or *air*, the interaction of titanium with nitrogen under intense thermal effects leads to the formation of a narrow ***sublayer ε-phase*** behind the dendrite layer.
(c) Below, in *two-phase* $\alpha + \beta$-alloys and pseudo-α alloys from the melt, under conditions of ultra-high laser heating rates and ultra-high heat transfer rates from the melt under the influence of hydrodynamic forces arising in the initial period of the radiation pulse, a ***fine-grained martensitic*** structure is formed. Martensitic areas of α' and α'' phases are randomly located in a layer with a thickness of 100–180 μm, depending on the radiation power density. The

presence of α′ and α″ phases increases the microhardness to 4200–5500 MPa. Moreover, the α′ phase is much harder than the α″ phase.

The **heat affected zone** during laser surface modification ***with fusion*** is weakly etched and has a multilayer structure.

When treated at low speeds in the medium of *air* or *nitrogen*, a needle-like martensitic structure is formed in the *upper* part of the heat affected zone, oriented within the α-phase grains (Fig. 4.32d).

The **heat affected zone** during laser ***hardening*** by continuous radiation ***without melting*** (on the example of high-strength heat-resistant α + β-alloy *BT3-1*: 5.5–6.5% Al, 2.0–3.0% Mo, 0.8–2.3% Cr, 0.2–0.7% Fe, 0.2–0.4% Si) has the following features:

- *microhardness* increases significantly with increasing power density and decreases with increasing processing speed (Fig. 4.33);
- The *depth* of the heat-affected zone *increases* with increasing power density (Fig. 4.33);
- the *width* of the heat-affected zone *decreases* with increasing processing speed;
- the *chemical composition does not change* significantly;
- the X-ray diffraction patterns show line widening, which indicates *crystal fragmentation* and an increase in micro distortions of the crystal lattice;
- when irradiated in air, the interaction of the sample with nitrogen can lead to the formation of *titanium nitride* TiN, which further increases microhardness;
- corrosion resistance of the surface increases.

The **neutral environment** (protective gases such as helium and argon) is favorable for laser surface modification compared to the active environment (air and nitrogen) because it provides:

- *Homogeneous structure* due to limited access of nitrogen and oxygen.
- *Redistribution of martensitic* phases towards increasing α″-phase and decreasing α′-phases. Accordingly, the *microhardness decreases*. Under pulsed laser radiation, the microhardness is evenly distributed and amounts to 3250–3400 MPa (the microhardness of the base metal is 3600 MPa).
- Formation of a *coarse-grained needle* structure under pulsed laser radiation due to the high rate of melt crystallization.

The **overlap** of the laser impact zones during pulsed laser radiation leads to the recrystallization of the structure formed under the influence of the previous pulse processing, characterized by an increase in microhardness in the overlapping zones. The microhardness in the overlapping areas increases to 3400–3500 MPa.

As the **power density** W_p increases, the depth and width of the laser impact zones for all titanium alloys increases.

Starting from the threshold power density $W_p = 1.2 \times 10^5$ W/cm^2, sufficient to melt the surface layer and form a modified structure in it, material vaporization and formation of a crater in the center of the zone are observed.

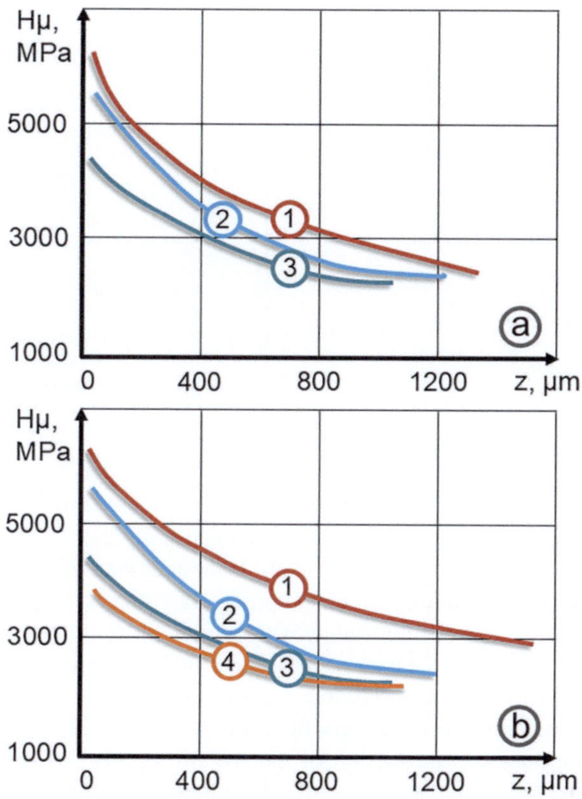

Fig. 4.33 Distribution of microhardness $H_{(\mu)}$ along the depth z of the heat-affected zone at a change in the power density Wp of laser radiation and at a change in the processing speed V in the titanium alloy BT3-1: a—Wp = 2.5.10^4 W/cm^2 (curve 1), Wp = 10^4 W/cm^2 (curve 2), Wp = 0.64 × 10^4 W/cm^2 (curve 3) at V = 2.0 m/min; **b**—V = 0.26 m/min (curves 1), V = 1.0 m/min (curves 2), V = 2.0 m/min (curves 3), V = 3.0 m/min (curves 4) at Wp = 0.64 × 10^4 W/cm^2

In the range of $1.2 \times 10^5 \le W_p \le 2.0 \times 10^5$ W/cm^2, the intensity of growth of the size of the laser impact zones slows down. In the laser impact zones, dissolution of alloying elements and formation of new compounds occur, which require additional energy consumption, as well as phase transitions and bulk changes in the structure caused by the formation of new phases.

With further growth of the power density $2.0 \times 10^5 \le W_p \le 2.9 \times 10^5$ W/cm^2, the growth of the width of the laser impact zones stops, since the specific melting energy is lower than the specific vaporization energy (by analogy with iron-based alloys).

Since the power density is a determining parameter in the formation of the size and structure of laser impact zones, the *optimal power density* should be determined when calculating the parameters of laser surface modification: $W_{p_1} < W_p < W_{p_2}$.. Pulsed laser treatment of titanium alloys at optimal modes provides a sufficient depth

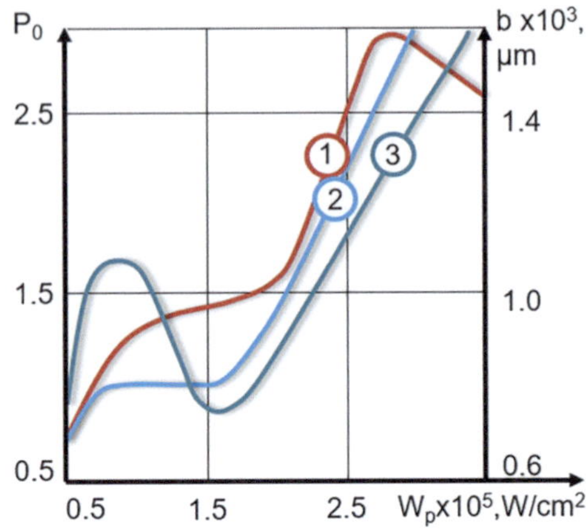

Fig. 4.34 Dependence of the width of the laser impact zone 'b' (curve 1) and the surface configuration parameter P_0 (curves 2 and 3) on the radiation power density W_p during laser surface treatment in neutral (2) and air (3) environments

of the modified layer (180–200 μm) and optimal geometric dimensions of the laser impact zones. The presence of a concavity (crater) usually increases wear resistance by retaining lubricant in the contact zone and localizing wear products.

The **geometry of the surface** after laser processing affects the performance characteristics of products and is determined by the depth of the craters (m) in the center of the laser impact zone and the excess of the solidified melt over the original surface (n) (the excess is formed due to volume redistribution). The ratio ($P_0 = m/n$) is a generalized parameter of the surface configuration after laser irradiation. The surface configuration parameter is a function of the radiation power density $P_0 = f(W_p)$ (Fig. 4.34), in the range $W_{p_1} < W_p < W_{p_2}$ during *pulsed* laser surface treatment of titanium alloys in *neutral* and *air* environments. The optimal value is $1.0 \leq P_0 \leq 1.5$, which is confirmed by the results of tribological tests.

Subsequent to laser processing, partial **tempering** (heating to 570 °C → holding for 2 h → cooling on air) or double tempering (heating to 880 °C → holding for 1 h → cooling to 650 °C → holding for 2 h → cooling on air) *does not change the microhardness* of the surface layer.

The subsequent **thermal hardening** after laser treatment (heating to 950 °C → holding for 1 h → cooling to 530 °C → holding for 6 h → cooling in air) *increases the microhardness* by 1000–2000 MPa at a thickness of the heat-affected zone of 0.2–0.4 mm, since it is performed at temperatures close to the temperature of polymorphic transformation 'α + β ↔ β' (950–980 °C).

4.5 Aluminum Alloys

Aluminum alloys (Table 4.3) have high *thermal conductivity* and high *reflectivity* to laser radiation.

Table 4.3 Chemical composition of the main types of aluminum alloys (Al—base)

Alloy grade	Analog	Fe	Si	Mn	Ni	Cu	Mg	Zn	Others
AK4	2618	0.8–1.3	0.5–1.2	≤ 0.2	0.8–1.3	1.9–2.5	1.4–1.8	≤ 0.3	Ti ≤ 0.1
ALL4	AlSi10Mg	≤ 1.0	8.0–10.5	0.2–0.5	≤ 0.1	≤ 0.3	0.2–0.3	≤ 0.3	Be ≤ 0.2
AK5M7	G-AlSi(Cu)	≤ 1.3	4.5–6.5	≤ 0.5	≤ 0.5	6.0–8.0	0.2–0.5	≤ 0.6	
AK12M2MgH	G-AlSi12Cu	≤ 0.8	11–13	0.3–0.6	0.8–1.3	1.5–3.0	0.8–1.3	≤ 0.5	Ti 0.1–0.2; Cr ≤ 0.2
AL26		≤ 0.7	20–22	0.4–0.8	1.0–2.0	1.5–2.5	0.4–0.7	≤ 0.3	

Laser treatment of aluminum alloys ***without*** surface ***melting does not change*** the microstructure and microhardness. The hardening zone from the solid phase and the transition zone are not detected.

The *difficulties* of laser processing of aluminum alloys ***with*** surface ***melting*** are due to the following reasons:

- The limited dissolution of impurities and intermediate phases due to the short laser heating time;
- The difficulty of hardening in the absence of martensitic transformations. Laser hardening with melting of aluminum and AK4-type alloys does not significantly increase their microhardness (only by 60–80 MPa).

The main *difference between the microstructure* of laser exposure zones in aluminum alloys (compared to steels and cast irons) is the formation of only *one layer*—a melted layer hardened from the liquid state. There is no heat-affected zone outside the melt (Fig. 4.35).

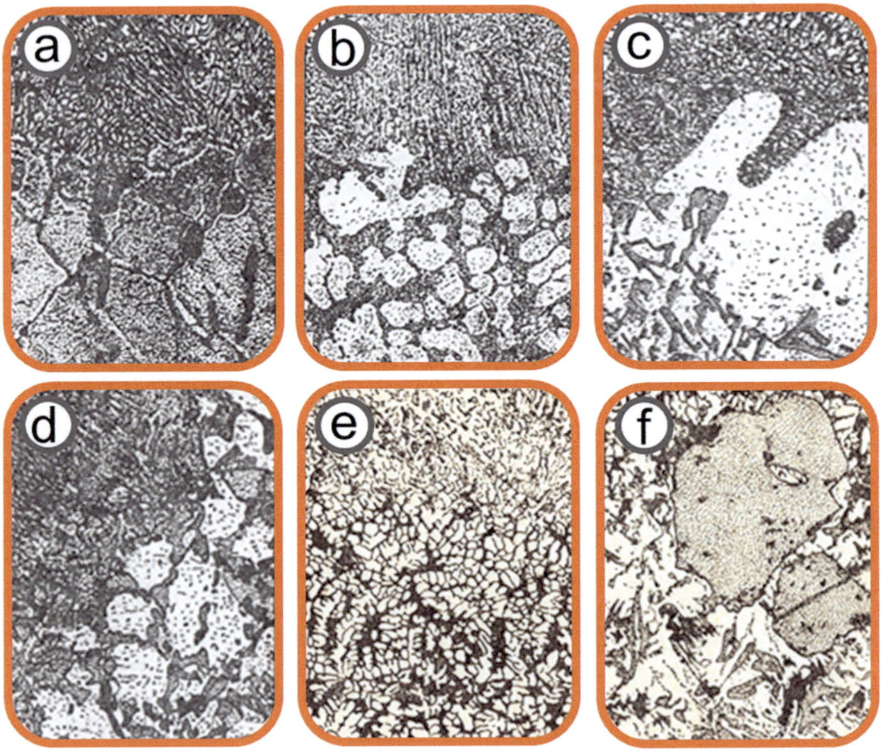

Fig. 4.35 Microstructure of surface layers of aluminum alloys after treatment with continuous laser radiation: **a**—AK4 alloy; **b**—AL4 alloy; **c**—AK5M7 alloy; **d**—AK12M2MgH alloy; **e**—AL26 alloy (top of the melting zone); **f**—AL26 alloy (bottom of the melting zone and initial structure)

The initial structure of ***non-eutectic aluminum alloys*** of the ***AK4*** type with a low silicon content is an α-solid solution with strengthening secondary phases of $CuAl_2$, Al_2CuMn, Mn_2Si, etc. The highest hardening is achieved during artificial aging with the formation of Guinier–Preston zones or secondary phases in a metastable state.

Laser treatment removes the hardening due to Guinier–Preston zones or due to the metastability of secondary phases. As a result of laser treatment, the α-solid solution *fine-grained structure* is formed, other results are solution supersaturation, disappearance of secondary phases, and an increase in the defectiveness of the structure (Fig. 4.35a). At the same time, the microhardness increases, but may not reach the initial state of the previously heat-strengthened alloy (Table 4.4).

A greater effect of increasing microhardness is achieved by laser treatment of ***silumin***, an aluminum alloy such as AL4; AK5M7; AK12M2MgH; AL26 with a high silicon content (in the initial state, silumin was subjected to hardening and aging to achieve maximum strength values).

The microstructure of the ***hypo-eutectic*** silumines of the Al-Si system (AL4 type—Fig. 4.35b) and the Al–Si–Cu–Mg system (AK5M7 type—Fig. 4.35c; AK12M2MgH type—Fig. 4.35d) consists of primary light grains of α-solid solution and α + Si eutectic.

The microstructure of AL26-type ***hyper-eutectic*** silumines consists of primary Si crystals up to 100 μm in size (Fig. 4.35f) and stable or metastable intermediate phases: Mg_2Si, AlMnFeSi, Al_4SiFe, $\Theta(CuAl_2)$, $W(Al_xMg_5Cu_4Si_4)$, $Al_5Mg_8Cu_2Si_6$, $Al_4CuMg_5Si_4$, Al_7Cu_2Me.

The *increase in microhardness* during laser processing is due to two *factors*:

(a) refining of the microstructure;
(b) supersaturation of the solid solution due to the dissolution of intermediate phases and silicon.

Microstructure refining

During laser processing of silumines with a composition close to eutectic, crystallization and *growth* of primary crystals or Si are *suppressed*. The microstructure in the melting zone has a quasi-eutectic structure characterized by *high* phase *dispersion*. The microhardness is higher than the eutectic microhardness in the original structure (Table 4.4). Large light gray primary Si crystals (Fig. 4.35f) are reduced in size by 18–40 times during laser treatment (Fig. 4.35e). This changes the morphology of the phases: in the original structure, the eutectics consist of *acute-angled* phases; after

Table 4.4 Microhardness H_{20} of aluminum alloys in the initial α-solution, in the eutectic and after laser melting

Condition	AK4	AL4	AK5M7	AK12M2MgH	AL26
Initial: α-solution eutectic	1110–1180 1600–1790	700–920 1010–1460	970–1140 1720–1950	1000–1330 1070–1890	700–1270 970–2200
After laser treatment	940–1790	970–1790	1460–2570	2350–2650	940–1790

laser treatment, the shape of the particles is close to the *globular* one. This improves the mechanical properties of the alloys.

Dissolution of intermediate phases and silicon

In addition to the grinding of phase components under laser hardening conditions, a number of phases pass from the liquid state to the *solid solution*. The diffractograms of the AL4 alloy in the initial state and after laser treatment show (111) Si lines and (220) Si lines with the superposition of Mg_2Si and AlMnFeSi. After laser treatment, the (311) Si line with an AlMnFeSi overlay and the (400) Al line with a Mg_2Si overlay almost completely disappear. Other Al lines do not change. Thus, the process of dissolution of silicon and Mg_2Si, AlMnFeSi phases in α-solid solution is underway.

Similar processes are observed in the AK12M2MgH alloy. After laser treatment, the lines of the $\Theta(CuAl_2)$ phase (251) and (350) completely disappear with the superposition of other phases: (240) and (402) with $Al_4CuMg_5Si_4$; (130) with Si, $Al_4CuMg_5Si_4$. The lines of $Al_4CuMg_5Si_4$ disappear at $2\theta = 38°18'$. The intensity of the (311) Si line decreases significantly.

The evidence of the supersaturation of the solid solution due to the dissolution of intermediate phases and silicon are:

- High increase in hardness of high-alloyed alloys such as AK12M2MgH compared to low-alloyed alloys such as AL4 (Fig. 4.36).
- Increasing the hardness of silumin during aging after laser treatment. Thus, at an aging temperature of 150 °C, the microhardness in the melting zone of the AK12M2MgH alloy continuously increases over 17 h from 2460 to 3050 MPa.

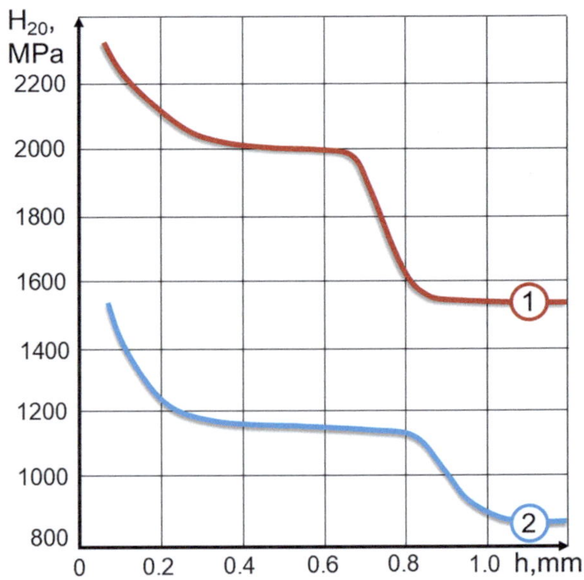

Fig. 4.36 Changes of microhardness by the depth of the heat affected zone in aluminum alloys AK12M2MgH (1) and AL4 (2)

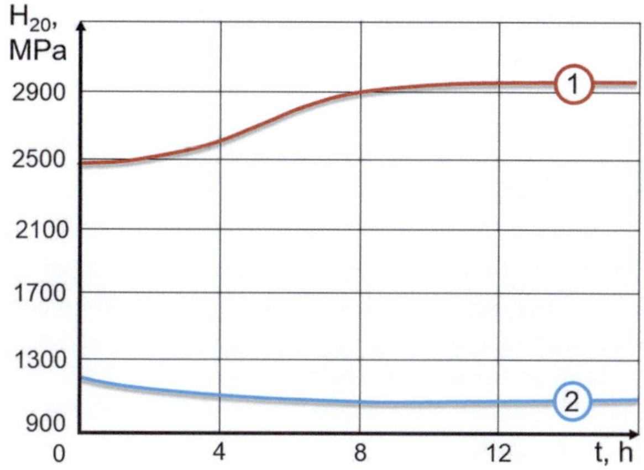

Fig. 4.37 Change in microhardness during aging at 150 °C of AK12M2MgH alloy: 1—in the melting zone; 2—in the non-irradiated area

At the same time, the microhardness of the structural components of the non-irradiated areas changes insignificantly (Fig. 4.37).

Aging after laser treatment

Aging after laser treatment leads to a significant increase in hardness.

Figure 4.38 shows an example of changes in the Vickers hardness of AK12M2MgH alloy during aging after laser treatment.

The laser treatment was performed by continuous irradiation with a power of P = 4 kW and a beam movement speed of V = 10 mm/s.

After laser treatment, the hardness of the fusion zone was HV90.

Air aging for 24 h at 150 °C and 200 °C resulted in an increase in hardness to HV146 and HV119, respectively.

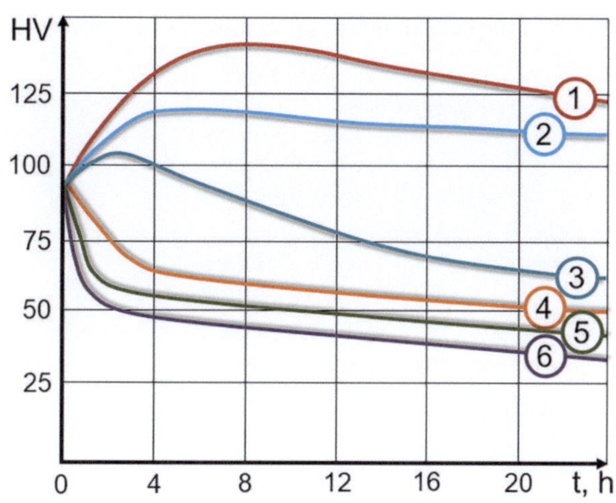

Fig. 4.38 Change in the Vickers hardness of AK12M2MgH alloy during aging after laser treatment: 1—at 150 °C; 2—at 200 °C; 3—at 250 °C; 4—at 300 °C; 5—at 350 °C; 6—at 400 °C

It should be noted that aging at temperatures of 250 °C, 300 °C, 350 °C and 400 °C leads to a decrease in hardness.

4.6 Copper Alloys

The most common are bronzes—alloys of copper with tin, as well as aluminum, silicon, beryllium, and lead. Less commonly used are: brass, an alloy of copper and zinc; melchior, an alloy of copper and nickel; and iron-copper alloys.

In ***dispersion-strengthened bronze*** (2.2–2.6% Ni + 0.5–1.0% Cr + 0.4–0.9% Si), in the initial state, there is no eutectic in the structure and the microhardness is 1520–1860 MPa.

During laser treatment, metastable intermediate phases are dissolved. Upon cooling, a *coarse-grained structure of the α-solid solution* is formed in the *melting zone* (Fig. 4.39a). Outside of the melting zone, the *heat-affected zone is not* metallographically *visible*.

As a result, after laser treatment, the microhardness on the surface of the melting zone does not increase significantly and is around 1460–1900 MPa.

The microhardness decreases *with depth* (Fig. 4.40a, curve 1). The minimum value of microhardness of the order of 1030–1230 MPa is observed in the heat-affected zone. This indicates the dissolution of the hardening phases in the solid state in a certain temperature range.

In ***tin bronze*** of the (3.0–5.0% Sn + 3.0–5.0% Zn + 1.5–3.5% Pb) type with a large crystallization temperature interval, a high cooling rate can cause the phenomenon of *interdendritic liquation* and a *change in phase composition*. In the initial state, the structure of bronze is a single-phase α-solid solution with a microhardness of 860–1070 MPa. In the melting zone, in addition to α-phase grains, chains of the secondary phase are observed (Fig. 4.39c). As a result of interdendritic liquation, the two-phase region of the state diagram shifts towards low concentrations and intermediate phases

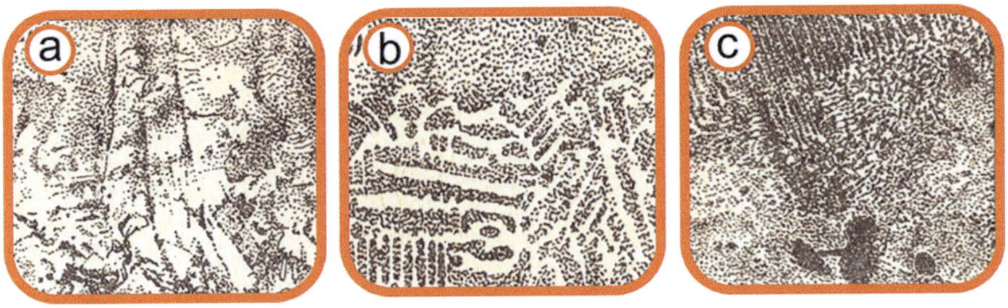

Fig. 4.39 Microstructure of surface layers of copper alloys after treatment with continuous laser radiation: **a**—bronze (2.2–2.6% Ni + 0.5–1.0% Cr + 0.4–0.9% Si); **b**—alloy (70% Fe + 30% Cu); **c**—bronze (3.0–5.0% Sn + 3.0–5.0% Zn + 1.5–3.5% Pb)

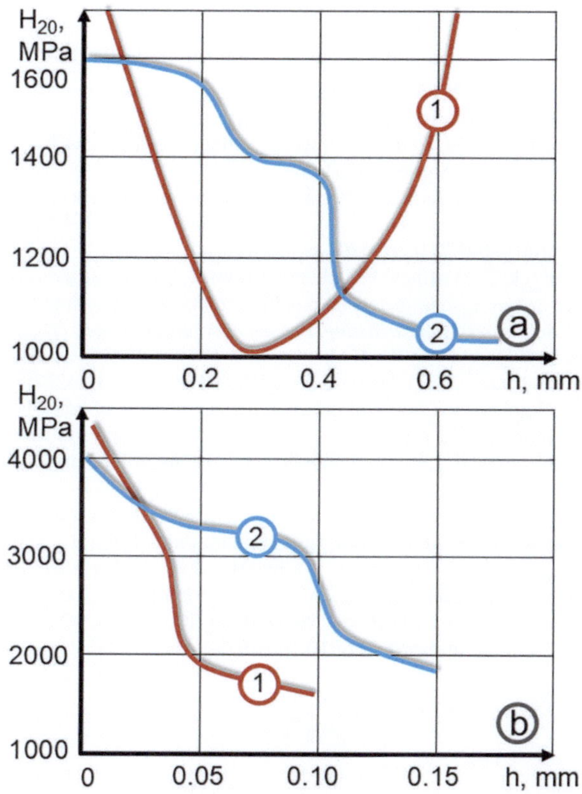

Fig. 4.40 Variation of microhardness with depth in copper alloys: **a**—laser treatment with continuous radiation of various alloys (1—bronze (2.2–2.6% Ni + 0.5–1.0% Cr + 0.4–0.9% Si); 2—bronze (3.0–5.0% Sn + 3.0–5.0% Zn + 1.5–3.5% Pb); **b**—laser processing of the alloy (70% Fe + 30% Cu) with different parameters (1—pulsed radiation; 2—continuous radiation)

β or δ are formed. As a result, the microhardness increases to 1600 MPa when treated with continuous laser radiation (Fig. 4.40a, curve 2).

In the ***alloy*** (70% Fe + 30% Cu), the initial microstructure consists of primary carbides of light crystals (Cu) with a microhardness in the range of 1180–1460 MPa and gray eutectoid α + (Cu) with a microhardness in the range of 2230–2450 MPa.

Laser treatment with melting leads to the formation of finely dispersed quasi-eutectics (Fig. 4.39b). The microhardness in the fusion zone increases to 2600–4000 MPa after laser treatment with continuous radiation and to 3300–4300 MPa after laser treatment with pulsed radiation (Fig. 4.40b).

References

1. Ł. Łach, Recent advances in laser surface hardening: techniques, modeling approaches, and industrial applications. Cryst. **14**(8), 726 (2024). https://doi.org/10.3390/cryst14080726
2. F. Frerichs, Y. Lu, T. Lubben, T. Radel, Process signature for laser hardening. Metals **11**, 465 (2021). https://doi.org/10.3390/met11030465
3. G. Billaud, A. El Ouafi, N. Barka et al., Thermal analysis of laser surface hardening of gears—3D simulation and experimental validation. Int. J. Adv. Manuf. Technol. **137**, 2355–2369 (2025). https://doi.org/10.1007/s00170-025-15292-6
4. D. Lesyk, B. Mordyuk, S. Martinez, V. Dzhemelinskyi, A. Lamikiz, Effects of optimized laser-ultrasonic surface hardening parameters on residual stress and structure-phase state of medium-carbon steel, in *Advanced Manufacturing Processes V. InterPartner 2023. Lecture Notes in Mechanical Engineering*, eds. by V. Tonkonogyi, V. Ivanov, J. Trojanowska, G. Oborskyi, I. Pavlenko (Springer, Cham, 2023). https://doi.org/10.1007/978-3-031-42778-7_27
5. V. Kovalenko, L. Golovko, V. Chernenko,Hardening and alloying of machine parts with a laser beam. V- Kyiv: “Tehnika” 192 (1990)
6. L. Golovko, S. Lukyanenko, Laser technologies and computer modeling. Kyiv: “Vistka” 296 (2009)
7. V. Kovalenko, A. Verhoturov, L. Golovko, I. Podchernyaeva, in *Laser Surface Hardening and Electric-Sparc Surface Hardening of Materials* (Plinum Publish. Corp.”, 1988), pp. 175
8. I. Ryabtsev, S. Fomichov, V. Kuznetsov, Y. Chvertko at al., in *Surfacing and Additive Technologies in Welded Fabrication: International Welding Engineers Textbook Series Under the Editorship of Borys Paton*, ed. by B. Paton (Switzerland AG: “Springer Nature”, 2023), pp. 229. ISBN 978-3-031-34389-6. https://doi.org/10.1007/978-3-031-34390-2

Chapter 5
Physical and Mechanical Characteristics of Surface Layers After Laser Treatment

Abstract A direction for increasing the service life and ensuring the geometric stability of machine parts and structures by improving the physical and mechanical characteristics of the metal surface layer, including wear resistance, the level and distribution of residual surface stresses, corrosion resistance, fatigue strength, and heat resistance, is proposed. The effect of laser treatment on the wear resistance of gas-thermal coatings, metal-cutting tools, and titanium alloys has been investigated. It has been determined that the general mechanism for increasing wear resistance after laser hardening consists of microstructure refinement, increased microhardness, high hardness of the crystalline phase, reducing the friction coefficient, forming a micro-relief with increased oil retention capacity formed by laser surface treatment, and reducing chemical activity. The features of the three-stage mechanism of increasing the wear resistance of titanium alloys after laser surface hardening are shown. The research results are summarized in the form of principles for forming the physical and mechanical characteristics of surface layers during laser processing. Based on the analysis of the mechanism of formation of thermal and phase residual surface stresses, the parameters of laser treatment that determine the stress level have been established and technological recommendations have been proposed to ensure the geometric stability of products. The dual effect of laser treatment on the corrosion resistance of metals has been determined. The positive effect consists in the transformation of selective and uneven corrosion on the surface into uniform continuous corrosion due to the formation of a fine-grained microstructure. The negative effect consists in the formation of a three-electrode electrochemical cell with different standard electrode potentials on the surface and an increase in the corrosion rate at the boundary between them. Technological recommendations for improving corrosion resistance are provided. The maximum heating temperatures of parts during operation have been established, above which a decrease in microhardness and significant structural transformations are observed.

I. Krivtsun et al., *Laser and Hybrid Technologies for Surface Engineering and Welding*, Springer Series in Materials Science 358, https://doi.org/10.1007/978-3-032-15038-7_5

5.1 Wear Resistance

The *purpose* of laser surface treatment is to increase the service life and ensure the geometric stability of machine parts and structures.

The *goal* is to improve the physical and mechanical characteristics of the metal, including wear resistance, level and distribution of residual surface stresses, corrosion resistance, fatigue strength, heat resistance.

Laser surface treatment significantly affects the properties of the materials being treated, positively changing their surface layers' physical and mechanical characteristics. Choosing the right treatment process can greatly increase the wear and corrosion resistance of surface layers [1, 2]. This effect is also achieved by forming a desired distribution of residual stresses [3]. This section presents studies demonstrating the feasibility of enhancing the physical and mechanical properties of surface layers of various materials through laser processing.

Principles of formation of physical and mechanical characteristics of surface layers during laser processing:

(1) the surface hardness and its distribution over the depth depends on the initial chemical composition and the parameters of the heating–cooling thermal cycle, which determines the formation of the microstructure;
(2) laser irradiation can effectively strengthen alloys whose hardness can be increased by conventional heat treatment (Tables 5.1, 5.2, 5.3, 5.4 and 5.5);
(3) alloys whose hardness cannot be changed by traditional heat treatment methods can be strengthened by laser irradiation due to the high cooling rate;
(4) *continuous* laser radiation provides *greater possibilities* of hardness control and a greater depth of the hardened layer (2.0–3.0 mm—with surface melting

Table 5.1 Hardness of carbon steels after laser hardening from the liquid state (LP) and in the solid phase (SP)

Metal	Chemical composition, %	Hardness in the initial state, HV	Hardness (LP), HV	Hardness (SP), HV
Low carbon steels	C 0.03; Mn 0.05; Si 0.01	100	230–320	170–180
	C 0.2; Mn 0.5; Si 0.27	200	450–500	400–650
Carbon steels	C 0.45; Mn 0.65; Si 0.27	250	700–750	650–850
	C 0.65; Mn 1.0; Si 0.27	180	900–1100	1000–1100
	C 0.8; Mn 0.25; Si 0.27	280	1000–1150	1050–1200
High carbon steels	C 1.2; Mn 0.25; Si 0.25	280	1000–1200	1100–1300

Table 5.2 Hardness of chromium steels after laser hardening from the liquid state (LP) and in the solid phase (SP)

Metal	Chemical composition, %	Hardness in the initial state, HV	Hardness (LP), HV	Hardness (SP), HV
Chromium steels	C 0.4; Cr 1.0; Mn 0.65; Si 0.27	280	800–900	800–850
	C 1.0; Cr 1.5; Mn 0.28; Si 0.25	300	1100–1200	950–1100
Nickel-chromium steels	C 0.4; Cr 0.6; Ni 1.2; Mn 0.7	320	600–650	700–750
	C 0.12; Cr 0.8; Ni 3.0; Mn 0.5	260	450–500	500–550
	C 0.14; Cr 1.0; Ni 3.0; Mn 1.0	280	700–800	800–900
Silicon-chromium steels	C 0.9; Cr 1.1; Si 1.4; Mn 0.5	700	700–900	950–1000
Manganese-chromium steels	C 1.0; Cr 1.1; W 1.4; Mn 0.9; Si 0.25	750	1000–1050	950–1100
High-alloy chromium steels	C 1.5; Cr 12.0; Mo 0.5; V 0.25	780	500–600	1000–1100
	C 0.85; W 6.0; Mo 5.0; Cr 4.1; V 1.9	780	700–1100	980–1200
	C 0.75; W 18.0; Mo 1.0; Cr 4.2; V 1.2	800	700–900	950–1100
High- chromium steels	C 0.2; Cr 13.0; Mn < 0.8	230	660–720	600–700
	C 0.4; Cr 13.0; Mn < 0.8	280	830–900	780–820

and 1.0–1.5 mm—without surface melting) compared to pulsed laser radiation (0.5–0.8 mm—with surface melting and 0.12–0.15 mm—without surface melting);

(5) the hardness of the hardened layer after laser treatment practically does not depend on the type of preliminary volume heat treatment—quenching, annealing or tempering (with preliminary volume quenching, the hardened layer has a 20–60% greater depth and a more homogeneous structure);

(6) laser treatment of ***cast irons*** (Table 5.3) is most effective at a low power density of 1–5 kW/cm^2 and a low processing speed of 0.8–1.5 m/min, which ensures maximum and evenly distributed hardness, as well as a reduction in the level of imperfections (cracks and pores);

Table 5.3 Hardness of cast irons after laser hardening from the liquid state (LP) and in the solid phase (SP)

Metal	Chemical composition, %	Hardness in the initial state, HV	Hardness (LP), HV	Hardness (SP), HV
Grey cast irons	C 2.6; Si 1.2; Mn 0.4; S 0.12	160	600–800	380–650
	C 2.9; Si 1.6; Mn 1.0; S 0.12	210	850–1000	850–890
	C 3.4; Si 2.0; Mn 0.8; S 0.02	250	900–1100	880–980
	C 3.5; Si 1.8; Mn 0.5; S 0.2	250	930	250–930
	C 3.5; Si 2.0; Mn 0.5; S 1.0	220	1180	220–1190
	C 3.4; Si 4.0; Mn 0.5; S 0.02	150	1100	150–1100
Alloy cast irons	C 2.7; Si 0.3; Mn 1.0; Cr 27.5	–	600	450–600
	C 2.6; Si 2.3; Mn 1.0; Cr 2.0; Ni 21.0; S 0.02; P 0.02	200	450	200–450
	C 3.0; Si 0.7; Mn 0.8; Cr 1.8; Ni 4.2; S 0.06; P 0.06	–	300	400
	C 3.3; Si 2.1; Mn 6.0; Ni 12.0	180	700	180–700

Table 5.4 Hardness of titanium alloys after laser hardening from the liquid state (LP) and in the solid phase (SP)

Metal	Chemical composition, %	Hardness in the initial state, HV	Hardness (LP), HV	Hardness (SP), HV
Titanium alloys (Ti—base)	Al 4.0; Mn 1.4	280	800–1600	350
	Al 1.5; Cr 2.5	180	780	200
	Al 5.1; Cr 2.0; Mo 1.9	340	750–890	360–500
	Al 5.0; Sn 2.5	300	800	340
	Al 6.0; V 4.6	340	860–890	490–510
	Al 6.5; Mo 3.3	280	600	400
	Al 2.7; V 4.5; Mo 5.0	380	490–510	300–430

Table 5.5 Hardness of aluminum alloys after laser hardening from the liquid state (LP) and in the solid phase (SP)

Metal	Chemical composition, %	Hardness in the initial state, HV	Hardness (LP), HV	Hardness (SP), HV
Aluminum alloys (Al—base)	Cu 2.2; Mg 1.6; Si 0.8; Ni 1.1	170	90–180	170
	Cu 4.2; Mg 1.5; Mn 0.8	150	100–170	150
	Cu 0.3; Si 9.5; Mg 0.3; Fe 0.64	60	90–100	60
	Si 7.0; Mg 0.3	90	100–120	90
	Si 5.0; Mg 0.4; Cu 7.0	90	120–260	90
	Si 12.0; Mg 1.0; Mn 0.5	120	240–270	120
	Si 13.0; Mg 0.9; Ni 1.0; Sn 0.01	110	180–200	110

(7) laser treatment of ***cast irons*** with a *pearlite* matrix provides a higher hardness of the hardened layer compared to a ferrite matrix;
(8) laser hardening of cast irons *without melting* the surface produces a *heterogeneous* microstructure with a wide range of microhardness;
(9) under the same laser treatment parameters, the *depth* of the hardened layer in *steels* (Tables 5.1 and 5.2) is 20–40% greater than in cast irons (Table 5.3);
(10) the hardness of ***aluminum*** alloys increases only during laser processing *with* surface *melting* (in eutectic alloys by 1.5–2 times);
(11) *aging* of aluminum alloys at 150 °C for 15 h after laser treatment increases the hardness of the hardened layer by 15–20%;
(12) deformed *dispersion-hardening aluminum* alloys *lose* their hardness and strength during laser processing;
(13) laser processing of ***titanium alloys*** (Table 5.4) is most efficiently performed with surface *melting* in an *air* environment (when laser processing in an inert environment, the hardness is 1.2–1.5 times lower);
(14) laser processing of titanium alloys *without melting* produces a heterogeneous microstructure and, as a result, a large hardness dispersion;
(15) with an increase in the degree of *alloying* of titanium alloys, the hardness of the hardened layer *decreases*.

When *hardening* surfaces of a large area, an *uneven distribution of hardness* occurs due to the formation of overlapping heat affected zones at the boundary:

- in steels and cast irons—tempering zones with lower hardness;
- in titanium alloys—recrystallization zones with high hardness.

In order to ensure a *uniform hardness distribution* when laser processing large surfaces the following is recommended:

(a) to perform the treatment with a beam *scanning* the surface. In one pass, a hardened zone with a uniformly distributed hardness up to 16–20 mm wide can be produced.
(b) for most steels and cast irons, a three-stage processing is to be used: preheating the workpiece to the temperature of increased resistance of supercooled austenite → laser treatment → further cooling of the workpiece in water.
(c) when hardening the outer and inner surfaces of cylindrical parts, focus laser beams into a ring with different cross-sectional sizes.

Wear resistance is the most important indicator of laser processing and is determined by a number of factors:

- surface hardness;
- dispersion of the surface layer structure;
- the roughness of the contacting surfaces and the coefficient of friction;
- the ratio of physical and mechanical properties of friction pairs;
- the nature of the friction;
- the speed of movement of friction pairs;
- specific pressure in the friction spot;
- the presence of lubrication;
- the ratio of the hardness levels of the contacting surfaces;
- distribution of hardened areas over the friction surface.

Laser treatment with continuous radiation significantly increases the *wear resistance of steels and cast irons* in sliding friction conditions.

For example, after laser hardening by continuous radiation of C45 steel (C 0.42–0.5; Si 0.17–0.37; Mn 0.5–0.8; Ni < 0.25; Cr < 0.25), the sliding friction coefficient decreases by 10% and the wear rate—by 6 times compared to the normalized or improved state (Table 5.6).

Laser hardening improves the wear resistance of parts operating in both *lubricated friction* (maximum efficiency is achieved) and *dry friction*. This increases the wear resistance not only of the hardened part, but also of the part in contact with it.

Laser irradiation *prevents* the formation of a network of *microcracks* on the friction surface of steels, which are typical for bulk hardening. During the friction process, the wear *rate* and friction *coefficient* remain *stable*.

Table 5.6 Friction coefficients and wear parameters of C45 steel under sliding friction on a hard alloy with different heat treatment methods

Type of treatment	Friction coefficient	Wear intensity
Improvement	0.42	3.1×10^{-6}
Normalizing	0.44	2.4×10^{-6}
Laser hardening	0.39	4.6×10^{-7}

Wear resistance of gas-thermal coatings

Gas-thermal powder coatings provide tribotechnical characteristics of carbon steel products. Powders made of iron-based eutectic alloys hardened by refractory borides (such as X10CrNiTi18-10 + TiB_2 + CrB_2) or carbides (such as X10CrNiTi18-10 + TiB_2 + VC) are widely used. Gas-thermal coatings are applied by plasma, detonation, or other methods [4].

In the initial state, the porosity of plasma coatings is 8–15%, and that of detonation coatings is 2–5%.

The microstructure of plasma coatings consists of white and gray layers with individual inclusions of round and irregular shapes, which have a eutectic structure of the cast state. White layers occupy from 60 to 80% of the cross-sectional area, have high microhardness (10,000–12,000 MPa) and therefore determine the wear resistance of the coating.

The parameters of laser irradiation of gas-thermal coatings is chosen so that the penetration depth approaches the coating thickness (for example, $P = 0.8$ kW, $d = 2$ mm, $V = 0.2$ m/min).

Four layers are formed in the surface layer by laser radiation:

(a) a surface molten layer with a fine-grained structure;
(b) molten layer with enlarged crystals of the strengthening phase;
(c) layer with the structure of the original sprayed material;
(d) layer with the structure of the substrate material.

The microstructure of the layers is determined by the temperature distribution and cooling rates along the depth. The cooling rate is maximum in the surface molten layer and is 5×10^4–10^5 °C/s. As you move away from the surface, the temperature gradient decreases, which causes the decomposition of metastable phases in the solid state.

Increasing the wear resistance of the friction pair as a result of laser treatment is due to:

- transferring part of the coating to the cast state;
- formation of oxide films on the melted areas, which are transferred to the adjacent non-melted areas during friction.

The wear pattern of the areas in the melted state is significantly different from that of the original coating: the surface of the original coating has traces of fracture in the form of cracks and chips; the surface of the melted coating is smooth, without visible traces of fracture. Moreover, the wear of a fully melted coating is two times less than that of the original coating.

As the area of the melted areas increases, the friction coefficient decreases.

Minimal overall wear of the friction pair is ensured by scanning the laser beam in zones:

- the smooth melt strip with a cast structure has high strength, low friction coefficient, and is not subject to chipping and peeling,

- a porous strip of outgoing powder coating (optimal width 0.5–3.0 mm) retains lubrication and acts as an oil accumulator.

Wear resistance of metal cutting tools

The friction process is one of the components of the fracture mechanism of metal cutting tools.

The friction of metal cutting tools is evaluated according to the cutting process diagram "prism (tool steel)—cylinder forming element (steel to be processed)".

The wear resistance of tool steel type X82WMOCRV6-5-4 after laser hardening exceeds the wear resistance of steel in the original state (quenching + tempering) by 1.5 times (Fig. 5.1). At the same time, the friction coefficient between the contacting surfaces is halved (Fig. 5.2).

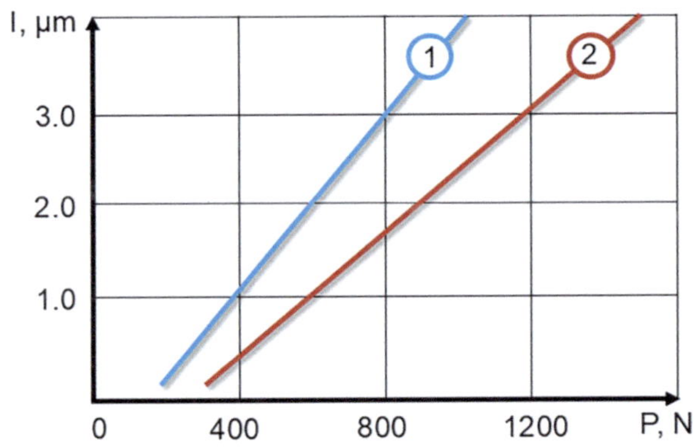

Fig. 5.1 Change in wear intensity (I) of the pair "X82WMOCRV6-5-4 steel—X10CrNiTi18-10 steel" at different pressures (P): 1—in the initial state; 2—after laser hardening

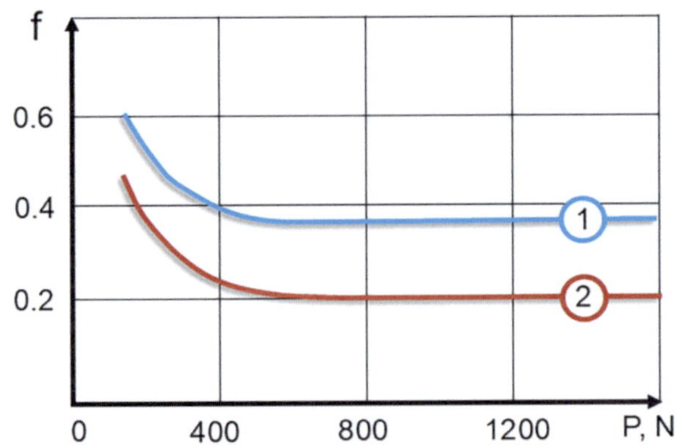

Fig. 5.2 Variation of the friction coefficient (f) of the pair "X82WMOCRV6-5-4 steel—X10CrNiTi18-10 steel" at different pressures (P): 1—in the initial state; 2—after laser hardening

The increase in wear resistance after laser hardening is the result of improving the interrelated physical properties of the surface layer:

- microstructure refinement;
- increase in microhardness;
- reduction of friction coefficient;
- reduction of chemical activity.

Reducing the friction coefficient and of chemical activity of the cutting element surfaces has a favorable effect on the temperature conditions of the tools and wear resistance, improves chip flow on the front surface and reduces the friction of the back surface of the cutting element with the workpiece.

Wear resistance of titanium alloys

The effectiveness of laser surface hardening depends on the loading conditions, the environment, the type of friction area and the geometry of the hardened areas on the contact surface.

The determination of the total linear wear at sliding friction in dry friction conditions and friction with lubrication of the tribo-couple "titanium alloy VT3-1—steel 41Cr4" without treatment and after laser surface hardening (Fig. 5.3) shows:

- For an *untreated* titanium alloy, the linear wear in dry friction (curve 1) is virtually indistinguishable from the linear wear in lubricated friction (curve 2).
- *Laser surface hardening* reduces the linear wear of the titanium alloy by a factor of 25 (curves 3, 4). During dry friction testing, linear wear increases sharply as the hardened layer wears away (curves 3).

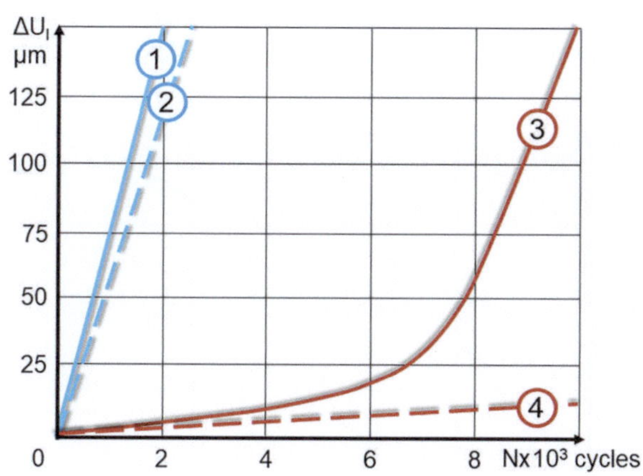

Fig. 5.3 Change in total linear wear ΔU_l during tests at P = 2 MPa of titanium alloy VT3-1 (5.5–6.5% Al, 2.0–3.0% Mo, 0.8–2.3% Cr, 0.2–0.7% Fe, 0.2–0.4% Si): 1—untreated (dry friction mode); 2—untreated (lubricated friction mode); 3—after laser surface hardening (dry friction mode); 4—after laser surface hardening (lubricated friction mode)

- *Lubrication* in combination with laser surface *hardening* increases the wear resistance of titanium alloy by several orders of magnitude (curves 4).

Laser surface treatment increases the wear resistance of titanium alloy-steel friction pairs under *fretting corrosion* by 1.6–3 times.

In laser surface hardening, the decisive *factors for increasing wear resistance* are:

(a) *modification* of the surface layer *structure*;
(b) formation of *microrelief* with increased lubricant *retention* capacity.

The *modification of* the surface layer *structure* in the air environment is formed at ultra-high laser heating and heat removal rates from the surface deep into the material, which determines the *features*:

- fine grain size;
- high hardness of the crystalline phase.

Modification of the surface layer *structure* is achieved by controlling the following *parameters*:

- power density,
- radiation pulse duration,
- coefficient of overlap of laser exposure zones.

Experiments show that the *effect* of increasing wear resistance by laser surface hardening is manifested at a total *treatment area of 20–30%* of the friction surface area.

Laser surface hardening in a neutral medium of the titanium alloy VT3-I working with bronze CuAl10Fe4Ni4 in highly loaded lubricated tribo-assemblies with an *overlap coefficient* $K_{ov} = 0.3–0.5$ increases the wear resistance of the friction pair by 2–6 times, reduces the friction coefficient in contact by 1.5–2.5 times (Fig. 5.4).

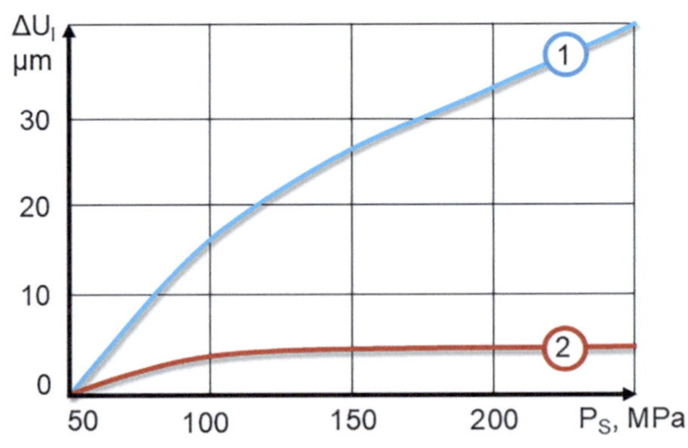

Fig. 5.4 Change in the total linear wear ΔU_l with a change in the specific load P_S of the friction pair "titanium alloy BT3-I—bronze CuAl10Fe4Ni4" under lubrication: 1—without treatment; 2—after laser surface hardening

The *mechanism* of increasing the wear resistance of titanium alloys after laser surface hardening is as follows. At the initial stage of the friction pair destruction process, sliding occurs on oxide adsorption films, which in titanium alloys have a fairly high adhesive bond with the base.

Next, the crystalline dendritic layer is cyclically loaded, which has a high microhardness due to the high level of hardening due to laser irradiation in the active medium. Accordingly, the probability of localized foci of oxide film destruction and subsequent adhesion of contact surfaces is reduced. The formation of hard metal wear particles, the accumulation of oxides and their abrasive effect are inhibited.

At the next stage of the friction process, a layer with a fine-grained structure and embedded crystalline phase areas on the surface comes into contact with the wearing surface. The decisive influence of increased hardness on wear resistance is complemented by the effect of strong cohesive bonds in the newly formed grains, which counteracts plastic deformation.

5.2 Residual Surface Stresses

Mechanism of residual surface stress formation

When parts are hardened using laser heating, the surface layers of the metal are subjected to high stresses. Residual surface stresses are formed from two components: thermal (σ_t) and phase (σ_p).

Thermal residual stresses occur during the cooling phase. The laser-heated zone tends to shorten. However, the adjacent zone of cold base metal prevents this shortening. As a result, an elastic tensile strain ε and, consequently, tensile stresses occur in the laser-affected zone:

$$\varepsilon = \frac{\Delta L}{L} = \frac{L\alpha \Delta T}{L} = \alpha \Delta T \qquad (5.1)$$

where α is the coefficient of thermal expansion;

ΔT is the heating temperature under laser radiation.

Thermal residual surface tensile stresses can reach the yield strength. To balance the stress diagram in the cross-section, the tensile stresses of the laser-affected zone are compensated by the compressive stresses that occur in the adjacent zones of the base metal (Fig. 5.5).

Phase residual stresses arise at the stage of crystallization of the laser-affected zone as a result of martensitic transformation. Martensite has a higher specific volume. When expanding in the free state, a martensite grain should have the volume shown by the red dashed line (Fig. 5.6b). However, there is a constraint from the adjacent grains, which causes compression stresses. To balance the stress diagram in the cross-section, the compression stresses of the martensitic transformation zone are

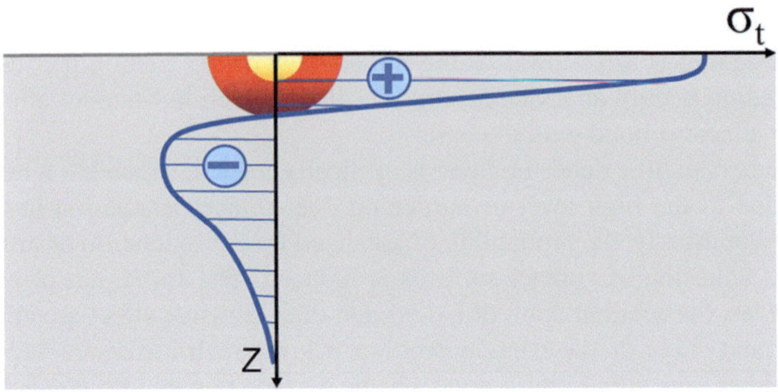

Fig. 5.5 Scheme of distribution of thermal residual stresses σ_t from the surface into the metal Z under laser radiation

compensated by the tensile stresses that occur in the adjacent zones of the base metal (Fig. 5.6a).

The magnitude and sign of the resulting residual stresses σ_Σ depend on the ratio of thermal σ_t and phase σ_p stresses, taking into account possible plastic effects. When $\sigma_t > \sigma_p$, tensile stresses are formed on the surface, and when $\sigma_t < \sigma_p$, compressive stresses are formed.

Note:

(1) *In this section, we consider the longitudinal residual stresses of the first kind directed along the laser processing tracks;*

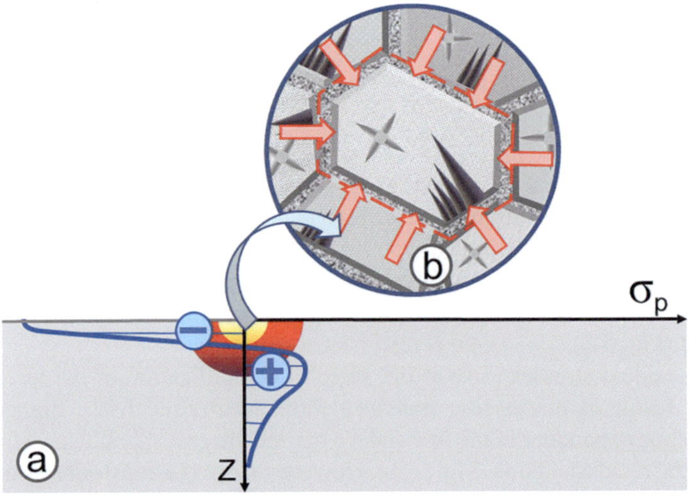

Fig. 5.6 Phase residual stresses σ_p under laser radiation: **a**—scheme of distribution of σ_p from the surface deep into the metal Z; **b**—scheme of formation

(2) *The magnitude and nature of the residual stress distribution were determined by the X-ray method by line displacement on the X-ray images;*
(3) *steel was studied in the initial annealed and hardened state;*
(4) *flat samples were irradiated on technological equipment with a solid-state laser in the pulsed mode and with a powerful* CO_2*laser in the continuous mode in the range of changes in radiation power density* $W_p = 10\text{–}0.200$ *kW/cm*^{2}*, at processing speeds* $V = 0.2\text{–}4$ *m/min, with an overlap coefficient* $K_{ov} = 0.7$ *(*$K_{ov} = S / d_0$*; where S is the processing step;* d_0*is the track width);*
(5) *To increase the absorption capacity, the irradiated surfaces were treated with an alcohol solution of picric acid, a colloidal solution of graphite, or a special ZnO-based coating.*

Geometric stability

Deformations associated with residual surface stresses can cause deviations in the geometry of the product.

The deformation after laser hardening depends on:

- the hardness of the alloys in the initial state—as the hardness increases, the deformation increases;
- the overlap ratio of laser heating zones or hardening tracks—with an increase in the degree of overlap, the deformation increases;
- sequence of applying reinforcement tracks.

To ensure the *geometric stability* of the product, it is optimal to apply spots or tracks of reinforcement:

- starting *from the center to the edges*;
- *symmetrically*;
- *uniformly* on the surface.

Effect of laser power density on the distribution of residual stresses [5]

Patterns of distribution of residual stresses along the depth of the hardened layer depending on the laser power density (Fig. 5.7):

(a) At *low power densities* (when the processed material is heated below the melting point), residual stresses are formed due to the thermal component. Large *tensile stresses* develop in the surface layer, and their distribution area is commensurate with the thermal impact zone. The stress distribution diagram is stretched.
(b) At *medium power densities* (when a thin layer of molten material appears on the surface of the processed material), the phase component begins to act in the formation of residual stresses and compensate for the thermal component. The tensile stresses are reduced and the stresses are *leveled* out in depth.
(c) At *high power densities* (when intense melting of the material occurs on the surface of the material being processed), the phase component is added to the formation of residual stresses. The residual stresses in the surface layer of the

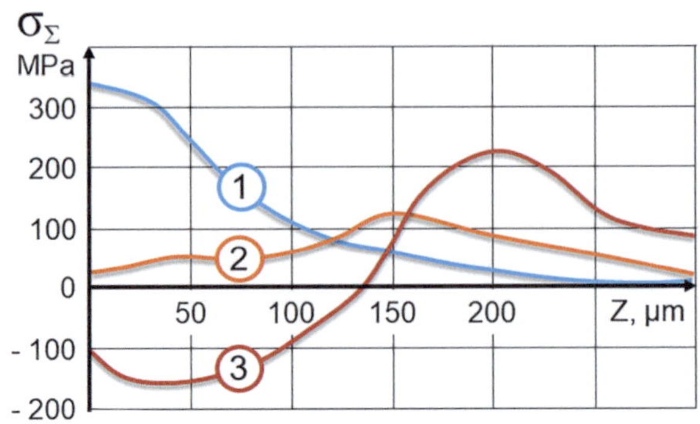

Fig. 5.7 Distribution of residual stresses σ_Σ along the depth z of the hardened layer in C45 steel (0.45% C, 0.3% Si, 0.8% Mn, 0.2% Cr): 1—power density $W_p = 23$ kW/cm^2; 2—power density $W_p = 37$ kW/cm^2; 3—power density $W_p = 48$ kW/cm^2

hardened material switch their sign—*compressive stresses* appear. Their distribution area is commensurate with the melting zone. Tensile stresses increase in the depth of the heat affected zone. The stress distribution diagram is compressed.

Effect of steel alloying on residual stress distribution

As the degree of alloying of martensitic steels increases, the *influence of the phase component* on the level and nature of the residual stress distribution *increases.*

Starting from low power, the influence of the compression stresses of the phase component is more pronounced in higher alloyed steels (curves 1 in Fig. 5.7) compared to lower alloyed steels (curves 2 in Fig. 5.7). With an increase in power, the total surface residual stresses change sign from tension to compression (Fig. 5.7c).

An exception is martensitic steels with a significant amount of residual austenite (curves 3 in Fig. 5.8), in which the initial surface thermal residual tensile stresses are reduced due to the compressive phase component, but they remain tensile even at high power densities (curves 3 in Fig. 5.8c).

The residual stresses extend to a depth commensurate with the depth of the heat affected zone. The total compressive residual stresses generated at high power densities gradually decrease with depth.

Influence of laser radiation parameters

The studies were carried out with laser surface treatment with an overlap coefficient $K_{ov} = 0.7$ ($K_{ov} = S/d_0$; where S is the processing step; d_0 is the track width).

In carbon steels, residual tensile stresses are formed in the surface layer under both pulsed and continuous irradiation (Fig. 5.9). During laser processing with melting, the level of tensile stresses is higher than without melting.

With *pulsed* radiation (curves 1, 2 in Fig. 5.9), residual tensile stresses are formed at a shallow depth and have a significant gradient. The zone of compensating residual

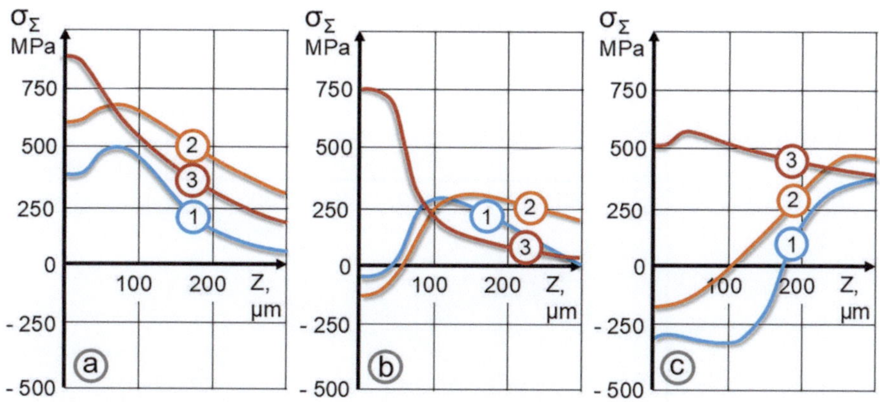

Fig. 5.8 Distribution of residual stresses σ_Σ along the depth z of the hardened layer in martensitic steels (**a**—power density $W_p = 20\ kW/cm^2$; **b**—power density $W_p = 60\ kW/cm^2$; **c**—power density $W_p = 200\ kW/cm^2$: 1—steel of 107WCR5 type (1.0% C, 0.3% Si, 1.0% Mn, 1.1% Cr, 1.4% W); 2—CT80 (0.8% C, 0.3% Si, 0.3% Mn); 3—steel of X165CRMOV12 type (1.5% C, 12.0% Cr, 0.5% Mo, 0.3% Si, 0.3% Mn, 0.3% V)

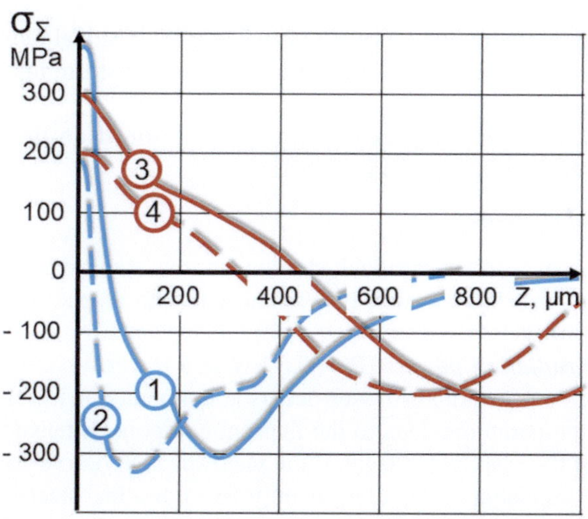

Fig. 5.9 Distribution of residual stresses σ_Σ along the depth z of the hardened layer in carbon steel (0.4% C) after laser treatment: 1—by pulsed radiation with surface melting; 2—by pulsed radiation without melting the surface; 3—by continuous radiation with surface melting; 4—by continuous radiation without melting the surface

compression stresses is characterized by a significant depth of propagation and stress level.

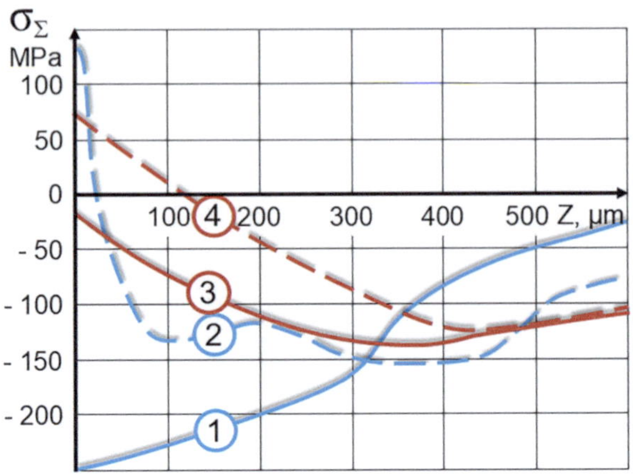

Fig. 5.10 Distribution of residual stresses σ_{Σ} along the depth z of the hardened layer in ferritic cast iron after laser treatment: 1—by pulsed radiation with surface melting; 2—by pulsed radiation without melting the surface; 3—by continuous radiation with surface melting; 4—by continuous radiation without melting the surface

With *continuous* irradiation (curves 3, 4 in Fig. 5.9), residual tensile stresses are formed at a great depth and with a small gradient they transform into lower residual compressive stresses.

In cast irons, residual compression stresses are formed on the surface during melting treatment (curves 1, 3 of Fig. 5.10).

Surface residual stresses

The distribution of surface residual stresses in the transverse (relative to the processing direction) cross-section is symmetrical with respect to the axis of the laser processing track.

At ***low laser radiation powers*** (P = 0.5 kW at a traverse speed of V = 2.0 m/min), the formation of residual stresses occurs at temperatures lower than the phase transformation temperatures. Due to the thermal component, tensile stresses occur on the surface of the track at the level of the *yield strength* (curve 1 in Fig. 5.11).

The phase component is manifested only in high-alloy steels in the form of "dips"—a decrease in the level of tensile stresses with peaks near the edges of the track (curve 2 in Fig. 5.11). The increase in the share of compressive stresses due to the increase in the volume effect during martensitic transformation occurs either with an increase in the concentration of alloying elements and carbon or with an increase in the cooling rate.

With an ***increase in laser radiation power***, the influence of the phase component changes the residual stresses in terms of their magnitude and distribution across the laser processing track.

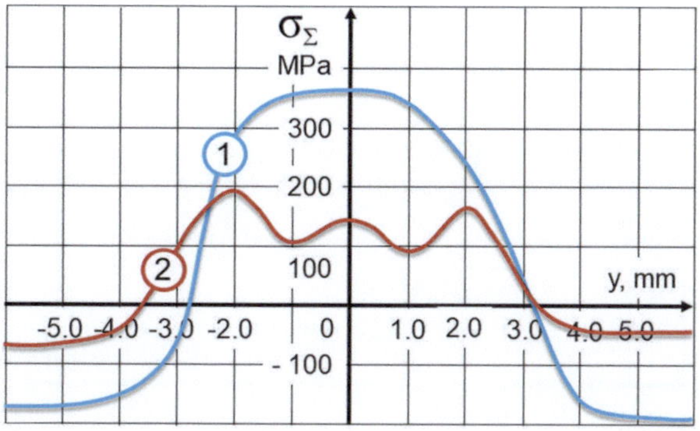

Fig. 5.11 Distribution of surface residual stresses σ_Σ in cross-section after laser treatment (power P = 0.5 kW, traverse speed V = 2.0 m/min, track width 4 mm) without initial hardening of steels: 1—steel of C45 type; 2—steel of X75WCrV18-4-1 type

At a laser radiation power of P = 1 kW, residual compression stresses are formed in the melting zone due to the volume effect during martensitic transformation (Fig. 5.12a).

Increasing the laser radiation power to P = 2 kW (Fig. 5.12b) leads to self-tempering in the central part and to an increase in the amount of residual austenite. As a result, in the center of the melting zone, the residual stress distribution curve rises to almost zero. Deep minima remain at the edges and small tensile stresses at the boundary of the laser impact zone.

At a laser radiation power of P = 3.5 kW (Fig. 5.12c), the width of the melt zone increases significantly, the compression stresses along the edges of the melt zone decrease significantly, and the residual stress distribution curve is smoothed.

The ***laser processing speed*** affects the magnitude and distribution of residual stresses depending on the degree of *alloying.*

In C45 structural carbon steels, residual tensile stresses are formed in the center of the melting zone at low machining speeds. With an increase in the processing speed, the tensile stresses in the center of the strip change sign from tensile to compressive.

In tool carbon steels of the CT80 type, the level of residual stresses in the center of the track and the nature of the transverse distribution do not change with speed.

In tool alloy steels of 107WCR5 type, at low speed, residual compressive stresses are formed in the center of the track (Fig. 5.13a). With an increase in speed, residual tensile stresses appear in the center of the track (Fig. 5.13b and c) due to the process of self-tempering of residual austenite.

The ***type of preliminary heat treatment*** affects the magnitude and nature of the residual stress distribution (Fig. 5.14).

In pre-hardened steel, the tensile stress peaks on the periphery of the track coincide exactly with the boundaries of the hardening and tempering zones.

In low-quenched steel, the tensile stress peaks are shifted to the tempering zone. As the tempering temperature increases, the width of the tensile stress zone increases.

Annealed steel has the widest tensile stress zone.

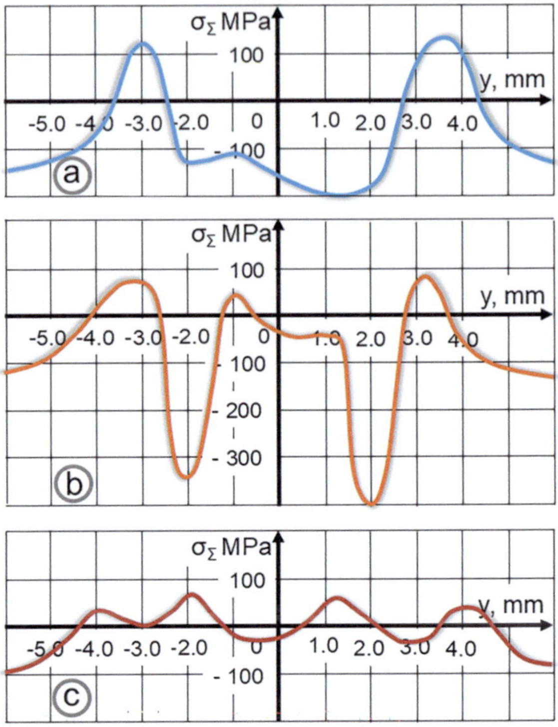

Fig. 5.12 Influence of laser radiation power on the distribution of surface residual stresses σ_Σ in the cross-section of CT80 after laser treatment at a speed of V = 2.0 m/min: **a**—power P = 1 kW; **b**—power P = 2 kW; **c**—power P = 3.5 kW

In pre-hardened steel, tensile stresses are as low as 100 MPa, in low-quenched steel they reach a maximum of 360 MPa, and in annealed steel they drop slightly to 270 MPa.

The ***pulse energy*** of pulsed radiation affects the magnitude and distribution of residual stresses. At all values of the radiation energy, tensile stresses are formed on the surface of the CT80 steel track.

With an increase in pulse energy E_i from 10 to 40 J, the transition from non-melting to significant melting occurs. The residual stresses in the center of the track decrease from 250 to 20 MPa. The magnitude of residual stresses is decisively influenced by the repeated thermal action during the imposition of stains.

The magnitude and sign of stresses switch after treatment with continuous laser radiation with overlapping tracks.

Thus, the effect of the pulse energy on residual stresses in the center of the tracks is the opposite of the effect of power during continuous laser processing.

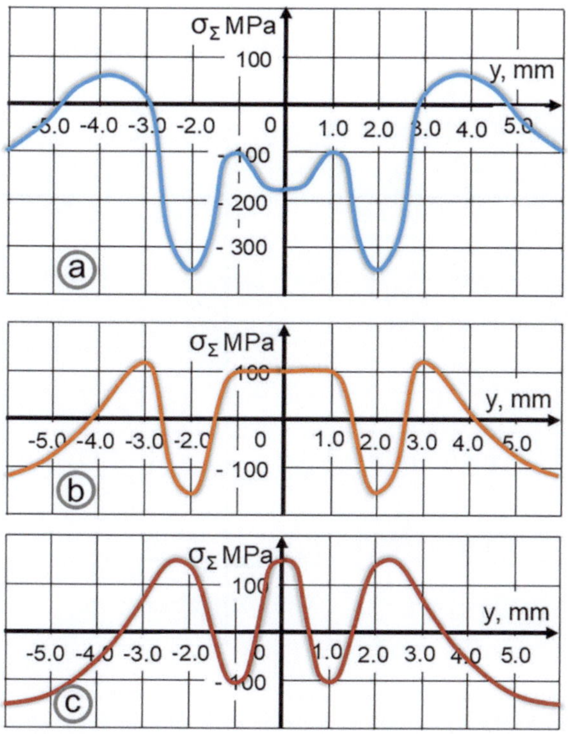

Fig. 5.13 Effect of the laser beam displacement speed on the distribution of surface residual stresses σ_Σ in the cross-section of 107WCR5 type steel after laser treatment at a power of P = 1 kW: **a**—speed V = 0.5 m/min; **b**—speed V = 1.0 m/min; **c**—speed V = 2.0 m/min

Residual deformations

During the cooling phase, the shrinkage force F_T (thermal component) occurs due to thermal contraction of the laser-heated zone. The shrinkage force is applied to the center of gravity of the zone of elastic–plastic deformation during heating–cooling and is equal to:

$$F_T = S_{ep}\sigma_y \quad (5.2)$$

where S_{ep} is the cross-sectional area of the zone in which elastic–plastic deformations occur during heating and cooling;

σ_y is the yield strength.

The phase component of the shrinkage force is the total effect of increasing the specific volume of the martensitic transformation at the stage of crystallization of the laser-affected zone. The phase component of the shrinkage force is opposite to the direction of the thermal component.

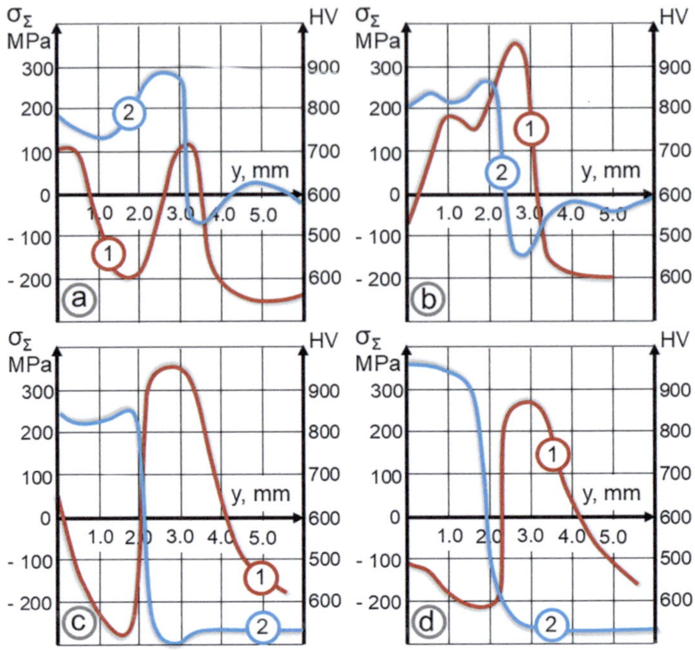

Fig. 5.14 Distribution of residual stresses (curves 1) and microhardness (curves 2) in the cross-section of steel of 100CR6 type (1.0% C, 0.3% Si, 0.3% Mn, 1.5% Cr) after laser treatment at P = 1.3 kW, V = 2.0 m/min with different preliminary heat treatment: **a**—quenching 840 °C; **b**—quenching + tempering 260 °C; **c**—quenching + tempering 600 °C; **d**—annealing

The shrinkage force causes a bending moment relative to the center of gravity of the cross-section of the product, which can lead to distortion of the geometric dimensions.

The ***dimensional stability*** of a product during laser processing depends on:

(a) ***Stiffness*** of the product. Laser processing of *high-stiffness* parts (including thick-walled plates) *does not cause* distortion of geometric dimensions. When laser processing *low-stiffness* parts, primarily thin plates, geometry changes can be quite *significant*. Increasing the length of the product reduces the stiffness and therefore increases the deflection.
(b) ***Shrinkage force values***. The shrinkage force increases with increasing width and depth of the laser impact zone or when laser processing surfaces over a large area. At lower laser power densities, which provide large values of the depth and width of the laser impact zone, there is a noticeable increase in the deflection value.
(c) The ***sign of the shrinkage force***, which is determined by the ratio of the influence of the thermal and phase components depending on the initial structure of the metal. The predominance of the thermal component causes deflection in the

direction of the laser beam, while the predominance of the phase component causes deflection in the opposite direction.

Technological recommendations for improving the dimensional stability of products during laser processing:

(a) to process symmetrically *from the middle towards the edge*;
(b) to place the laser heating zones evenly and symmetrically relative to the *center of gravity* of the product cross-section (to create a compensating shrinkage force, similar laser heating zones can be applied from the opposite side);
(c) provide a *time interval* between the application of successive laser heating zones;
(d) create a preliminary reverse bend before laser processing.

5.3 Corrosion Resistance

Metal ***corrosion*** is the process of chemical, electrochemical or biochemical interaction of a metal with the environment, which results in partial or complete destruction of the metal or changes in its properties.

The most dangerous types of corrosion are caused by the presence of micro- and macro-inhomogeneities on the metal surface (Fig. 5.15a):

(a) *Microstructure heterogeneity* (1). Coarse grains and the presence of texture (directional orientation of grains due to rolling) accelerate corrosion processes. The most favorable microstructure in terms of corrosion resistance is fine-grained. Microstructure heterogeneity results from uneven heating and unequal cooling rates of products, which is typical for welding and related processes. Three zones with different microstructures occur in welded joints: the weld (2), the heat-affected zone (3), and the base metal (4). In solutions, the weld zones form a three-electrode electrochemical cell. Typically, anodic accelerated dissolution processes are localized in the heat-affected zone close to the fusion line.
(b) *Anomalies* in chemical composition: either at *grain boundaries* caused by the accumulation of impurities or in grain zones adjacent to the boundary caused by depletion of alloying elements due to diffusion. The depletion of alloying elements that provide corrosion resistance occurs as a result of the formation of carbides at the grain boundary. Grain boundaries and adjacent zones (5) become anodic sites of metal dissolution. Intergranular corrosion develops (6).
(c) The *release of dislocations* to the surface (7) in the plastic deformation zones reduces the work of ions leaving the metal for the solution, and thus the zones become anode areas.
(d) *Microcracks* (8) and destruction of *surface* passivation films (9) due to the mechanical factor. Anodic processes of metal dissolution are localized at the tops of defects due to the formation of conditions for the release of ions into solution, stress concentration and differential aeration (the appearance of a corrosive

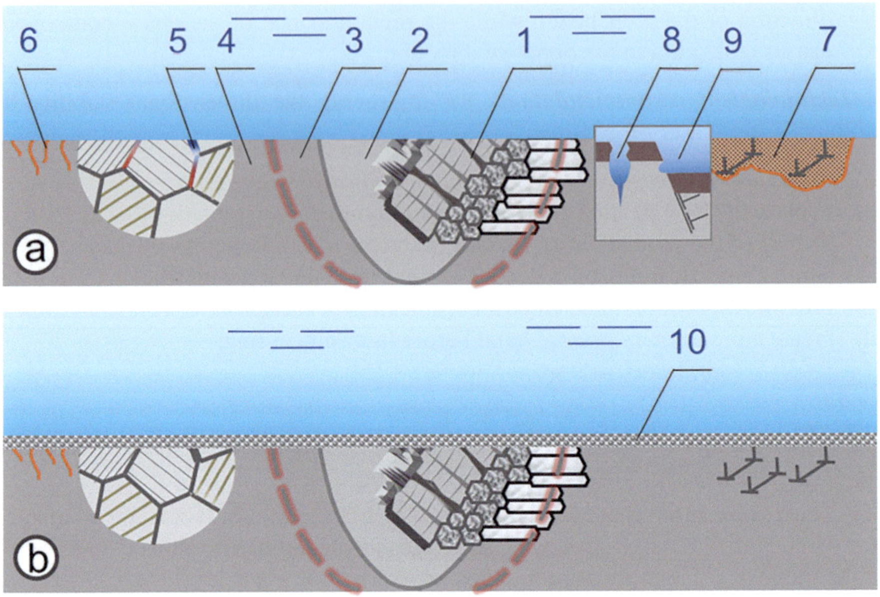

Fig. 5.15 Scheme of corrosion destruction of metals: **a**—in the initial state; **b**—laser treatment (1—microstructure heterogeneity; 2—weld; 3—heat affected zone; 4—base metal; 5—grain boundaries and adjacent zones; 6—intercrystalline corrosion; 7—dislocations coming to the surface in the zones of plastic deformation; 8—microcracks; 9—destruction of surface passivation films; 10—melted layer by laser radiation)

electrochemical element due to the lack of mixing of the electrolyte with a further decrease in the concentration of dissolved oxygen).

The ***positive effect*** of laser treatment on the corrosion resistance of metals is the transformation of selective and uneven corrosion on the surface into uniform continuous one due to (Fig. 5.15b):

(a) obtaining a *fine-grained* microstructure (Fig. 5.16);
(b) uniform distribution of impurities, reduction of anomalies at *grain boundaries* as places of the initial stage of corrosion damage;
(c) dissolution and *uniform distribution of chromium* in solid solution, hence increasing the resistance against intercrystalline corrosion of stainless steels;
(d) obtaining a *cast* structure;
(e) *remelting* of surface *microdamages*, including dislocation exits and microcracks.

The ***negative effect*** of laser treatment on the corrosion resistance of metals is the formation of a three-electrode electrochemical cell on the surface (similar to welds and surfaced beads [6]):

(a) the melting zone;
(b) heat affected zone;

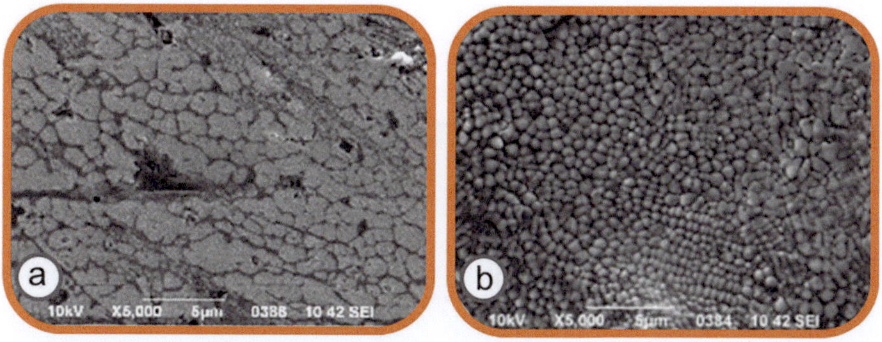

Fig. 5.16 Surface microstructure of samples made of C45 steel: **a**—in the initial state; **b**—after treatment with pulsed laser radiation

(c) the base metal.

Different microstructures of the zones determine different standard electrode potentials (Fig. 5.17). The difference in the electrode potentials of neighboring zones determines the corrosion rate at the boundary between them. Typically, the maximum corrosion rate is observed in the heat affected zone at the boundary with the melting zone.

Therefore, to ensure the corrosion resistance of the product, laser processing should be carried out with ***full coverage*** of the surface in contact with the aggressive environment. The overlap with a coefficient $\boldsymbol{K_{ov} = 0.7\text{–}0.8}$ is optimal.

The corrosion resistance of structural chromium steels [7, 8] in a high humidity atmosphere at a temperature of 30 ± 2 °C after laser treatment with an overlap of $K_{ov} = 0.7$ increases by 20% (Fig. 5.18).

Laser treatment increases the corrosion fatigue strength of carbon steels (Fig. 5.19). Moreover, with an *increase* in the amplitude of cyclic *deformation*, the protective effect of laser treatment *decreases*.

A slight increase in corrosion resistance is achieved at *low* laser beam travel speeds (Fig. 5.20), which ensure the maximum volume of melted metal.

5.4 Fatigue Strength

Material ***fatigue*** is the weakening of a material as a result of the gradual *accumulation of damage* under the influence of alternating (cyclic) stresses due to changes in its properties and the formation and development of cracks, which causes failure.

The ***positive effect*** of laser treatment on fatigue strength is determined by:

(a) obtaining a *fine-grained* microstructure;
(b) uniform distribution of impurities, reduction of anomalies at *grain boundaries* as places of the initial stage of fatigue fracture;

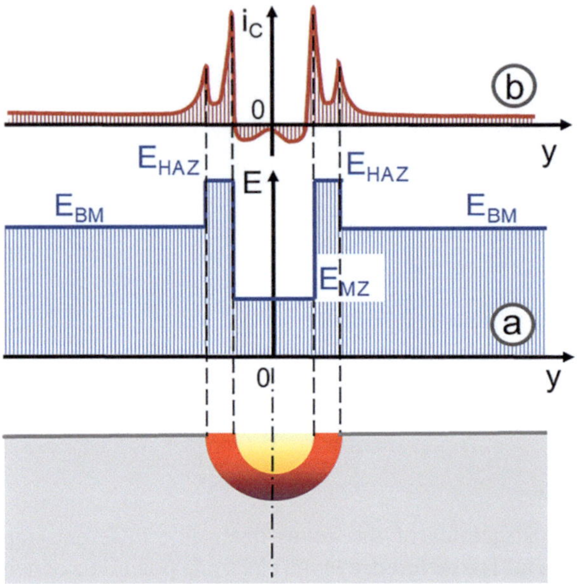

Fig. 5.17 Scheme of cross-sectional distribution of electrochemical corrosion characteristics on the surface of C45 carbon steel in seawater after laser treatment: **a**—standard electrode potential E (averaged over the melting zone—E_{MZ}, heat affected zone—E_{HAZ}, base metal—E_{BM}); **b**—corrosion rate i_c

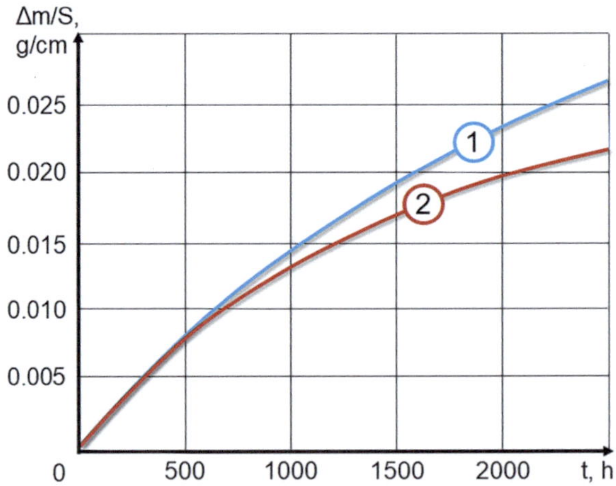

Fig. 5.18 Dependence of the specific change in mass Δm/S (S is the corroded surface area) of samples on time t in an atmosphere with high humidity of 41Cr4 steel (0.4% C, 1.0% Cr, 0.8% Mn, 0.3% Si): 1—in the initial state; 2—after laser treatment

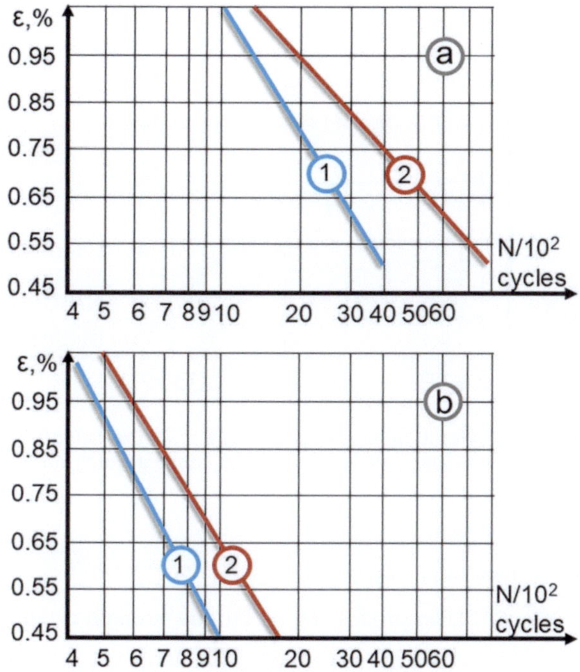

Fig. 5.19 Dependence of cycles to failure N (frequency f = 15 cycles/min) on the amplitude of cyclic strain ε of C30 steel (1—in the initial state; 2—after laser hardening at P = 1 kW; V = 0.2 m/min; d = 4.0 mm; K_{ov} = 0.7): **a**—3.5% aqueous solution of NaCl; **b**—25% aqueous solution of H–COOH

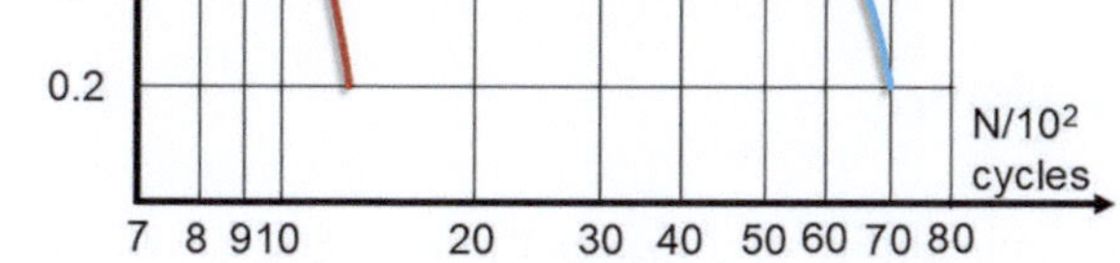

Fig. 5.20 Dependence of cycles to failure N (frequency f = 15 cycles/min, cyclic strain amplitude ε = 0.5%) on the laser processing speed V (P = 1 kW; d = 4.0 mm; K_{ov} = 0.7) of C30 steel: 1—3.5% aqueous solution of NaCl; 2—25% aqueous solution of H–COOH

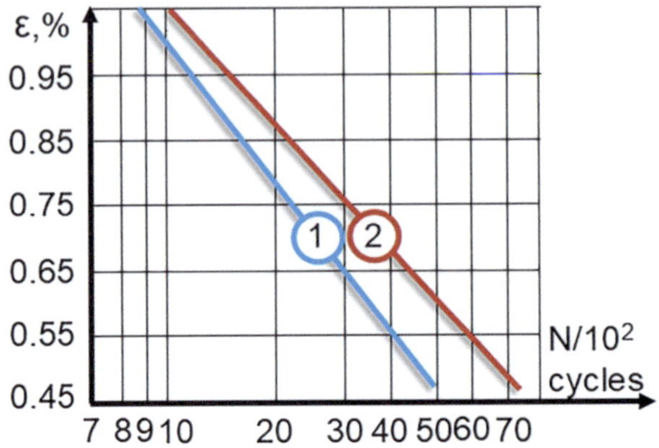

Fig. 5.21 Dependence of cycles to failure N (frequency f = 15 cycles/min) on the amplitude of cyclic strain ε of C30 steel: 1—in the initial state; 2—after laser hardening at P = 1 kW; V = 0.2 m/min; d = 4.0 mm; $K_{ov} = 0.7$

(c) *remelting* of surface micro-defects as *stress concentrators*.

The ***negative effect*** of laser processing on fatigue strength is determined by:

(a) *hardening* as a result of laser treatment, *reduced plasticity* and, consequently, *embrittlement* of the metal surface;
(b) the presence of *residual stresses* and plastic *deformation* on the surface.

The resulting effect of laser treatment on fatigue strength is determined by a combination of the above factors and is regulated by the laser power density.

Laser treatment increases the fatigue strength of carbon steels at a low level of deformation (Fig. 5.21). With an *increase* in the amplitude of cyclic *deformation*, the effect of laser treatment *decreases* (Fig. 5.22).

A slight increase in fatigue strength is achieved at *low* laser beam travel *speeds* (Fig. 5.22), which ensure the maximum volume of molten metal.

5.5 Heat Resistance

Heat resistance is the ability of a material to maintain its operational properties, structural and phase state at elevated temperatures.

Heating of the product during operation can lead to the *tempering* of the laser-hardened surface layer. Therefore, heat resistance during laser processing can be determined by the maximum heating temperature that does not cause changes in the microstructure of the surface layer.

Heat resistance was evaluated on low-carbon steels of C22 type (0.22% C, 0.3% Si, 0.5% Mn), carbon steels of C45 type (0.45% C, 0.3% Si, 0.65% Mn), and high-carbon steels of CT80 type (0.8% C, 0.3% Si, 0.3% Mn). The laser treatment was

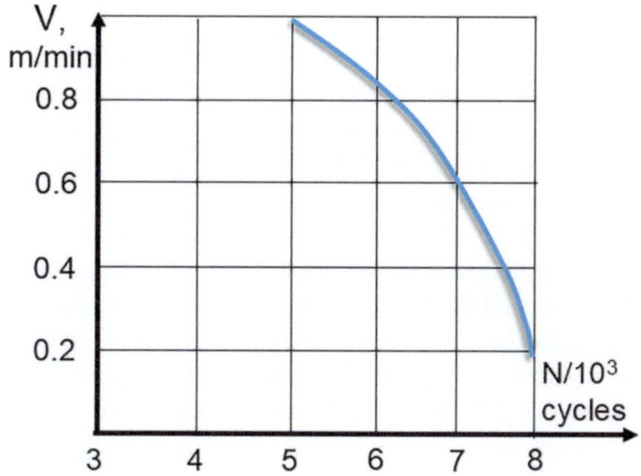

Fig. 5.22 Dependence of cycles to failure N (frequency f = 15 cycles/min, cyclic strain amplitude $\varepsilon = 0.5\%$) on the laser processing speed V (P = 1 kW; d = 4.0 mm; $K_{ov} = 0.7$) of C30 steel

performed by continuous irradiation with a CO_2 gas laser at a power density of $W_p = 25$ kW/cm^2 at a speed of V = 0.25 m/min. The processing parameters ensured the maximum depth of the laser heating zone. The parameters of tempering of the laser-hardened surface layer: heating to a given temperature (100, 200, 300, 400, 500, and 600 °C), holding for 1 h, and subsequent cooling in air.

The *microstructure* of the laser heating zone consists of two layers:

(a) The *upper molten* layer consists of finely dispersed martensite (system of (110) α planes corresponding to angle 2θ = 67°10′) and residual austenite (line (111) γ corresponding to angle 2θ = 68°46′). In low-carbon steels of the C22 type, the diffraction maximum corresponding to the (111) γ-austenite line is not detected due to the low carbon content.

(b) The *heat affected zone* is a transition zone from the upper molten layer to the parent material. The microstructure of the heat affected zone depends on the steel grade. In C22 and C45 steels, the microstructure consists of ferrite grains and martensite formed in place of pearlite grains. In CT80 steels, the microstructure is based on primary martensite.

No changes occur in the laser heating zone during subsequent tempering at ***100 °C***.

An increase in the heating temperature to ***200 °C*** leads to a slight decrease in the microhardness of the laser heating zone due to the decomposition of quenching martensite into tempering martensite with the release of ε-carbides. In addition, in C45 and CT80 steels, the decomposition of residual austenite into α-phase and dispersed carbides begins. Diffraction maxima corresponding to the (110) α-martensite plane system are recorded.

With an increase in the tempering temperature to ***300*** °C, further decomposition of martensite in all types of steels and residual austenite in CT80 steels is observed. In addition, the carbide transformation of ε-carbides into cementite begins, which leads to a violation of the lattice coherence of α-solid solution and carbide. The microhardness of the laser heating zone is significantly reduced.

Martensite decomposition and carbide transformation are completed at a heating temperature of ***400*** °C. All supersaturated carbon is released from martensite in the form of cementite. The tetragonality of the α-solid solution lattice is eliminated and martensite is completely transformed into ferrite.

At tempering temperatures of ***500–600*** °C, C22 and C45 steels change their ferrite structure: the density of dislocations decreases, the grain enlarges, the grain shape approaches equilibrium, and the phase scaling that occurs during laser hardening is removed. In addition, further coagulation of cementite occurs in CT80 steels.

Thus, the heat resistance of steels after laser treatment is maintained when the parts are heated up to ***150–200 °C*** during operation. An increase in temperature leads to a decrease in microhardness and significant structural changes.

References

1. J. Yao, Q. Zhang, R. Liu, G. Wu, in *Laser applications in surface modification* (Springer Singapore Hardcover, 2022) ISBN 978–981–16–8921–5. https://doi.org/10.1007/978-981-16-8922-2
2. D.A. Lesyk, B.N. Mordyuk, S. Martinez, M.O. Iefimov, V.V. Dzhemelinskyi, A. Lamikiz, Influence of combined laser heat treatment and ultrasonic impact treatment on microstructure and corrosion behavior of AISI 1045 steel. Surf. Coat. Technol. **401**, 126275 (2020). https://doi.org/10.1016/j.surfcoat.2020.126275
3. G. Muthukumaran, D. Babu, Analysis of residual stress distribution and corrosion in laser surface hardened low alloy steel with a flat top-hat laser beam, using a high-power diode laser. Arab. J. Sci. Eng. **47**, 8785–8803 (2022). https://doi.org/10.1007/s13369-021-06350-8
4. I. Ryabtsev, S. Fomichov, V. Kuznetsov, Y. Chvertko at al., *Surfacing and Additive Technologies in Welded Fabrication: International Welding Engineers Textbook Series Under the Editorship of Borys Paton*, ed. by P. Borys, (Switzerland AG: "Springer Nature", 2023), pp. 229 P. ISBN 978–3–031–34389–6. https://doi.org/10.1007/978-3-031-34390-2
5. V. Kovalenko, A. Verhoturov, L. Golovko, I. Podchernyaeva, in *Laser Surface Hardening and Electric-Sparc Surface Hardening of Materials* (Plinum Publish. Corp., 1988), pp. 175
6. S. Fomichov, O. Linyucheva, G. Vasyliev, E. Chvertko, in *Corrosion of Welded Structures* (Switzerland AG, "Springer Nature", 2024), p. 222. ISBN 978–3–031–68992–5. https://doi.org/10.1007/978-3-031-68993-2
7. I. Antropov, V. Murga, in *Surface Microstructure and Corrosion Resistance of Alloy Steels after Laser Treatment*. Bulletin of the East Ukrainian National (University named after V. Dahl., 2009), p. 105–106
8. V. Murga, I. Antropov, I. Zhikharev, J. Omeman, The influence of the surface phase composition on the corrosion properties of steel. Issues Nucl. Sci. Technol. Series: Phys. Radiat. Damage Radiat. Mater. Sci. **4**, 136–139 (2011)

Chapter 6
Technological Processes of Laser Surface Engineering

Abstract The physical phenomena, calculations of thermal processes, and technological know-how of the authors on methods of laser surface engineering are presented: increasing the rigidity of thin-sheet metal products; forming sheet metal products; thermo-deformation strengthening of metal products; hardening of cutting tools; thermal cycling of plasma coatings; laser thermal deformation synthesis of diamond-containing composite coatings; laser powder cladding of wear-resistant coatings; laser casting and projection casting technologies for the production of bimetals. To increase the rigidity of thin-sheet metal products, laser forming of structural stiffening ribs is proposed, which consists in directed linear melting of the surface of a thin-sheet product and the occurrence of two effects: the effect of compensation of working stresses due to the formation of residual stresses equal in magnitude and opposite in sign to the maximum working stresses; the effect of reinforcement due to the creation of a pattern of local areas with increased strength and hardness as a result of structural and phase transformations. Based on calculations of the stress–strain state of a thin-sheet disc under various loading conditions, optimal shapes of structural stiffening ribs have been developed. The mechanisms of longitudinal and transverse bending of the sheet during laser processing are considered. Directions for preventing warping during laser processing of thin-sheet parts are proposed by means of: double-sided laser processing; multi-pass laser processing; ensuring a rectangular HAZ shape. Technological solutions for ensuring a rectangular HAZ shape are provided. The mechanisms of laser forming of sheet products are analyzed: temperature gradient, bulging, and polymorphic transformations. Based on calculations of temperature fields and deformations and experimental studies of the influence of the number of passes, the diameter of the focusing spot, and the speed of the laser beam, comprehensive parameters are proposed that ensure controlled laser forming. The features of laser forming of unhardened and pre-hardened carbon steels are analyzed. Technological schemes for combined laser thermal deformation strengthening are proposed, which consists of combining laser ultra-fast local heating of metals with mechanical plastic deformation by a roller. It has been experimentally proven that high-temperature plastic deformation of the surface layer suppresses the development of recrystallization processes, increases the density and directionality

I. Krivtsun et al., *Laser and Hybrid Technologies for Surface Engineering and Welding*, Springer Series in Materials Science 358, https://doi.org/10.1007/978-3-032-15038-7_6

of dislocations, forms branched sub-boundaries, refines grains, increases microhardness and the depth of the hardened layer, and forms compressive residual stresses. Based on the range of favorable temperatures, the optimal distance for applying the deforming load and the force that ensures guaranteed plastic deformation have been determined. Examples of industrial applications of laser thermal deformation strengthening are provided. The adequacy of the mathematical model of the heat transfer process during the strengthening of blade-shaped elements (cutting tools) with a laser beam has been developed and experimentally verified. The influence of the following parameters on the temperature field of the blade has been analyzed: laser beam speed, distance between the laser beam axis and the cutting edge, heating spot radius, and blade sharpening angle. The types of defects and the limiting conditions for their formation have been determined. Based on the analysis of critical values of the laser processing mode, a diagram of the range of technological modes of guaranteed quality of cutting tools was developed using CT80 steel as an example. The scientific and technological principles of combined plasma-laser coating are considered. It consists of two sequential processes: plasma spraying for controlled surface doping and subsequent laser thermal cycling to distribute the alloying elements in depth and ensure a favorable microstructure and mechanical properties of the surface layer. A mechanism of laser thermal cycling has been proposed, which consists in the occurrence of a thermal shock during ultra-fast heating and excitation of an elastic wave, followed by tunneling of the alloying elements, grain refinement, and recrystallization in the deeper layers of the metal. Based on calculations of the temperature field of a moving laser beam in a two-layer medium and studies of mass transfer of alloying elements and changes in the mechanical properties of the surface layer, technological recommendations for laser thermal cycling of plasma coatings have been developed. Examples of combined plasma-laser coating application in industry are given. A new technology has been developed for laser thermal deformation synthesis of diamond-containing composite coatings, which includes laser melting of the product surface with simultaneous dosed addition of metal powder binder and artificial diamonds, followed by plastic deformation of the composite layer during cooling to 650–500 °C. Based on studies of the effect of heating temperatures and laser irradiation conditions on the mechanical properties of synthetic and natural diamonds, quality assurance requirements for high-temperature diamond processing have been established. Based on mathematical modeling of laser thermodeformation synthesis of diamond-implanted composite coatings, optimal technological schemes, laser irradiation parameters, and plastic deformation modes of the composite layer in the production of diamond cutting wheels and tool drums with replaceable cutting inserts with diamond-implanted coatings have been determined. Technological schemes of laser-powder surfacing are considered, which ensure a minimum depth of penetration and a higher proportion of base metal in the deposit compared to electric arc, plasma, and gas surfacing. Based on mathematical modeling of the laser-powder surfacing process, it has been proven that it is possible to control the maximum heating temperature, heating rate, surface layer cooling rate, lifetime, and shape of the surfacing pool by changing the heat flux density distribution function in the focusing zone. It has been substantiated that the optimal distribution of heat flux density is uneven, with the maximum shifted to the edge of the focus spot in the direction of the laser beam. The technological parameters of laser surfacing and the physical and mechanical properties of surfaced layers from nickel-based, iron-based, and boron carbide powders were analyzed. It has been proven that the level and nature of residual stress distribution is determined by two components: the composition of the deposited powder and the nature of the heat flux density distribution of the laser radiation. The mechanisms and factors contributing to crack formation have been analyzed. The optimal parameters of the laser surfacing (power density level, heat flux density distribution, and laser beam travel speed) have been determined, at which the number of cracks is reduced by 2–3 times. Examples of laser surfacing technologies are proposed to increase the service life of fast-rotating parts: gas turbine blades of aircraft engines (by applying a wear-resistant coating to the contact shelves); drill bits (first, by increasing the hardness of the interchangeable bore surface by laser irradiation after normalization and carburizing with quenching, and second, by solid lubrication of the interchangeable bore - cutter friction unit by surfacing with a mixture of bronze and iron powders); diesel engine turbines (by creating a multilayer coating on the working surfaces of the sliding bearing rings). Two laser technologies for producing bimetals have been developed: laser casting and projection casting. Laser casting technology involves melting a rectangular section of the surface layer of the bimetal base to a depth of 30–50 μm with a laser beam, while simultaneously feeding the molten functional component of the bimetal from a foundry induction ladle onto the melted surface. Projection casting technology consists of two operations: forming a macro-relief on the surface of the bimetal base (by laser cutting, milling, or plastic deformation) and pouring the melt of the functional component of the bimetal from a foundry induction ladle over the entire length of the workpiece with a surface macro-relief. Macro-relief can be formed on the surface of the functional component of the bimetal in order to preserve its metallurgical properties (for example, when using heat-resistant alloys). Controlling the configuration of the protrusions allows you to control the temperature distribution, the strength of the bond with the bimetal base, and productivity.

6.1 Laser Stiffening of Thin Sheet Metal Products

6.1.1 Structural Stiffeners

Stiffness is the ability of a structure or its elements to resist deformation (change in shape and/or size) from an applied force *along a chosen direction.*

Thin-sheet metal structures significantly reduce material consumption, energy consumption, and labor intensity. The main task in designing thin-sheet metal structures is to ensure rigidity.

The problems associated with increasing the rigidity of thin sheet metal products, as well as the methods for solving these problems, are described in research [1, 2].

Let's address highly effective methods for controlling the stress state of products using laser irradiation.

The laser creation of ***structural stiffeners*** *involves* directed linear melting of the surface of a thin sheet product and the creation of two effects:

(a) *the effect of compensation of working stresses*—due to the formation of residual stresses equal in magnitude and opposite in sign to the maximum working stresses;
(b) *the reinforcement effect*—by creating a pattern of local zones with increased strength and hardness as a result of structural and phase transformations.

Control of creation of structural stiffeners is performed by:

- laser radiation mode—the size of the focus area, pulsed or continuous radiation, power, speed;
- topography—the location, shape and size of the melting zones;
- the choice of materials that are subject to structural and phase transformations or are subject to thermal deformation processes without transformations.

The stress state of the laser heating zone (Sect. 5.2) is determined by the laser processing parameters and material properties (Table 6.1).

Additional opportunities are provided by combined effects—laser heating followed by plastic deformation.

Table 6.1 Character of residual stresses in the core during laser heating of metals

Laser heating to temperature: $T \leq T_{melt}$	Laser heating to temperature: $T < T_{A1}$
Austenitic corrosion-resistant steels, titanium, aluminum and copper alloys 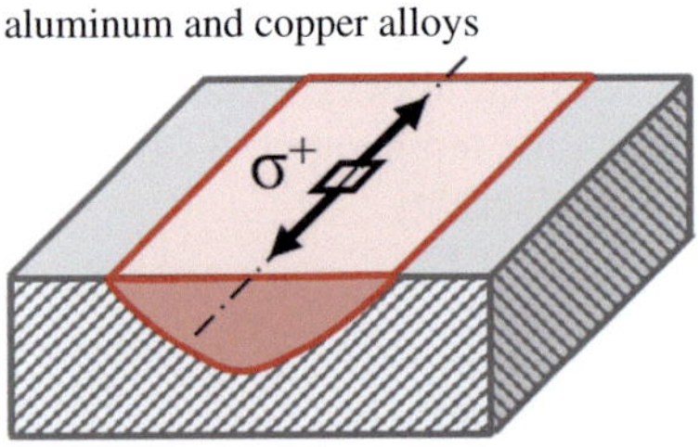	Iron-carbon alloys in which polymorphic transformations do not occur 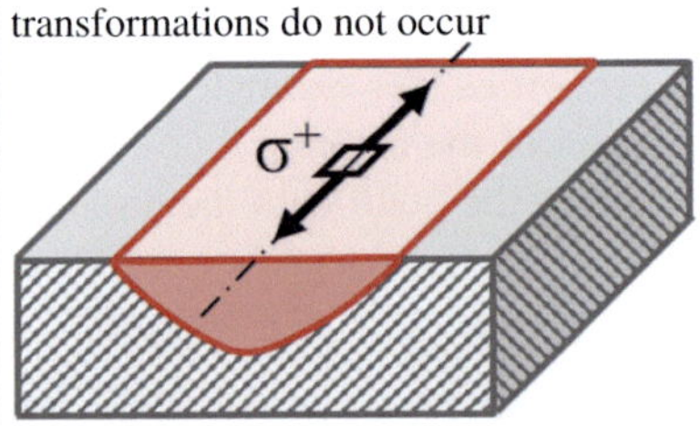
Laser heating to temperature: $T \geq T_{A1}$	Laser heating to temperature: $T \geq T_{A3}$
Iron-carbon alloys in which polymorphic transformations occur 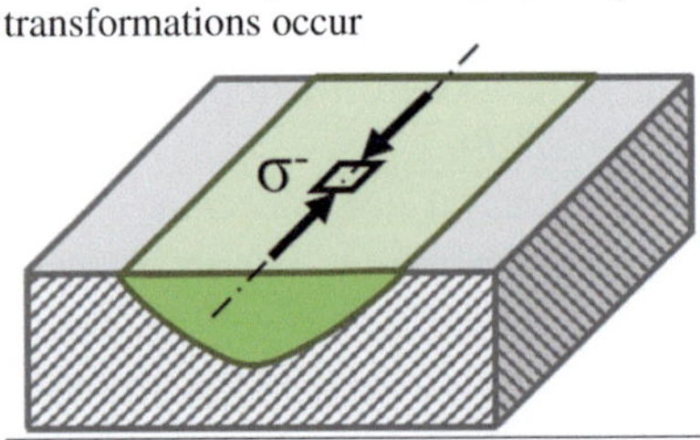	Iron-carbon alloys in which polymorphic transformations occur 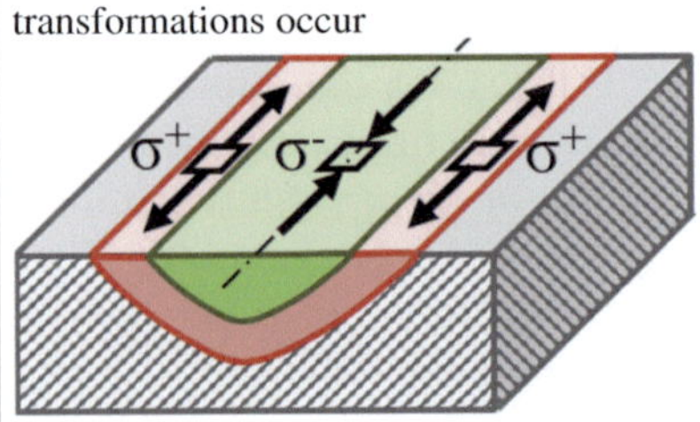

Mathematical modeling of the stress–strain state

The *purpose of the modeling* is to determine the optimal shape of structural stiffeners based on the nature of the distribution, magnitude and sign of stresses that occur in thin-sheet structures under external loading.

The *object of modeling* is the body of a diamond cutting blade in the form of a thin-sheet disk subjected to cutting forces. The stress–strain state under different loading conditions was modeled by the finite element method using the MSC Nastran software [2].

A three-dimensional geometric model of the disk with a part of the shaft is shown in Fig. 6.1. A disk of 200 mm dia and 0.5 mm thick made of chromium steel (1.3% C, 1% Cr) is rigidly mounted on a shaft of 30 mm dia. A torque M is applied to the shaft.

During rotation, the disk is subjected to radial R_1 and tangential R_2 reactions of the material being processed. Calculations were performed in a wide range of R_1:R_2 ratios (20N $\leq R_1 \leq$ 300N; 10N $\leq R_2 \leq$ 420N). Distributions of normal, tangential, maximum tangential stresses and strains in different planes were calculated.

In the local zone of load application, normal compressive stresses act, and in the peripheral zones normal tensile stresses act (Fig. 6.2a, b). The distribution of stresses significantly depends on the ratio of the radial and tangential load components.

With a large radial component ($R_1 \gg R_2$), the zone of action of normal compressive stresses is extended in the radial direction (Fig. 6.2a).

With the increase of the tangential component R_2, the zone of action of normal compressive stresses bends *in the form of a spiral* in the direction opposite to the direction of rotation of the disk (Fig. 6.2b).

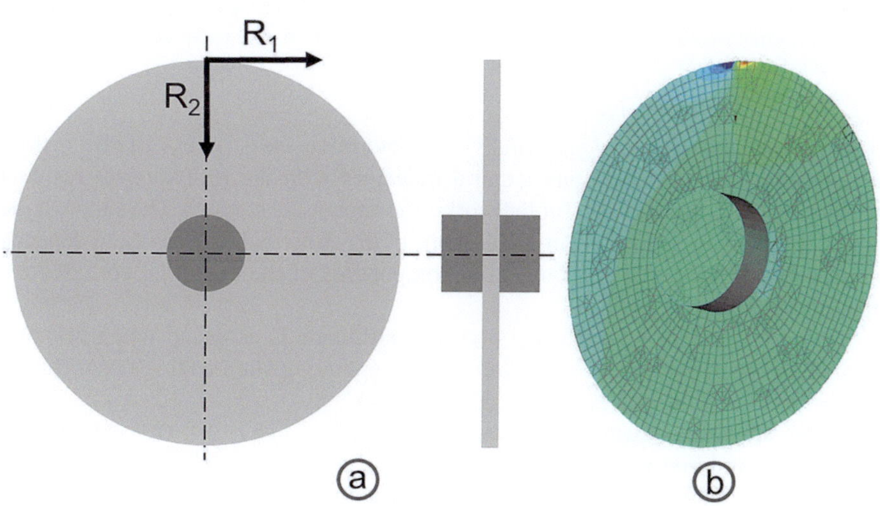

Fig. 6.1 The disk model: **a** calculation scheme; **b** scheme of breakdown into elements

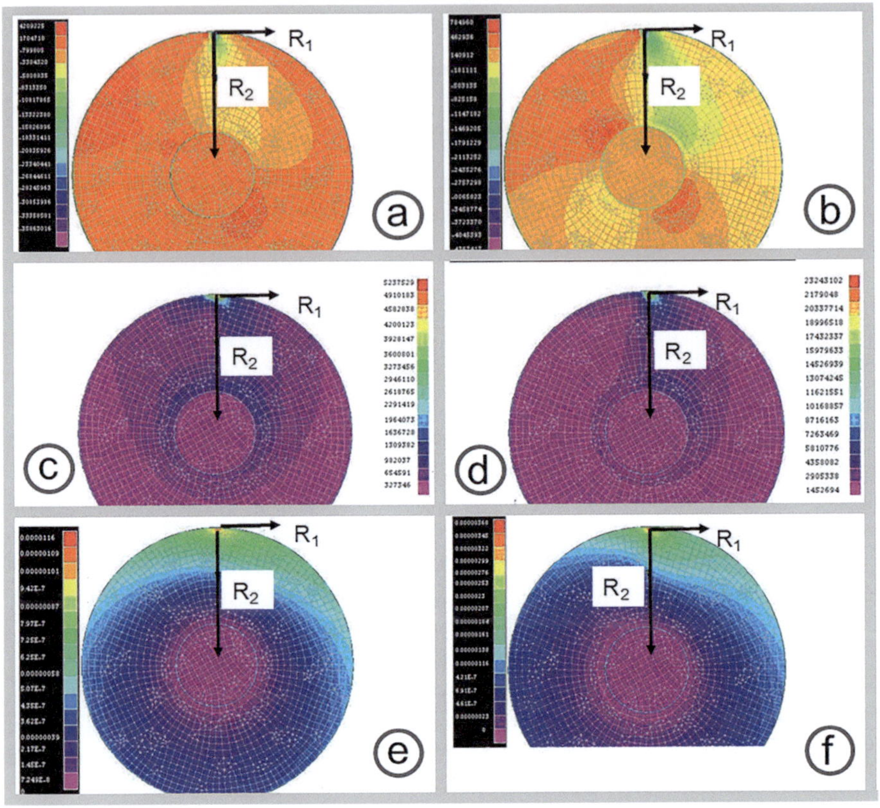

Fig. 6.2 Stress–strain state of a thin-sheet disk under the action of cutting forces R_1, R_2: **a**, **b** distribution of normal stresses; **c**, **d** distribution of tangential stresses; **e**, **f** distribution of deformations (**a**, **c**, **e** $R_1 = 10$ N, $R_2 = 50$ N; **b**, **d**, **f** $R_1 = 20$ N, $R_2 = 80$ N)

Increasing the critical load values that determine the stiffness of the disk is achieved by laser heating along a spiral trajectory with the formation of residual tensile stresses to compensate for the working compressive stresses. The shape of the spiral laser heating path determines the shape of the structural stiffeners and depends on the ratio of the radial and tangential components of the operating compressive stresses (Fig. 6.3).

The highest efficiency of using structural stiffeners is achieved when the total thermal and phase stresses compensate for the operating stresses that occur during the operation of a thin sheet product (Fig. 6.4a, b).

At the same time, structural stiffeners are reinforcement of increased strength due to structural and phase transformations, including the formation of hardened martensitic structures. The creation of a branched pattern (Fig. 6.4c) ensures high stiffness of the thin-sheet product regardless of external loads.

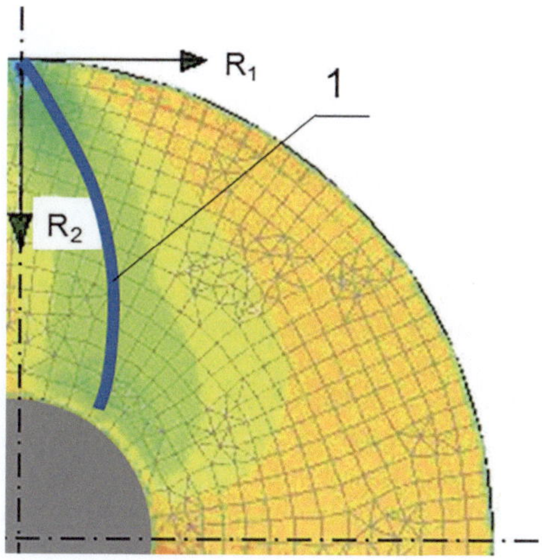

Fig. 6.3 The shape of the spiral trajectory (1) of laser heating that forms a structural stiffener

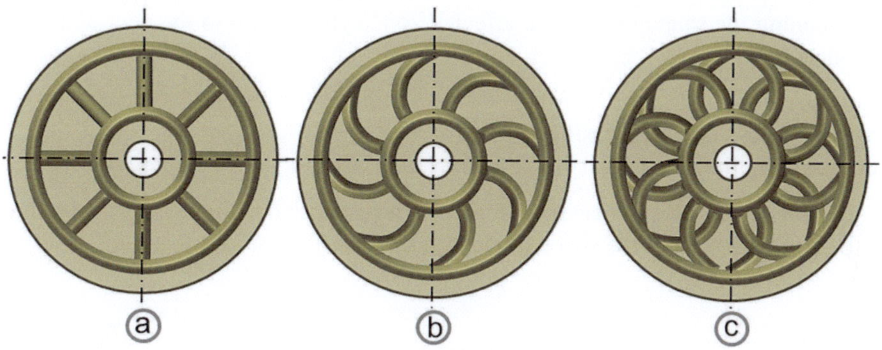

Fig. 6.4 Shapes of patterns of structural stiffeners: **a** for working loads directed along the radius; **b** for working loads directed in the tangential and radial directions; **c** universal

6.1.2 Preventing Excessive Distortion During Laser Processing

Manufacturing of structural stiffeners by local heating with laser irradiation is associated with the occurrence of a mold defect—excessive distortion.

Excessive distortion is a deviation in the shape and dimensions of a product that occurs as a result of residual stresses and strains. Excessive distortion can be considered as a combination of longitudinal bending and transverse bending.

Excessive distortion is a common imperfection in welded joints of thin-sheet structures. The predominant factor in distortion of welded joints is the loss of stability under the influence of longitudinal and transverse shrinkage forces.

In laser processing, the magnitude of shrinkage forces (Sect. 5.2) is small due to the small area of the heat affected zone. The predominant factor in excessive distortion is the occurrence of a bending moment due to the asymmetry of the heat affected zone relative to the neutral axis of the sheet.

The longitudinal bending is caused by the action of the bending moment M_L from the longitudinal shrinkage force P_L on the shoulder δ_c (Fig. 6.5). The resultant longitudinal shrinkage force is applied to the center of gravity of the shape, which limits the heat affected zone (with single-pass laser radiation, the shape of the heat affected zone is usually a triangle). The bending arm is the distance from the center of gravity C of the heat affected zone to the neutral axis of the sheet.

The **transverse bending** is caused by the bending moment M_T due to the uneven distribution of the transverse shrinkage force p_T over the thickness of the sheet (Fig. 6.6). The transverse shrinkage force acting in the layer Δh (where the shape of the heat affected zone can be considered rectangular) is proportional to the linear thermal expansion of the heat affected zone in the plastic zone (zone width Δe). With a triangular shape of the heat affected zone, the maximum transverse shrinkage force acts on the side of the laser irradiation, and the minimum—on the opposite side.

Prevention excessive distortion during laser processing of thin sheet parts is based on obtaining the shape of the heat affected zone symmetrical to the neutral axis of the sheet by:

(a) double-sided laser processing;
(b) multi-pass laser processing;
(c) ensuring the rectangular shape of the HAZ.

Double-sided laser processing (Fig. 6.7a) is performed with the same parameters with strict repetition of the shape of the structural stiffeners. The center of gravity C

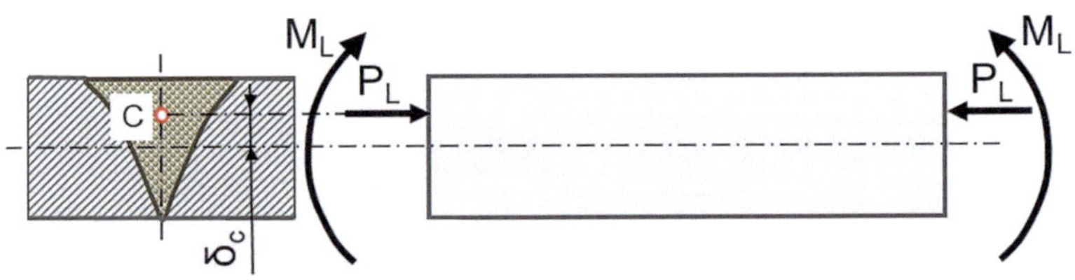

Fig. 6.5 Scheme of longitudinal bending of a sheet during laser processing

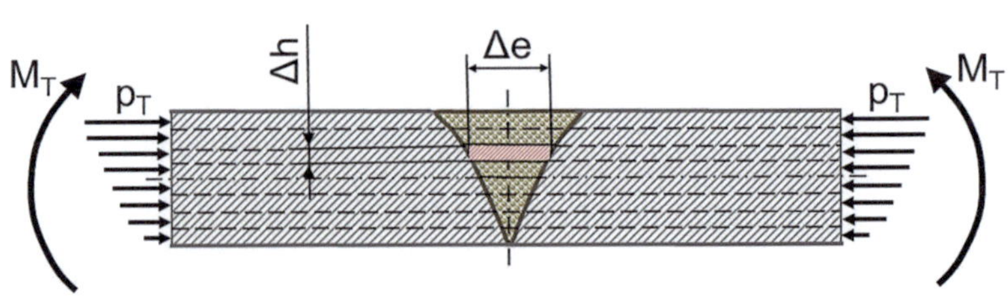

Fig. 6.6 Schematic of sheet transverse bending during laser processing

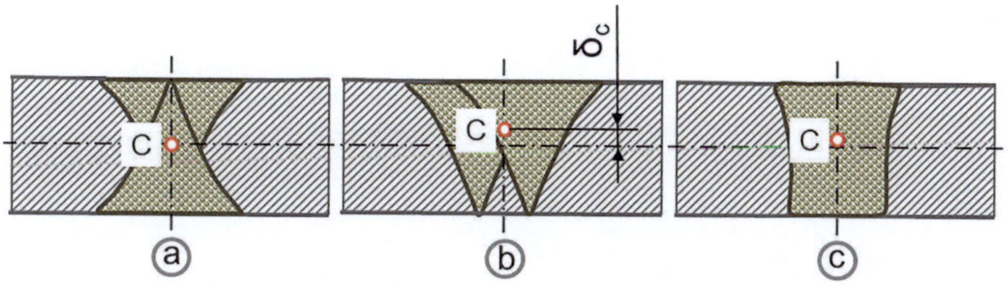

Fig. 6.7 Scheme of methods for preventing excessive distortion during laser processing of thin sheet parts: **a** double-sided laser processing; **b** multi-pass laser processing; **c** ensuring a rectangular shape of the HAZ

of the shape, which limits the total heat affected zone, is on the neutral axis of the sheet, hence the bending moment is zero.

To avoid deformation of the sheet, two-sided laser processing must be carried out in a bonded state. An example of the device is shown in Fig. 6.8. A thin-sheet disk 1 is installed in the heat-removing copper lower matrix 2 and clamped by the upper matrix 3. The dies have grooves through which laser heating is carried out. The shape and size of the grooves correspond to the configuration of the structural stiffeners. After processing, the dies with the clamped disk are turned over and the second side is laser processed. After complete processing, the disk is released.

Multipass laser processing (Fig. 6.7b) is performed on one side with the same parameters in 2–3 passes. With each pass, the center of gravity C of the shape, which limits the total heat affected zone, approaches the neutral axis of the sheet. The bending arm and, consequently, the bending moment are reduced.

The rectangular shape of the HAZ (Fig. 6.7c) is provided by:

(a) controlling the intensity distribution of laser radiation in the cross-section;
(b) processing with a defocused laser beam.

The center of gravity C of the shape, which limits the overall heat affected zone, practically coincides with the neutral axis of the sheet, so the bending moment is close to zero.

The shape of the heat affected zone corresponds to the shape of the temperature isotherms of structural phase transformations. With the Gaussian distribution of laser radiation intensity (Sect. 3.2), the shape of the heat affected zone is close to triangular (Fig. 6.9)—the largest width is on the irradiated surface, the smaller one is on the opposite one.

The intensity distribution in the cross-section of the laser radiation is controlled by focusing (this changes the radiation power density in the focusing plane) and using an optical gradient filter (Fig. 6.10). To ensure that the part is heated to the full depth, the beam power density distribution must be heterogeneous.

When the laser radiation power is unevenly distributed over the heating zone (100% of the radiation intensity in the front and sides of the focusing zone, 25% in

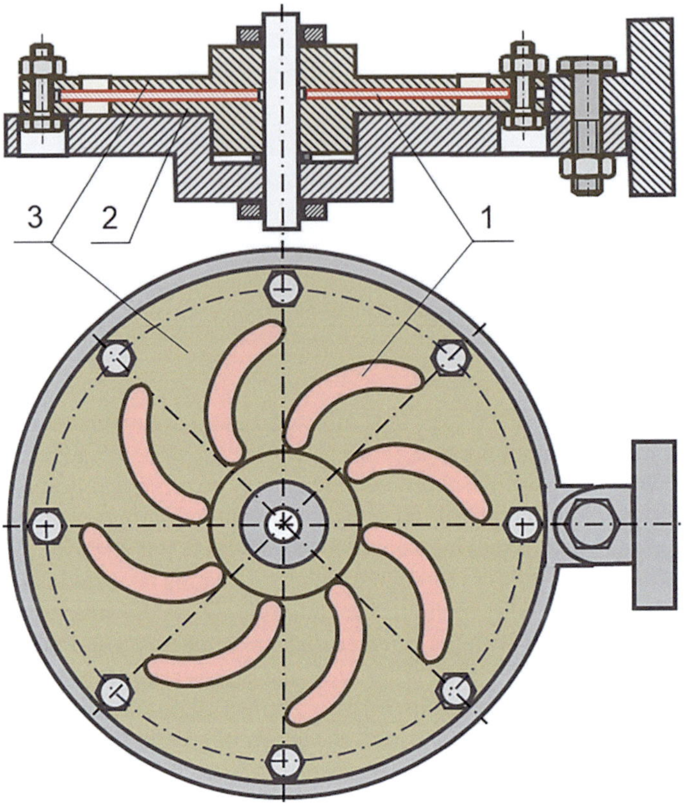

Fig. 6.8 Device for double-sided laser processing of a thin sheet disk: 1—thin sheet disk; 2—lower matrix; 3—upper matrix

the center, and 1% in the tail), the isotherms change their configuration and limit the region of structural and phase transformations to an almost rectangular shape.

6.2 Laser Forming of Sheet Metal Products

6.2.1 Laser Forming Mechanisms

The use of laser radiation for shaping metal products is a well-known process, which is being researched and applied in industry by a large number of scientists and specialists in various industrial fields [3–7]. This section is devoted to highlighting many issues related to the forming with a certain degree of precision of a large number of products from structural metals and alloys, including pre-hardened ones,

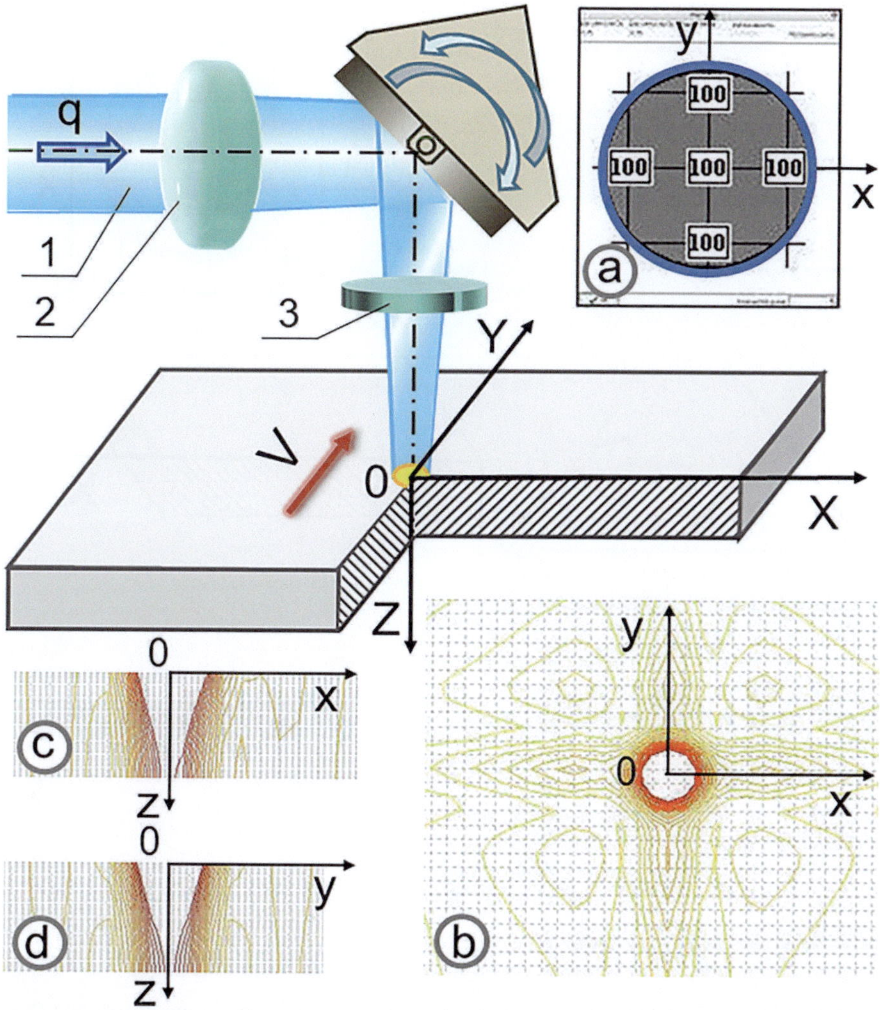

Fig. 6.9 Uniform distribution of laser radiation intensity (**a**) and temperature distribution (**b** on the surface of the sheet; **c** in cross section; **d** in longitudinal section): 1—laser beam; 2—focusing system; 3—optical gradient filter

the forming of three-dimensional cylindrical products with arbitrary generatrices, and so on.

Laser forming consists of the <u>*controlled*</u> formation of a stressed-strained state due to the laser irradiation <u>*parameters*</u> and the <u>*location*</u> of the tracks and, as a result, the formation of the desired product geometry.

Laser forming is used to manufacture shaped parts and allows to:

- use a wide range of materials, including elastic or brittle (e.g. after hardening);

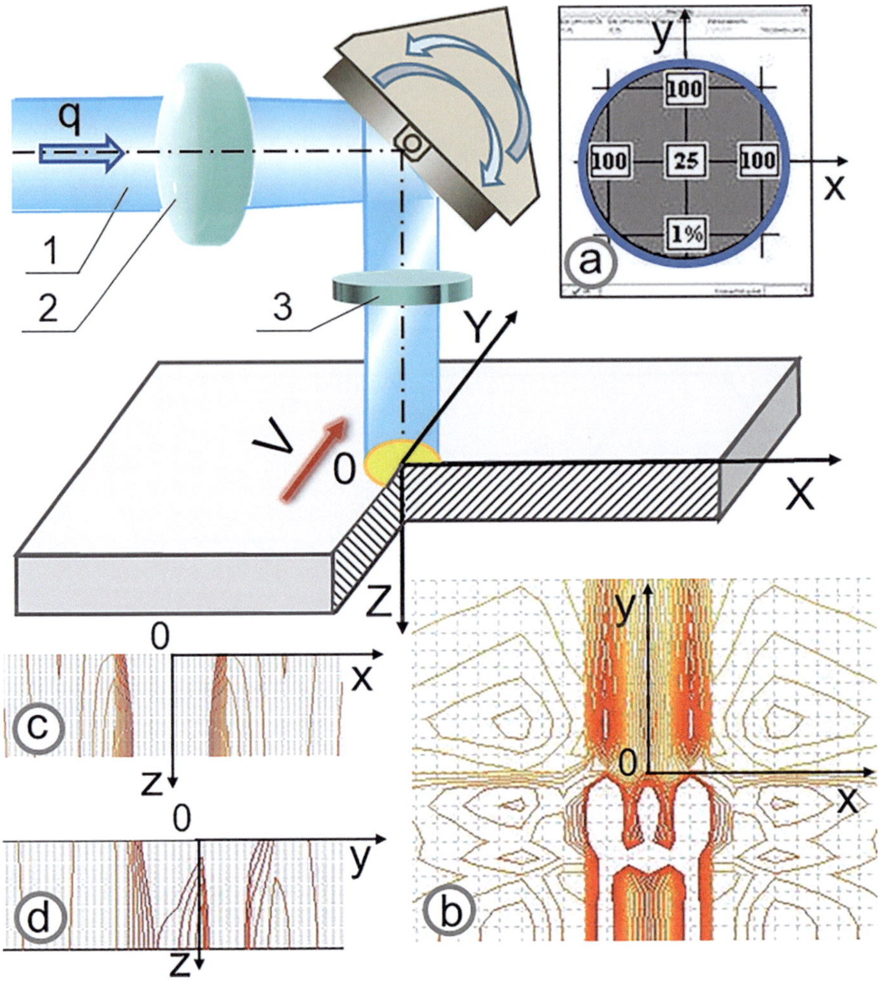

Fig. 6.10 Uneven distribution of laser radiation intensity (**a**) and temperature distribution (**b** on the surface of the sheet; **c** in cross section; **d** in longitudinal section): 1—laser beam; 2—focusing system; 3—optical gradient filter

- increase the accuracy of providing bending angles to 1–2 min by eliminating the reverse elastic effect after removing the working load during stamping;
- bend large-sized products and create local shaped elements in them;
- form products of more complex configuration;
- ensure high rigidity by creating structural stiffeners (see Sect. 6.1);
- eliminate distortion of the cross-section of the workpiece when bending narrow profiles, which consists in reducing the thickness at the bending point, expanding inside the corner and narrowing on the outside;
- eliminate expensive stamping equipment;

- form small bent elements (shelves smaller than 6 mm).

Laser forming is achieved through the manifestation of three complementary physical phenomena (*mechanisms*):

(1) the temperature gradient mechanism—prevails in the presence of unevenly heated layers in depth (when the diameter of the heating spot is commensurate with or greater than the sheet thickness) [8];
(2) the bulge mechanism—prevails when there is a possibility of local loss of stability due to thermal expansion (when the diameter of the heating spot is significantly larger than the sheet thickness) [9];
(3) the mechanism of polymorphic transformations is predominant in the processing of carbon alloyed martensitic steels [10].

The temperature gradient mechanism consists in the different thermal expansion of the upper layer of metal, which is intensively heated by the laser beam, and the lower layer, which remains cold due to the locality of the process, which leads to the emergence of a transverse shrinkage force, the formation of a bending moment and causes bending of the sheet. The mechanism works when the diameter of the laser beam is commensurate with or greater than the thickness of the sheet.

During the heating stage, the thermal expansion of the upper irradiated layer and the lack of expansion of the lower cold layer lead to the following sequential phenomena (Fig. 6.11a):

(a) The temporary transverse force p_h arises in the irradiated layer.

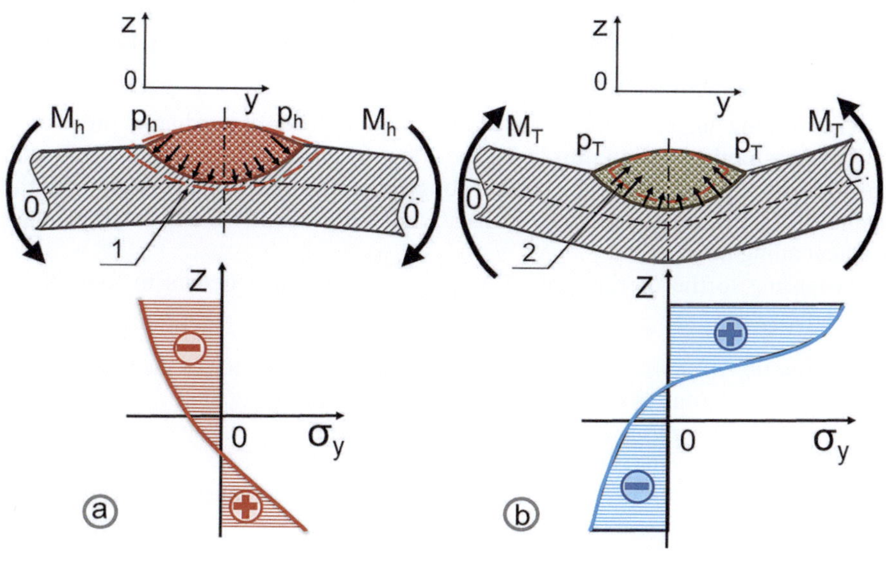

Fig. 6.11 Scheme of laser forming by the temperature gradient mechanism (1—contour of the laser heating zone during free thermal expansion; 2—contour of the laser heating zone during free thermal contraction): **a** heating stage; **b** cooling stage

(b) The force p_h creates a bending moment M_h relative to the neutral plane of the section. There is a slight transverse bending of the plate towards the cold layer.
(c) Since the free thermal expansion of the irradiated layer is prevented by the restraint from the cold layer, a temporary compression strain occurs in the irradiated layer. The compression strain is proportional to the heating temperature (Sect. 5.2) and initially develops in the elastic region (elastic strain). Since the yield strength of metals drops sharply upon heating, a larger proportion of the compression strain occurs in the plastic region—plastic strain is formed. Further thermal expansion is converted into plastic compression strain.
(d) Simultaneously with the occurrence of elastic compressive deformations in the irradiated layer, temporary compressive stresses arise. Initially, the compressive stresses increase in proportion to the magnitude of the elastic deformations, reach the yield point (for the layer heating temperature) and stabilize.
(e) Transverse tensile stresses arise in the cold layer, which balance the compressive stresses of the irradiated layer.

During the cooling stage, thermal contraction of the irradiated layer, which has undergone plastic compression deformation, occurs under conditions of deformation restraint from the lower cold layer. This leads to the following sequential phenomena (Fig. 6.11b):

(a) Residual transverse shrinkage force p_T arises in the irradiated layer.
(b) The force p_T creates a bending moment M_T relative to the neutral plane of the section. There is a significant transverse bending of the plate towards the irradiated layer.
(c) Residual transverse tensile stresses at the yield point arise in the irradiated layer.
(d) Transverse residual compressive stresses arise in the cold layer, which balance the tensile stresses of the irradiated layer.

The bending angle can be increased by repeating the laser irradiation process.

The mechanism of bulging is the local loss of stability of the thin sheet metal due to thermal expansion of the laser heating zone. The mechanism works when the diameter of the heating spot is about ten sheet thicknesses and there is no temperature gradient along the depth.

According to the mechanism of bulging, laser forming occurs in four stages (Fig. 6.12):

(a) *Formation of compressive stresses* due to thermal expansion of the heating zone under conditions of restriction from the cold surrounding area.
(b) *Local loss of stability* and formation of a thermo-plastic-elastic bulge. The bulge is formed along the direction of the laser beam.
(c) The *growth of the bulge* along the entire length of the sheet as the laser beam moves.
(d) *Forming the bending angle* at the end of sheet processing.

The bending angle can be increased by repeating the laser irradiation process.

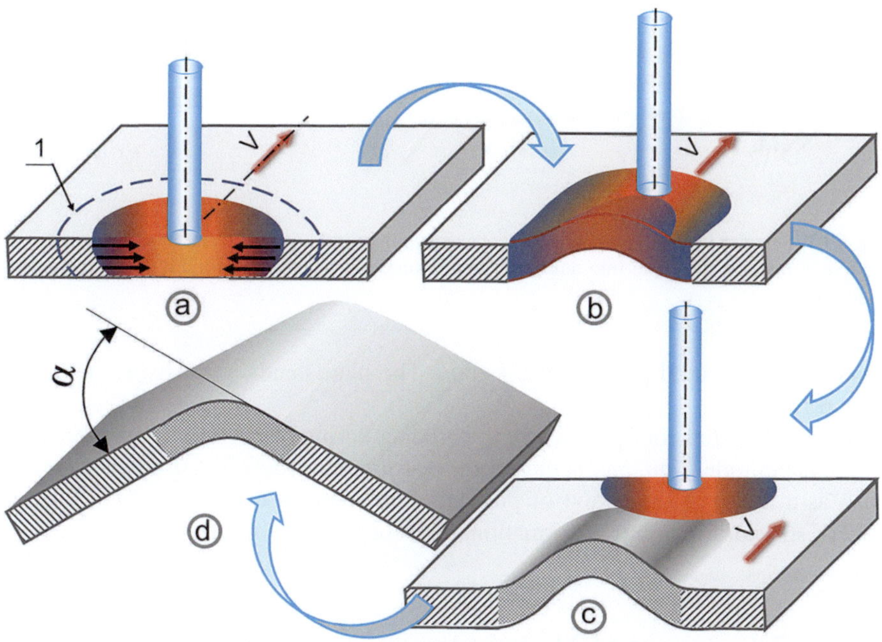

Fig. 6.12 Scheme of laser forming by the bulging mechanism: **a** formation of compressive stresses; **b** local loss of stability; **c** growth of bulge; **d** formation of the bending angle

The mechanism of polymorphic transformations involves the simultaneous expansion of the heat-affected zone and the creation of transverse forces and compressive stresses. This leads to the formation of a bending moment, causing the sheet to bend. This mechanism operates alongside the temperature gradient and bulging mechanisms during the processing of martensitic carbon alloy steels.

During heating at a temperature of A_1 (under equilibrium conditions, the PSK line of the iron-cementite diagram) in the heat-affected zone, *pearlite is transformed into austenite* with subsequent transformation of ferrite either into *austenite* in hypoeutectoid steels in the temperature range $A_1 - A_3$ (under equilibrium conditions, A_3 is the GS line), or into a solution of *secondary cementite* in hypereutectoid steels in the temperature range $A_1 - A_{CM}$ (under equilibrium conditions, A_{CM} is the SE line). The total effect leads to the occurrence of temporary deformation. The direction of the slight transverse bending of the plate depends on the ratio of the change in the volume of the metal during phase transformation and thermal expansion (Fig. 6.13a).

After the laser heating is stopped, as a result of ultra-rapid cooling, the auto-hardening process occurs with the formation of a hardening structure (martensite + residual austenite). The features of auto-hardening are that the specific volume of martensite is higher than the specific volume of austenite and the volumetric expansion occur under conditions of strict spatial limitation. As a result, residual transverse force and compressive stress arise with the formation of a bending moment.

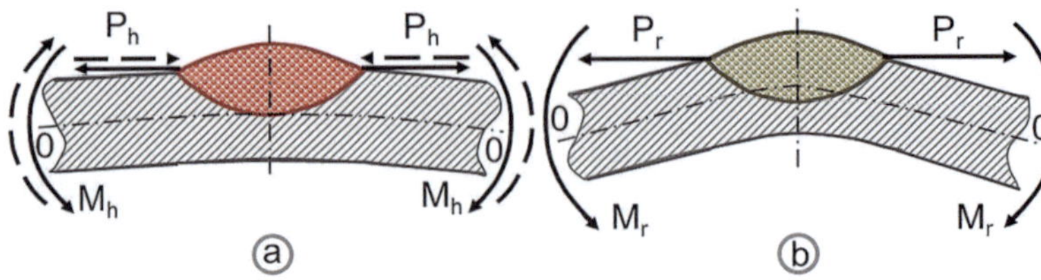

Fig. 6.13 Scheme of laser forming by mechanism polymorphic transformations: **a** heating with the transformation of ferrite into austenite; **b** auto-hardening with the formation of a martensite structure

relative to the neutral plane of the section. The plate bends towards the cold layer (Fig. 6.13b).

The final spatial configuration of the product during laser forming is determined by the magnitude, sign, and nature of the distribution of *residual deformations* that arise as a result of the action of possible combinations of three factors: thermal effects, structural-phase transformations, and forced force action.

6.2.2 Calculations of Thermal Processes During Laser Forming

The mechanism of laser forming is determined primarily by the temperature and cooling rate of the processing zone, as well as the temperature distribution in the adjacent zones of the workpiece.

Calculations of temperature fields when heating a plate with pulsed and continuous laser radiation depending on the speed of movement of the laser beam, power and distribution of radiation intensity, radius of the focusing spot, thickness of the workpiece, and thermophysical properties of the metal are given in Sects. 3.2 and 3. 3.

The software can be developed in the Delphi 7 visual programming environment, using OpenGL to display graphical information and Express Bars Suite 7 to develop the interface of the main program window.

Examples of temperature distribution calculations in the processing zone of 66Mn4 steel (composition: C $=$ 0.66%; Si $=$ 0.17–0.37%; Mn $=$ 0.9–1.2%; Ni $\leq$ 0.25%; Cr $\leq$ 0.25%; Cu $\leq$ 0. 2%) shown in Fig. 6.14.

Increasing the power of the laser beam leads to an increase in the temperature in the processing area.

Increasing the speed of the laser beam reduces the energy input and, as a result, leads to a decrease in the temperature in the processing zone.

When changing the cross-sectional shape of the laser beam to a rectangular or "horseshoe" shape, the heat distribution configuration in the treatment zone repeats

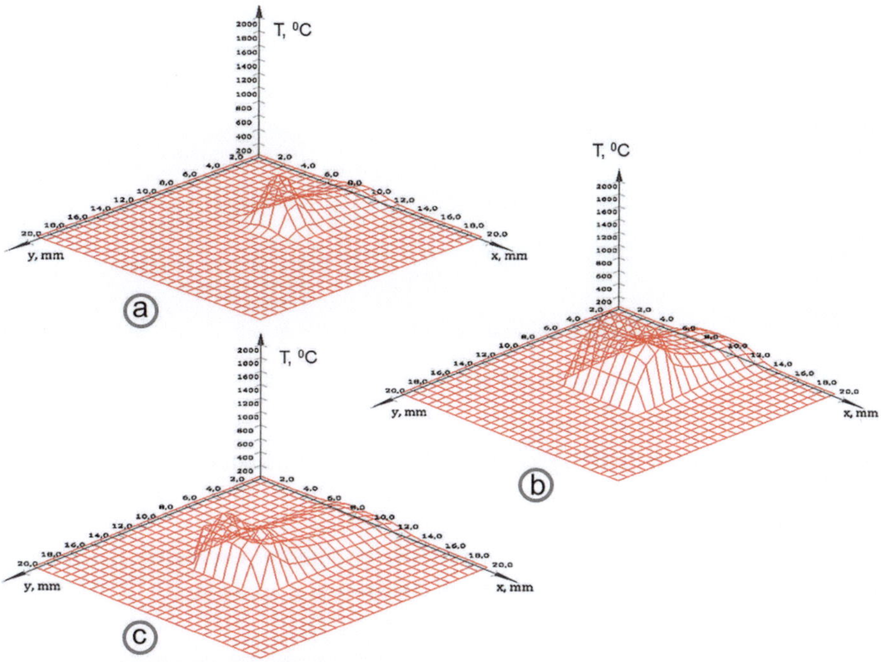

Fig. 6.14 Temperature distribution T in a plate with a thickness of $\delta = 1$ mm made of 66Mn4 steel for different cross-sectional shapes of the laser beam: **a** Gaussian intensity distribution (V = 1 m/min, P = 500 W, d = 3 mm); **b** square beam 3 × 3 mm (V = 3 m/min, P = 1500 W); **c** horseshoe beam d = 3 mm (V = 3 m/min, P = 1500 W)

the shape of the beam, the size of the thermally affected zone and the overall temperature level in the treatment zone increase.

Calculations of sheet metal deformation during laser forming can be performed in the Visual Weld software package.

The solution of the non-stationary problem of thermo-elasticity under the influence of laser irradiation is carried out using the finite element method.

Examples of calculations of temperature fields and deformations when modeling the laser irradiation process are considered on models of plates made of steels of 66Mn4 type and X10CrNiTi18-9 type (composition: C = 0.12%; Si ≤ 0.8%; Mn ≤ 2%; Ni = 8–9.5%; Cr = 18%; C u ≤ 0.3%; Ti = 0.8%) dimensions 100 × 50 × 0.5 mm and 200 × 50 × 0.5 mm with different steps between the heating tracks. In numerical modeling, the change in the thermophysical and mechanical properties of the metal with increasing heating temperature was taken into account. The grid size in the heating track area is 0.25 × 0.25 mm with a gradual increase to 1 × 1 mm (Fig. 6.15). The simulation of the plate clamp during laser irradiation was performed by rigidly fixing the selected nodes on the surface of the finite element model.

The calculated data on the distribution of temperatures and stresses by coordinates along the *Z* and X axes, and further calculations of bending angles for different

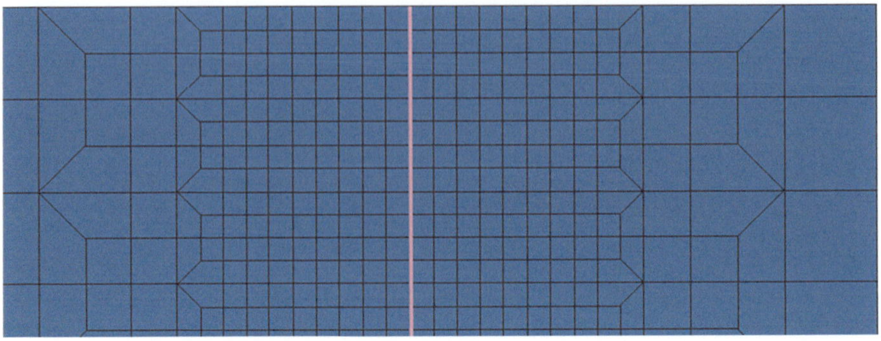

Fig. 6.15 Grid of the finite element model of the plate

heating schemes show good agreement with the experimental results (Fig. 6.16). The difference between the average values of the results of numerical simulation and the experiment does not exceed 1 degree, which is 8% of the discrepancy.

The *step size* between the irradiated tracks determines the curvature of the formed surface and is one of the main parameters of laser forming. Reducing the step between

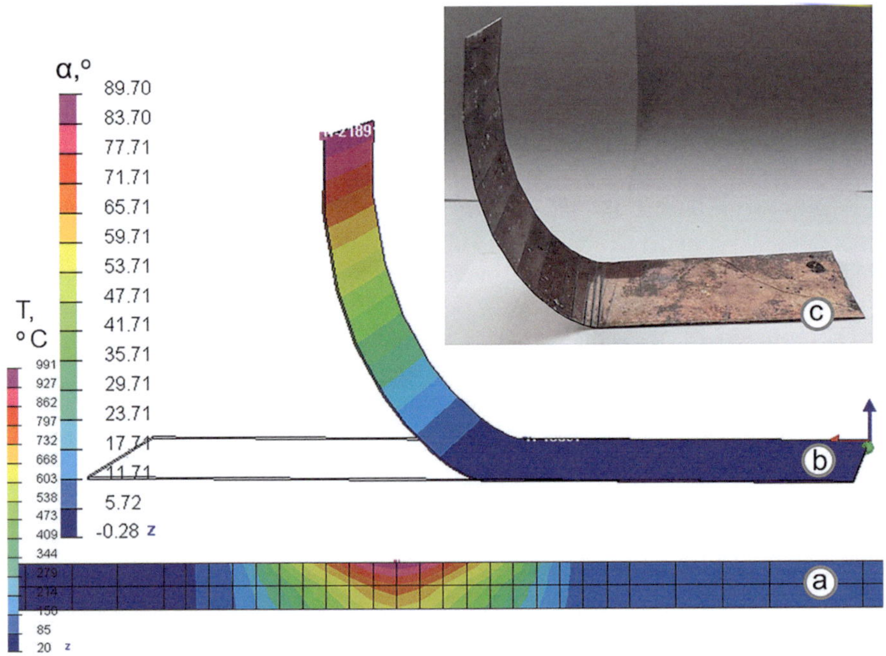

Fig. 6.16 **a** Calculated temperature distribution in the cross section after the 1st pass; **b** calculated changes in the bending angle α during laser irradiation with a variable step, starting from the edge; **c** experimental sample after laser irradiation with a variable step

the irradiated tracks leads to a decrease in the radius of curvature of the formed surface, while maintaining all other parameters of the process. For example, with a step of 5 mm, the radius of curvature of the surface is 29 mm, and with a step of 2.5–13 mm.

6.2.3 Laser Forming Parameters

The factors that determine the formation of residual deformation and the efficiency of laser forming consist of *three* groups.

(1) *Metal properties and workpiece geometry*:

- absorption capacity, which determines the proportion of laser energy used to heat the metal;
- thermal conductivity and heat capacity, which determine the distribution of heat in the volume;
- the initial microstructure and final phase composition of the metal determine the mechanism of laser forming and affect the nature and magnitude of structural residual stresses;
- the geometry of the workpiece, primarily thickness, determines the moment of inertia and the conditions for heat withdrawal from the processing zone to the cold layers of the metal.

(2) Laser forming *parameters*:

- *power* determines the amount of laser energy delivered to the processing area;
- the *diameter* of the focusing spot determines the concentration of laser energy in the processing area;
- the *number of passes* determines the accumulation of plastic deformation;
- the *speed* of movement of the laser beam determines the time it takes for the laser beam to pass through a single point and, accordingly, the amount of energy input into the processing area.

(3) Laser irradiation *conditions*:

- the *trajectory* of the laser beam determines the distribution of residual stresses and strains;
- the *number of passes* (number of repeated heating and cooling cycles) increases the deformation values to the desired level;
- the presence of a forced force action creates the required magnitude and distribution of residual deformations.

The metal properties and workpiece geometry are determined at the design stage based on the functional purpose of the product. However, the metal properties and the ratio of the sample thickness to the focusing spot diameter determine the laser forming **mechanism**, which must be established when determining the parameters:

(a) In austenitic stainless steels and metals where *structural-phase transformations do not occur* during the heating–cooling cycle, the *temperature gradient mechanism prevails.*
(b) In carbon alloyed steels and metals, where *structural-phase transformations occur* during the heating–cooling cycle, *two mechanisms operate simultaneously*: *temperature gradient* and *polymorphic transformations.* The influence of structural-phase transformations on the final stress–strain state of the product *is much smaller* than that of thermal influence.

The **number of passes** determines the total absolute deformation (measured by the bending angle α of the workpiece) in an almost directly proportional relationship for all steel grades (Fig. 6.17).

In steels of the 66Mn4 type, where structural-phase transformations occur, the absolute deformation of the workpiece in one pass is usually lower compared to steels of the X10CrNiTi18-9 type, where structural-phase transformations do not occur (Fig. 6.17). This is explained by the simultaneous action of mechanisms of temperature gradient and of polymorphic transformations and, as a result, the possibility of weakening (or strengthening) the total deformation.

Changing the **diameter of the focusing spot** at $d \leq 6$ mm (with other parameters unchanged) does not change the nature of the dependence of the total absolute deformation on the number of passes.

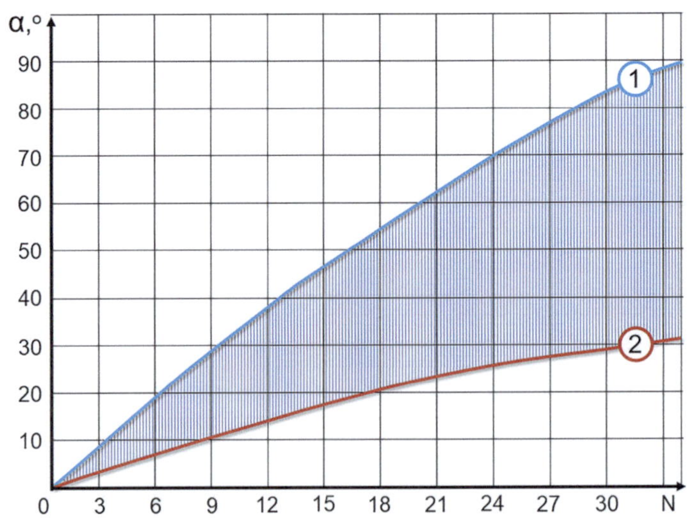

Fig. 6.17 Change in the bending angle α of the workpiece depending on the number of passes N during laser forming (P = 0.8 kW; V = 1.2 m/min.; d = 2 mm; δ = 0.5 mm): **a** steel X10CrNiTi18-9; **b** steel 66Mn4

For steels of 66Mn type, with an increase in the focal spot d ≥ 8 mm, the volume of the heat-affected zone increases. The increase in deformations due to structural-phase transformations accumulates and, starting from a certain pass, can completely compensate for the increase in deformations due to the temperature gradient. This changes the dependence of the total absolute deformation on the number of passes: during the first 10 passes, a proportional dependence is observed, then the increase in the bending angle stops (Fig. 6.18).

The efficiency of laser forming is higher when using diode-pumped solid-state lasers compared to CO_2 lasers due to better absorption by metals of short-wavelength (an order of magnitude shorter) radiation from a solid-state laser. The nature of the dependence of the bending angle on the number of passes is the same.

Laser beam speed is a key parameter of the processing and has a certain optimal range of controlled formation (Figs. 6.19 and 6.20).

Low speed of laser beam movement (V ≤ 1/5 m/s) leads to ***melting of the surface*** due to the high energy input, which is extremely *undesirable* during laser forming.

In the ***range of controlled forming***, increasing the speed of the laser beam increases the efficiency of the process until a certain maximum of efficiency is reached—the minimum number of passes required to form a 90° bend angle. Conditions are created for maximum plastic deformation. After passing the maximum, further increasing the speed increases the required number of passes to form a 90° bend angle, but the process remains controlled. The presence of a wide range of controlled forming allows to vary the laser forming parameters.

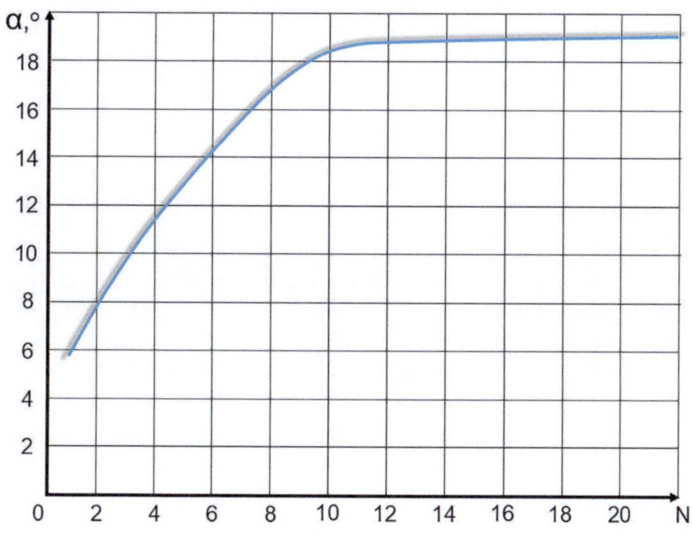

Fig. 6.18 Change in the bending angle α of the workpiece depending on the number of passes N during laser forming of 66Mn4 type steel with a focusing spot d = 8 mm dia (P = 0.8 kW; V = 1.0 m/min; δ = 0.5 mm)

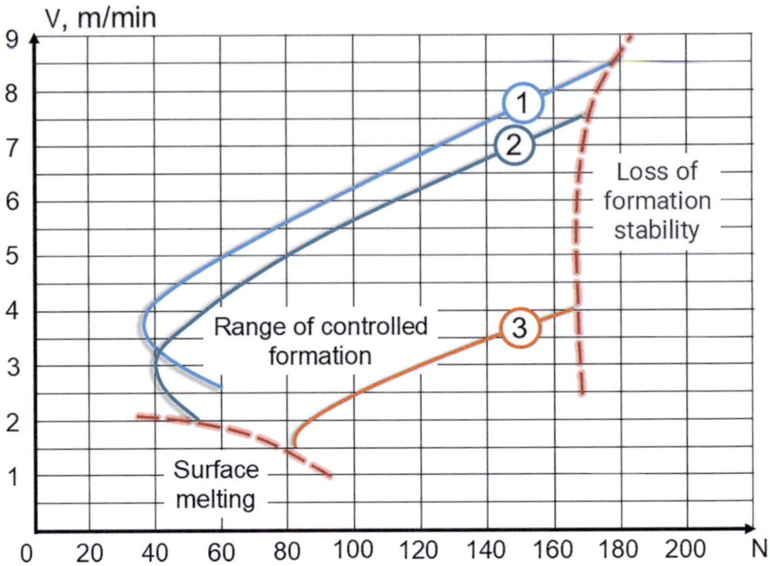

Fig. 6.19 Diagram of the range of controlled forming speeds at which a bending angle of $\alpha = 90°$ is ensured for samples made of type X10CrNiTi18-9 steel (P = 1 kW; d = 4 mm): 1—at $\delta = 0.5$ mm; 2—at $\delta = 1.0$ mm; 3—at $\delta = 1.5$ mm

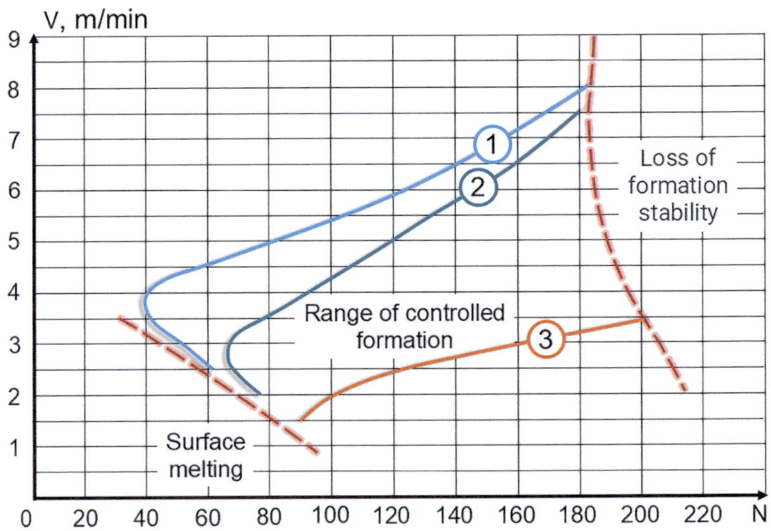

Fig. 6.20 Diagram of the range of speeds of controlled forming, at which the bending angle $\alpha = 90°$ is ensured for samples made of type 66Mn4 steel (P = 1 kW; d = 4 mm): 1—at $\delta = 0.5$ mm; 2—at $\delta = 1.0$ mm; 3—at $\delta = 1.5$ mm

An <u>*increase in the speed*</u> of the laser beam leads to a ***loss of forming stability*** (makes it impossible to form a 90° bend angle) due to the following phenomena:

(a) The efficiency of laser forming decreases. As the speed of the laser beam increases, the number of passes required to form a 90° bend angle increases significantly.
(b) The share of elastic deformations in compensation of local thermal expansion increases due to the reduction of energy input and the decrease in the intensity of heating of the processing zone. This leads to a decrease in residual deformations and a decrease in the controllability of the forming process.
(c) The rate of heating and cooling of the heat-affected zone increases, which affects the course of polymorphic transformations and leads to the formation of a significant amount of residual austenite.

Comprehensive parameters of laser forming

The guaranteed forming result is ensured by the optimal combination of parameters (laser radiation power P, W; laser beam movement speed V, mm/s; focusing spot diameter d, mm; workpiece thickness δ, mm) into ***complex parameters***:

- $p_e = P/(Vd)$—energy input parameter;
- $p_g = d/\delta$—geometric parameter;
- $p_c = p_e\ p_g$—complex forming parameter.

It has been experimentally confirmed that the same complex parameters provide the same result (the deviation of the bending angle does not exceed 6%) in different laser irradiation modes. For example, with complex parameters $p_e = 20$; $p_g = 4$; $p_c = 80$, the same result can be obtained with the following parameters:

- laser radiation power—0.8 kW, laser beam movement speed—1.2 m/min. (20 mm/s), focusing spot diameter—2 mm, sample thickness—0.5 mm;
- laser radiation power—0.6 kW, laser beam movement speed—0.9 m/min. (15 mm/s), focusing spot diameter—2 mm, sample thickness—0.5 mm.;
- laser radiation power—1 kW, laser beam movement speed—1.5 m/min (25 mm/s), focusing spot diameter—2 mm, sample thickness 0.5 mm;
- laser radiation power—1.6 kW, laser beam movement speed—1.2 m/min. (20 mm/s), focusing spot diameter—4 mm, sample thickness 1 mm.

The **spatial configuration** of the product is ensured by choosing the trajectory of movement of the laser beam (step and angle) and the number of passes on each irradiation track.

The ***step*** between the irradiation tracks forms the required curvature of the surface. <u>*Reducing the step*</u> between the irradiation tracks leads to a <u>*decrease in the radius of curvature*</u> of the formed surface, while maintaining all other parameters. For example, with steps of 15 mm, 10 mm, 5 mm and 2.5 mm, the radii of curvature of the resulting surface are 95.5 mm, 44 mm, 29 mm, and 13 mm, respectively.

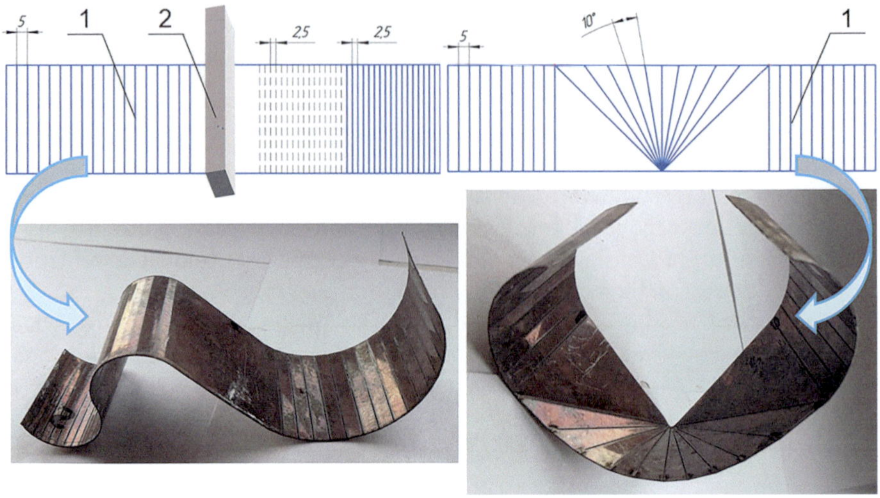

Fig. 6.21 Examples of trajectories of laser beam movement (type X10CrNiTi18-9 steel): 1—irradiation track; 2—clamp

To manufacture products of complex shape, the irradiation tracks are arranged at a certain ***angle*** to each other (Fig. 6.21). In this case, the nature of the reproduction of the deformation value (bending angle) will remain constant.

6.2.4 Laser Forming of Non-quenched and Pre-quenched Carbon Steels

Conducting heat treatment in the as-formed state will cause deformation and distortion of the geometry of the products. Forming pre-hardened products by stamping is associated with the risks of brittle cracking of hardened steels. There are also risks of cracking when stamping hard elastic steels. In contrast, laser forming allows forming all classes of steels without cracking.

Studies of changes in structure and properties during repeated cyclic thermal exposure were carried out on samples of 66Mn4 steel using the laser forming parameters: laser radiation power 2 kW, laser beam movement speed 5 m/min, focus spot diameter 6 mm.

The study found that both pre-quenched and non-quenched samples deform *in the direction of* the laser *treatment*. The amount of deformation increases with the number of passes. *Pre-quenched* samples deform with a *larger* bending angle compared to non-quenched ones. The average difference in deformation between pre-quenched and non-quenched samples is 40%.

No signs of melting, cracks, or other defects were found in either type of steel.

The irradiation zone, after the production of microsections (across the axis of the laser pass), was investigated by means of optical and electron microscopy, the microhardness of the laser pass area was determined in two mutually perpendicular directions with a fixed step (Fig. 3.30), to determine the distribution of microhardness in the cross section of the laser pass.

The initial microhardness values before laser treatment are 973.4 HV for pre-quenched steel and 233 HV for non-quenched steel.

After the first laser pass on the *pre-quenched steel*, a plateau of high microhardness values at 870 HV is observed in the central part of the pass. At the periphery of the pass, a tempering zone with a minimum value of 485 HV and a further increase to the initial value occurs.

After the first laser pass on the *non-quenched steel*, a high microhardness plateau of 820 HV also appears across the entire width of the laser pass (50 HV lower than in the pre-hardened steel). Beyond the plateau, a sharp decrease to the microhardness values of the base metal is observed. The microhardness distribution in the vertical direction reaches maximum values of 884 HV near the surface in the laser pass zone with a decrease in the transition zone to the values of the base metal.

The *microstructure* of pre-quenched and non-quenched steels consists of three zones (typical for surface laser hardening): a light zone of laser treatment, a narrow transition zone, and the microstructure of the base metal.

With an *increase in the number of passes*, an expansion of the transition zone is observed for both steel states. Due to the increase in the transition zone, the microhardness distribution values change: the width of the plateau with high hardness values narrows, and the transitions from maximum to minimum values become flatter.

Increasing the number of passes can lead to surface melting due to heat accumulation (the sample does not have time to cool down to the initial temperature during the time between cycles). The microstructure of the surface melting zone has the appearance of a columnar dendritic structure. Since melting is an undesirable phenomenon in laser forming, the time between passes for cooling should be increased or heat-removing clamps should be used.

6.3 Combined Laser Thermo-Deformation Hardening of Metal Products

Laser thermo-deformation hardening of metal products is aimed at increasing the wear resistance, fatigue resistance and fatigue wear of parts operating under extreme conditions (abrasive wear, corrosive media, cyclic alternating loads of critical parts of aircraft gas turbine engines, compressors, gear pumps, hydraulic drives of automotive and agricultural machinery). The metal microstructure must have at least high dispersion and hardness at the guaranteed depth.

The combination of *laser* ultra-high-speed and localized heating of metals with the process of mechanical *plastic deformation* allows forming hardened layers with structures that have all the advantages of laser hardening structures combined with plastic hardening structures [11, 12].

Combined laser thermo-deformation hardening is carried out in two stages:

(1) Laser radiation 1, defocused in the form of a circular spot with radius r (Fig. 6.22a), or obtained by high-frequency scanning of the beam at a width A (Fig. 6.22b), moves relative to the workpiece 2 with a speed V and *heats* the surface layer to the austenitizing temperature.
(2) With the distance from the heat source 3, the heated metal cools rapidly. At the distance l_f, where the surface temperature is 650–500 °C, the surface metal *is deformed* by the roller 4, to which a force F is applied. Martensite formed in the process of auto hardening inherits the structure of deformed supercooled austenite and thus provides the hardened layer 5 with a set of required physical and mechanical characteristics.

Determination of the technological *scheme* of thermo-deformation hardening includes:

(a) determination of the *distance* l_f before the deforming load is applied, where the metal is heated in the range of favorable austenitizing temperatures to avoid the recrystallization process;

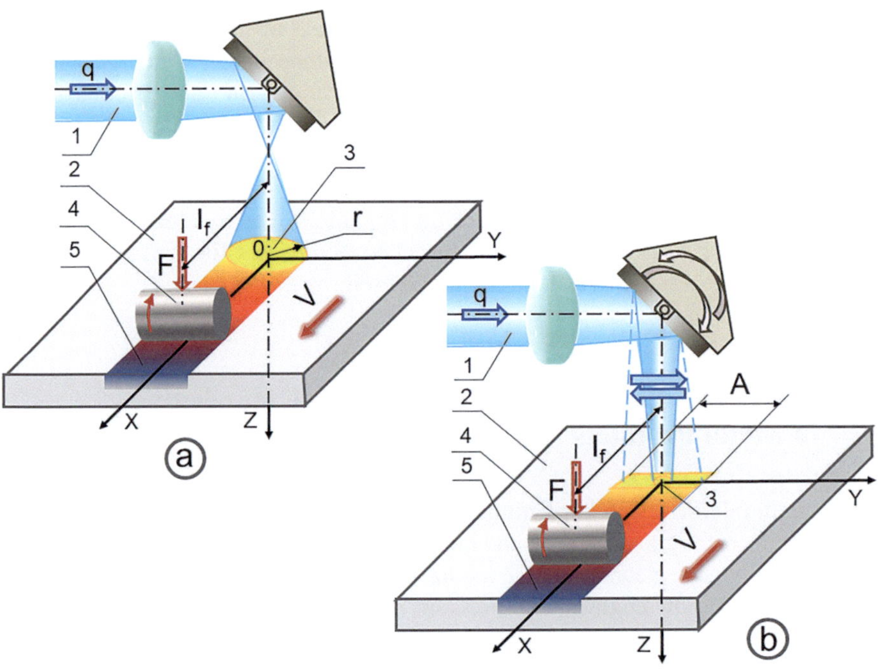

Fig. 6.22 Schemes of combined laser thermo-deformation hardening (1—laser beam; 2—workpiece; 3—heat source area; 4—deforming roller; 5—hardened layer): **a** defocused laser beam; **b** scanning laser beam

(b) determination of the minimum *force* F applied to the deforming roller, which ensures guaranteed plastic deformation of the surface layer in the range of favorable temperatures.

Determination of the distance l_f to the application of a deforming load where the metal is heated in the range of favorable temperatures is based on *calculations of the thermal field* of the laser heating source.

The design scheme of the laser thermo-deformation hardening process is shown in Fig. 6.23. A surface normally distributed laser heating source moves with a constant velocity V in the direction of the OY axis along the surface of a metal parallelepiped with dimensions $L_x \times L_y \times L_z$.

The power density of laser radiation on the surface is distributed according to the Gaussian law:

$$q(x, y, t) = q_{\max} e^{-5\frac{(x-x_C)^2+(y-y_C)^2}{r^2}} \tag{6.1}$$

where r is the effective beam radius;

t is time;

x_c, y_c are coordinates of the center of the focus spot on the irradiated surface.

The mathematical model is based on the three-dimensional heat conduction equation:

$$c\gamma\frac{\partial T}{\partial t} = \frac{\partial}{\partial x}\left(\lambda\frac{\partial T}{\partial x}\right) + \frac{\partial}{\partial y}\left(\lambda\frac{\partial T}{\partial y}\right) + \frac{\partial}{\partial z}\left(\lambda\frac{\partial T}{\partial z}\right) \tag{6.2}$$

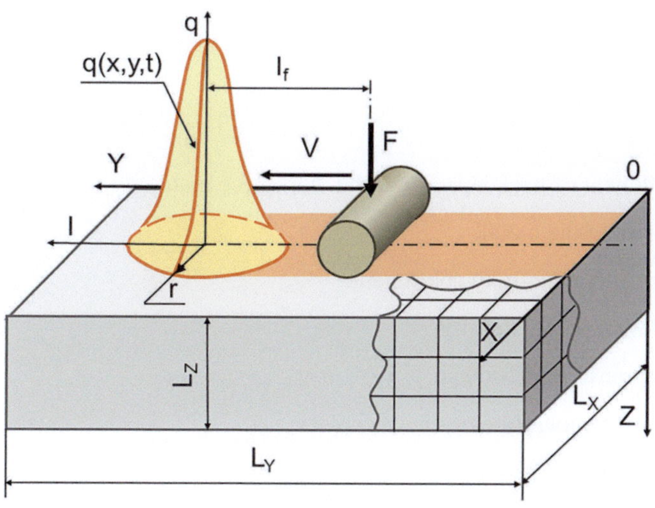

Fig. 6.23 Schematic diagram of the laser thermo-deformation hardening process

relative to an unknown temperature T = T(x,y,z,t),

with the initial condition: T(x,y,z,0) = T_0.

Boundary conditions on the surface to be treated in the laser irradiation zone:

$$\lambda \frac{\partial T(x, y, 0, t)}{\partial z} + q(x, y, t) = 0 \tag{6.3}$$

Outside the irradiation zone:

$$\lambda \frac{\partial T(x, y, 0, t)}{\partial z} + \alpha(T(x, y, 0, t) - T_c) = 0 \tag{6.4}$$

Boundary conditions on the other boundaries of the computational domain model heat exchange with the environment:

$$\lambda \frac{\partial T(0, y, z, t)}{\partial x} + \alpha(T(0, y, z, t) - T_c) = 0 \tag{6.5}$$

$$\lambda \frac{\partial T(x, 0, z, t)}{\partial y} + \alpha(T(x, 0, z, t) - T_c) = 0 \tag{6.6}$$

$$\lambda \frac{\partial T(L_x, y, z, t)}{\partial x} + \alpha(T(L_x, y, z, t) - T_c) = 0 \tag{6.7}$$

$$\lambda \frac{\partial T\left(x, L_y, z, t\right)}{\partial y} + \alpha\left(T\left(x, L_y, z, t\right) - T_c\right) = 0 \tag{6.8}$$

$$\lambda \frac{\partial T(x, y, L_z, t)}{\partial z} + \alpha(T(x, y, L_z, t) - T_c) = 0 \tag{6.9}$$

where x, y, z are spatial coordinates;

t is time;

c is the coefficient of heat capacity of the workpiece material;

ρ is the density of the metal;

λ is the thermal conductivity coefficient;

q is the radiation power density;

α is the heat transfer coefficient;

T_0 is the initial temperature of the workpiece;

T_c is the ambient temperature.

The studies were carried out on normalized steel C45 (composition: C = 0.45%; Si = 0.17–0.37%; Mn = 0.5–0.8%; Ni ≤ 0.3%; Cr ≤ 0.25%; Cu ≤ 0.3%) with the

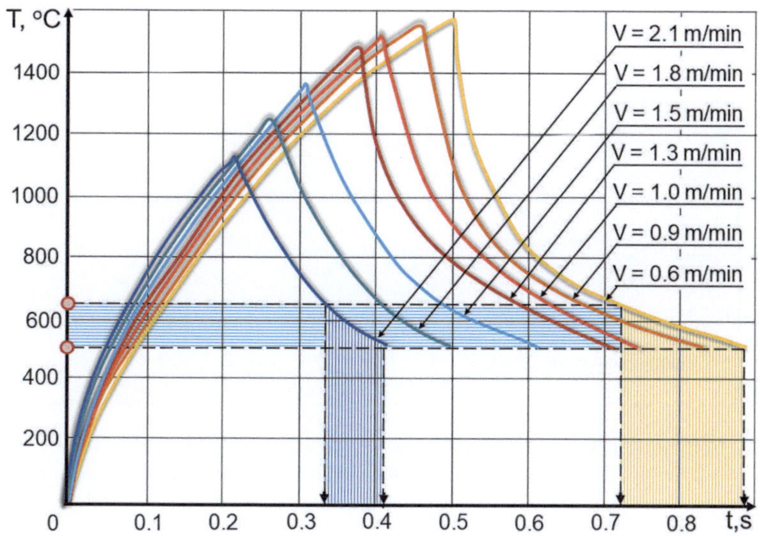

Fig. 6.24 Calculated changes in temperature T over time t at a surface point on the l-axis at different speeds V of the laser heating source (radiation power P = 1000 W; diameter of the laser focusing zone d_0 = 6.5 mm; absorption capacity A = 0.8)

following thermophysical characteristics: $\lambda = 0.314$ W/cm grad; a = 0.07 cm^2/s; c = 4.465 J/(cm^3 grad); $T_m = 1300$ °C; hardening temperature $T_h = 950$ °C.

As the machining speed increases, the maximum surface temperature of the material decreases (Fig. 6.24). At the same time, the depth of penetration of the isotherm of the desired temperature also decreases.

For C45 steel, the range of favorable temperatures is 650 °C $\geq T_f \geq$ 500 °C (Fig. 6.24). At the movement speed of the laser heating source V = 2.1 m/min, favorable temperatures will be established in the time interval of 0.34 s $\leq$ t $\leq$ 0.41 s; at a speed of V = 0.6 m/min—in the time interval of 0.72 s $\leq$ t $\leq$ 0.88 s. Consequently, the distance to the application of the deforming load is in the range of 11.9 mm $\leq l_f \leq$ 14.4 mm at the speed of movement of the laser heating source V = 2.1 m/min and in the range of 7.2 mm $\leq l_f \leq$ 8.8 mm—at the speed of movement of the laser heating source V = 0.6 m/min.

The second way to determine the distance l_f before the deforming load is applied is to analyze the distribution of isotherms (Fig. 6.25). The range of favorable temperatures of 650–500 °C is formed at a distance of 12.2 mm $\leq l_f \leq$ 17.0 mm from the center of the beam.

Therefore, for C45 steel, a distance of 10 mm $\leq l_f \leq$ 16 mm before the application of a deforming load can be recommended, where the metal is heated in the range of favorable temperatures for a wide range of laser thermal strain hardening modes.

The initial parameters for **determining the minimum force F** applied to the deforming roller, which ensures guaranteed plastic deformation of the surface layer, are:

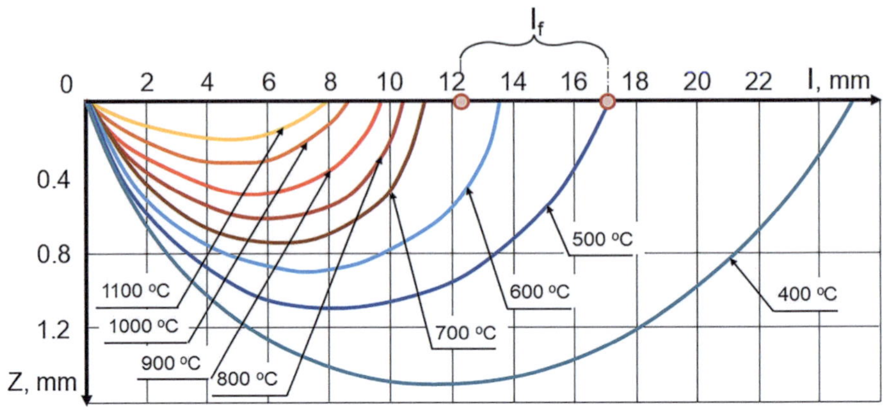

Fig. 6.25 Isotherms in depth z (radiation power P = 1000 W; laser heating source displacement speed V = 1.2 m/min; laser focusing zone diameter d_0 = 8 mm; absorption coefficient A = 0.8)

(a) deformation temperature in the range of favorable temperatures (650 °C ≥ T_f ≥ 500 °C) for C45 steel;

(b) yield strength σ_y at temperature T_f (Fig. 6.26a);

(c) the thickness of the layer δ with the changed structure to be deformed is determined either by isotherms (Fig. 6.25) or by the velocity V of the laser heating source (Fig. 6.26b).

The force F is determined by the experimental dependence:

$$F = 2\delta^2\sigma_y \tag{6.10}$$

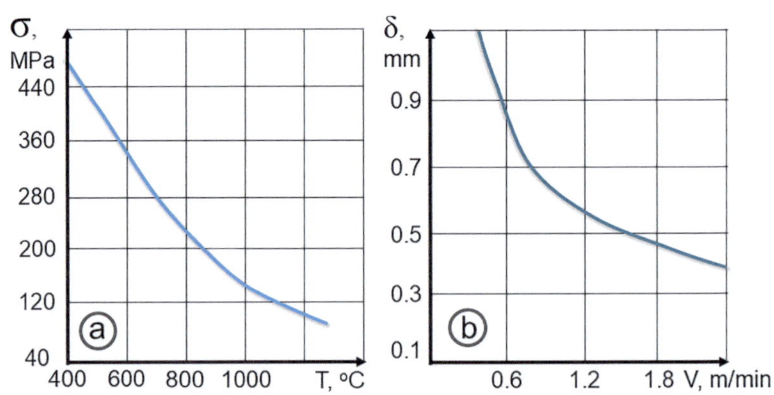

Fig. 6.26 Initial parameters for determining the force F applied to the deforming roller: **a** dependence of the yield strength σ_y of C45 steel on the heating temperature; **b** dependence of the thickness of the layer δ with a changed structure on the speed V of the laser heating source movement

For example, a C45 steel part is being machined at a speed of V = 0.8 m/min. Assign $T_f = 500$ °C from the range of favorable temperatures (650 °C ≥ T_f ≥ 500 °C). Yield strength $\sigma_y = 410$ MPa (at a temperature of $T_f = 500$ °C—Fig. 6.26a). The thickness of the layer to be deformed is $\delta = 0.7$ mm (at a speed V = of 0.8 m/min—Fig. 6.26b). Thus, the force to the deforming roller, which ensures guaranteed plastic deformation of the layer with a changed structure, is F = 402 N.

To adjust the load on the deforming roller, either a compression spring or a pneumatic actuator is used.

A compression spring has a number of disadvantages: low accuracy of load setting, dependence of force on strain level, and a small range of force adjustment.

The device for thermo-deformation hardening with a pneumatic drive (Fig. 6.27) consists of a series of hollow cylindrical parts with a laser beam 1 on the axis. Pressurized air 2 enters the pneumatic cylinder 4 through a pressure reducing valve 3. The pressure reducing valve regulates the air pressure and maintains it at the required level. The pressure is transferred to the piston 5, which is a hollow cylinder. A deforming roller 6 is mounted in the lower part of the piston. The pressure provides a deforming force in the range of F = 80–650 N. The distance to the application of the deforming load is regulated by screw 7 within the range of $l_f = 10$–20 mm. Spring 8 returns the piston 4 to its original position after processing and pressure release.

The **effectiveness** of laser thermo-strain hardening was evaluated by comparative tests of structural and phase transformations, the level and nature of residual stress distribution, wear resistance, and fatigue strength afterwards:

- laser hardening;
- laser thermo-deformation hardening.

Parameters of laser thermo-deformation hardening during the tests: radiation power P = 1000 W; diameter of the focusing zone $d_0 = 4; 5; 6; 7; 8$ mm; speed of movement of the laser heating source V = 0.6–2.1 m/min; force F = 430; 500; 570 N; distance to the application of the deforming load $I_f = 8; 10; 12$ mm.

In its original state, C45 steel has a heterogeneous coarse-grained ferrite-pearlite microstructure: lamellar pearlite and fine-grained ferrite along grain boundaries (Fig. 6.28a).

During laser hardening, due to the large temperature gradient, the sequence of phase transformations is clearly visible.

In the surface layer, at a heating temperature much higher than the point A_3, a microstructure is formed from areas of high-carbon (~ 0.8% C) austenite (formed at the site of pearlite colonies) and low-carbon austenite (formed at the site of ferrite). During high-speed cooling, austenite transforms into a mixture of high-carbon and low-carbon martensite with different hardnesses (Fig. 6.28b). The microhardness in the surface layer to a depth of 50 μm is $H\mu = 8000$–9000 MPa.

In the transition layer, pearlite colonies are partially transformed into austenite during the heating stage, which is transformed into martensite of high microhardness during the cooling stage. The microhardness in the transition layer at a depth of up to 650 μm is $H\mu = 6000$–7900 MPa (Fig. 6.28c).

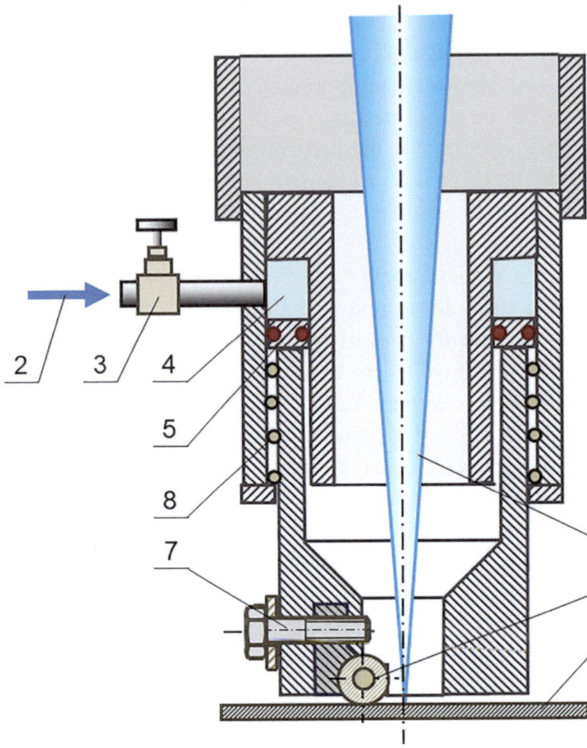

Fig. 6.27 Device for laser thermo-deformation hardening of metal products with a pneumatic drive: 1—laser beam; 2—pressurized air; 3—pressure reducing valve; 4—pneumatic cylinder; 5—piston; 6—deforming roller; 7—screw for adjusting the distance to the deforming load; 8—piston return spring; 9—workpiece

During laser thermo-deformation hardening, high-temperature plastic deformation inhibits the development of recrystallization processes, increases the density and orientation of dislocations, and forms branched subboundaries. A stronger grain refinement occurs (Fig. 6.28d, e).

The dislocation structure of *austenite* deformed at high temperature is characterized by the formation of three-dimensional dislocation networks of a stable fragmented, polygonized substructure. After high-temperature deformation, only stable dislocations remain. Recrystallization does not develop intensively.

Martensite inherits the dislocation structure of austenite. The microstructure of martensite corresponds in size and disorientation to the fragmented structure of the original austenite. Martensite has a lower total carbon content, which determines its increased ductility (compared to martensite obtained by conventional quenching).

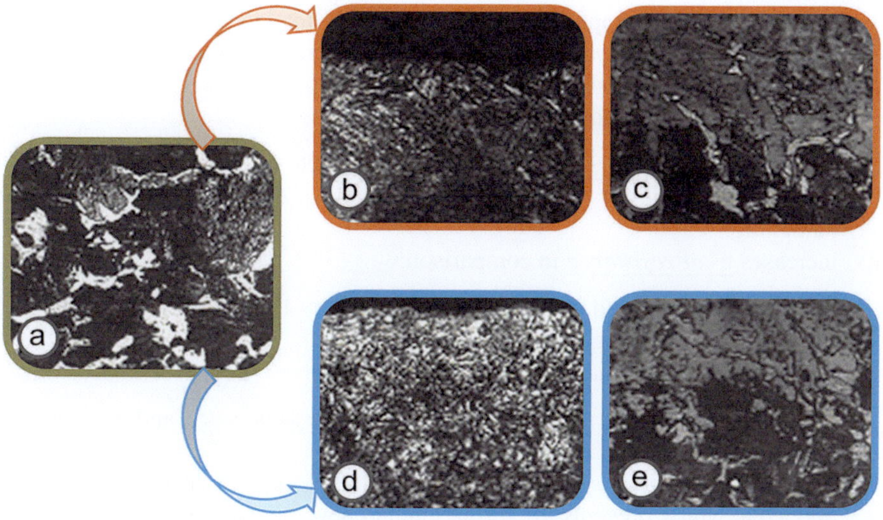

Fig. 6.28 Microstructure of C45 steel (P = 1000 W, d_0) = 4.5 mm, V = 1.0 m/min, l_f = 12 mm, F = 570 N) (×400): **a** in the initial state; **b** surface layer after laser hardening; **c** transition layer after laser hardening; **d** surface layer after laser thermal deformation hardening; **e** transition layer after laser thermo-strain hardening

An *increase in* the deforming *load* F leads to:

- grain grinding;
- increase in microhardness (Fig. 6.29).

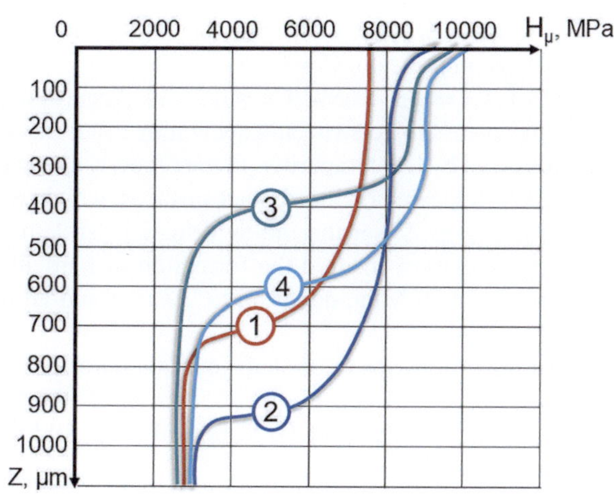

Fig. 6.29 Distribution of microhardness along the depth in C45 steel after laser treatment (P = 1000 W, d_0 = 4.5 mm, V = 1.0 m/min): 1—quenching; 2—at F = 430 N (l_f = 12 mm); 3—thermo-deformation hardening at F = 500 N (l_f = 12 mm); 4—thermo-deformation hardening at F = 570 N (l_f = 12 mm)

The *distance* l_f to the application of the deforming load in the range of favorable temperatures has virtually *no effect* on the level of microhardness on the surface and the distribution over the depth.

Thus, **laser thermo-deformation hardening**:

(a) increases *microhardness* by 1500 MPa compared to laser hardening;
(b) increases the *depth of the hardened layer* by 100–200 μm) compared to laser hardening;
(c) increases *wear resistance* in comparison:

- with the normalized one in the initial state—by 10 times,
- with bulk hardening in a furnace—by 2 times,
- with surface laser hardening—by 30%;

(d) reduces the *coefficient of friction*, improves *wear uniformity*, and prevents the formation of chips, swirls, and seizures;
(e) generates *compressive residual stresses* in the surface layer within 400–500 MPa, which prevents the formation and development of surface microdefects (initial stage of fatigue fracture, stress corrosion, corrosion fatigue);
(f) increases *fatigue strength* by 30–80% under low-cycle alternating loads.

Examples of laser thermo-strain hardening **applications**.

(a) Friction pair shaft-gear for plain bearings, end and involute surfaces of the gear teeth of an aircraft hydraulic unit made of high-strength cast iron with spheroidal graphite type EN-GJS-600-3 (composition: C = 3.0–3.6%; Si = 2.4–2.8%; Mn = 0.4–0.7%; Ni ≤ 0.4%; Cr ≤ 0.15%; Cu ≤ 0.3%). Increasing the microhardness to 8500–9000 MPa with a hardened layer depth of 0.5 mm ensured a reduction in wear of friction surfaces, hermetic stability of parts, stability of lubricant pressure, and an increase in the overall service life of aircraft engines.
(b) The surfaces of the distributor rings of a tractor hydraulic unit made of 100Cr6 steel (composition: C = 0.95–1.05%; Si = 0.17–0.37%; Mn = 0.2–0.4%; Ni ≤ 0.3%; Cr = 1.3–1.65%; Cu ≤ 0.25%). According to the original technology, the rings were subjected to bulk quenching in a furnace to a hardness of 650–700 HV, followed by low tempering. In the process, a network of radially directed microcracks appeared on the contact surfaces, which eventually led to the destruction of the rings. Laser thermo-deformation hardening by both pulsed and continuous radiation ensured the formation of hardening zones with microhardness up to 12,000 MPa, depth up to 0.7 mm and an increase in service life by 2.1 times.
(c) Turbocompressor rotor shaft journals made of 41Cr4 steel (composition: C = 0.36–0.44%; Si = 0.17–0.37%; Mn = 0.5–0.8%; Ni ≤ 0.3%; Cr = 0.8–1.1%; Cu ≤ 0.3%). Tests for 60 h showed a 2.5-fold reduction in wear compared to the standard.

6.4 Laser Hardening of Blade Tools

Blade tools include a wide range of products whose working elements have a wedge-shaped cutting edge, such as knives, blades, scissors, and wedges. The shape of the working body is a wedge with angle of sharpening varying from 90° to 2–3°. A number of works [13, 14] are devoted to the specific issue of laser hardening of tool steels.

To determine the limit values of the parameters of the laser processing that ensure the optimal structure in the surface layer, it is necessary to develop a mathematical model of heat transfer processes during laser hardening of the wedge and analyze possible ways to influence the thermal state of the wedge [15].

6.4.1 Mathematical Model

The design scheme of the heat transfer process during the hardening of wedge-shaped elements by a laser beam is shown in Fig. 6.30.

The object under study is a wedge-shaped body. A surface normally distributed laser heating source moves with a constant velocity V, parallel to the cutting edge of the wedge at a distance a^*. The conditions of heat exchange between the wedge and the environment are described by the Newton-Richman law. The ambient temperature T_m and heat transfer coefficient α are assumed to be constant and the same for all bounding surfaces.

The mathematical model of the heat transfer process during the strengthening of wedge-shaped elements by a laser beam for the quasi-stationary state is as follows:

$$\frac{1}{r}\cdot\frac{\partial}{\partial r}\left(rK\frac{\partial T}{\partial r}\right)+\frac{1}{r^2}\cdot\frac{\partial}{\partial\varphi^2}\left(K\frac{\partial T}{\partial\varphi}\right)+\frac{\partial}{\partial z}\left(K\frac{\partial T}{\partial z}\right)+VC_V\frac{\partial T}{\partial z}=0 \qquad (6.11)$$

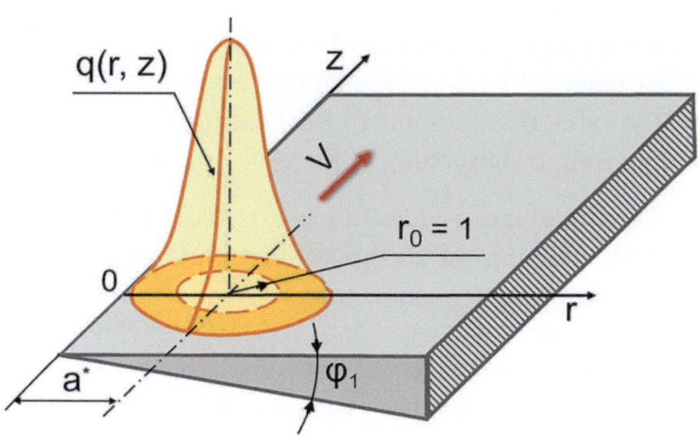

Fig. 6.30 Schematic diagram of the wedge laser hardening process ($\varphi = 0$—treated surface; $\varphi = \varphi_1$—untreated surface)

$$\left[K\frac{\partial T}{\partial r} - \alpha(T - T_m)\right]_{r=r_i} = 0 \tag{6.12}$$

$$\left[K\frac{\partial T}{\partial r} - \alpha(T - T_m)\right]_{r=r_e} = 0 \tag{6.13}$$

$$\left[K\frac{1}{r}\cdot\frac{\partial T}{\partial r} - \alpha(T - T_m)\right]_{r=r_0} = -q_{\max}\exp\left[-\left(\frac{z}{r_0}\right)^2 - \left(\frac{r-a^*}{r_0}\right)^2\right] \tag{6.14}$$

$$\left[K\frac{1}{r}\cdot\frac{\partial T}{\partial r} - \alpha(T - T_m)\right]_{\varphi=\varphi_1} = 0 \tag{6.15}$$

$$K\frac{\partial t}{\partial z}\bigg|_{z=\pm\infty} = 0 \tag{6.16}$$

where q_{max} is the maximum value of the heat supply distribution q(r, z);

T_m is the temperature of the medium;

V is the speed of the laser beam;

r_0 is the radius of the heating zone;

r_i, r_e are the inner and outer radii of the workpiece surface;

a^* is distance from the cutting edge to the laser beam axis;

φ_1 is the wedge sharpening angle;

α, K, C_v are the coefficients of thermal conductivity, thermal diffusivity, and volumetric specific heat of the material.

The choice of optimal technological parameters of laser hardening of wedge-shaped elements consists in determining the temperature conditions at which the necessary structural changes in the material are ensured.

The *temperature regime* of the wedge during hardening can be *controlled* by varying:

- the speed of movement of the laser beam V;
- distance a^* between the cutting edge and the center of the heating zone;
- the value of the radius of the heating zone r_0;
- the wedge sharpening angle φ_1.

The calculations according to Eqs. (6.11)–(6.16), given further in this section, were carried out using the method of complete polyargument systems with the *following initial parameters*: product material—tool carbon CT80 type steel (composition: C = 0.8%; Si = 0.17–0.33%; Mn = 0.17–0.33%; Ni ≤ 0.25%; Cr ≤ 0.2%); r_e = 100 mm; r_i = 0.001 mm; K = 1; C_v = 1.

The **effect of** the laser beam **velocity V** (Peclet number Pe = V (r/α)) on the temperature regime of the wedge.

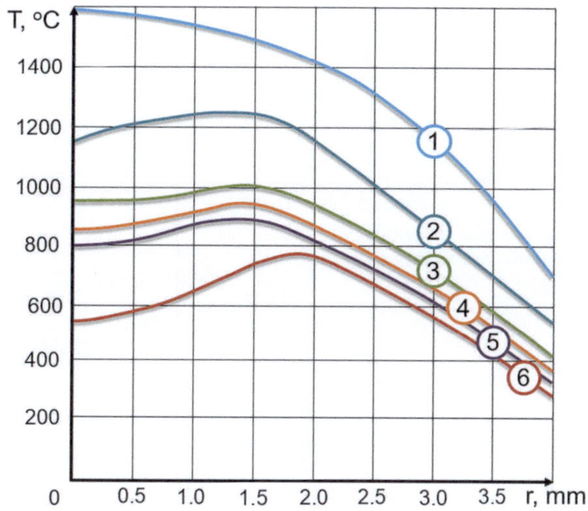

Fig. 6.31 Effect of laser processing speed V on the temperature distribution T on the surface (z = 2 mm; $\varphi_1 = \pi/3$; a* = 2 mm): 1—at V = 0.6 m/min; 2—at V = 1.0 m/min; 3—at V = 1.4 m/min; 4—at V = 1.6 m/min; 5—at V = 2.0 m/min; 6—at V = 3.0 m/min

In general, the temperature level in a wedge decreases with increasing velocity. The speed can significantly affect the nature of the temperature distribution in the radial direction (Fig. 6.31) (in contrast to the semi-infinite body model, where the shape of the isotherms remains unchanged by a semi-ellipsoid of rotation over the entire range of speed values).

At high speeds, the temperature change along the radius on the processed surface has an extreme—curve 6 in Fig. 6.31. As the speed decreases, the extremum becomes less pronounced and shifts to the cutting edge. At low speeds, a monotonic temperature drop along the radial coordinate is observed (curve 1 in Fig. 6.31).

The effect of velocity on the character of the *temperature field* of the wedge is associated with the manifestation of the edge effect:

- At high speeds, the zone of noticeable temperature changes is localized near the surface to be treated. The nature of the temperature field is determined by the conditions of heat dissipation and practically does not depend on the configuration of the object.
- At low speeds, the size of the zone of perceptible temperature changes increases and covers the cutting edge. The nature of the temperature field depends on both the heat dissipation conditions and the specifics of the wedge configuration.

As the speed increases, *the maximum* surface heating temperature decreases (Fig. 6.32) over the entire range of values of r and z.

The **effect of the distance *a**** of the laser beam axis to the cutting edge on the temperature regime of the wedge.

The temperature distribution on the wedge surface at different distances *a** of the laser beam axis to the cutting edge is shown in Fig. 6.33.

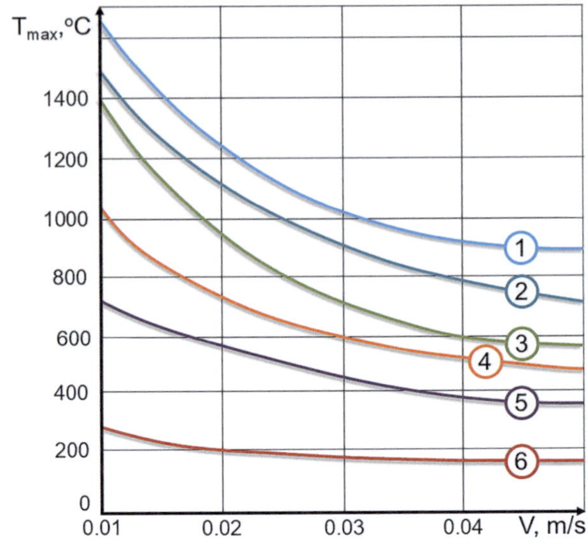

Fig. 6.32 Influence of laser processing speed V on the maximum temperature T_{max} of the surface ($\varphi_1 = \pi/3$; a* = 2 mm): 1—at z = 0, r = 0.1 mm; 2—at z = 0, r = 2 mm; 3—at z = 0, r = 4 mm; 4—at z = 3 mm, r = 2 mm; 5—at z = 1 mm, r = 2 mm; 6—at z = 3 mm, r = 2 mm

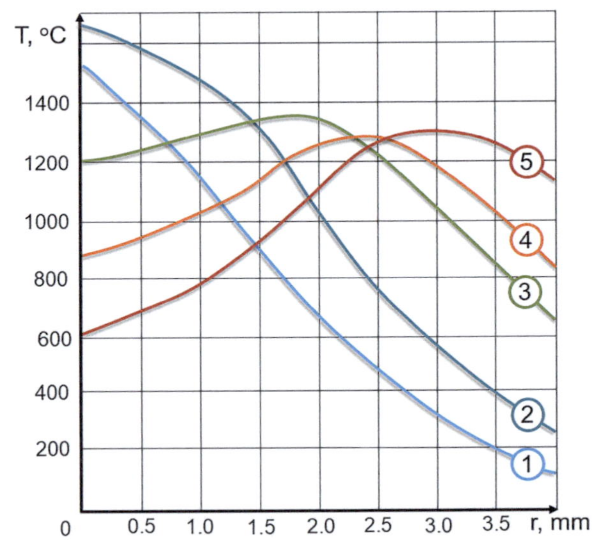

Fig. 6.33 Temperature distribution T on the wedge surface at different distances a* of the laser beam axis to the cutting edge (z = 1 mm; $\varphi_1 = \pi/3$; r = 2 mm; V = 1 m/min): 1—at a* = 0; 2—at a* = 1 mm; 3—at a* = 2 mm; 4—at a* = 2.5 mm; 5—at a* = 3 mm

At low values of a^* (curves 1, 2), the maximum temperature value is observed at the edge of the wedge. As the temperature moves away from the edge, it decreases monotonically.

At large values of a^* (curves 3, 4, 5), the dependence T = f (r) has an extreme character. In this case, the temperature extremum shifts away from the edge with increasing a^*.

The peculiarities of the change in the maximum temperature in the wedge cross-section depending on the distance a^* are determined by two factors:

(a) the value of the total heat flux Q supplied to the wedge;
(b) heat transfer conditions in the wedge.

With the growth of a^*, the following is observed:

- increase in the total heat flux Q, which causes a tendency to increase the maximum temperature of the wedge;
- Improved conditions for heat outflow from the heating zone, which causes a tendency to reduction of the maximum wedge temperature.

As a result, in the zone of small a^* values (curves 1, 2), the dominant factor is the influence of the total heat flux Q. In the zone of large a^* values (curves 3, 4, 5), the dominant factor is the effect of improving the conditions for heat outflow from the heating zone.

Starting from a certain distance a^*, the maximum temperature T_{max} in the transverse cross-section remains constant. The back surface of the wedge, which is not processed, has less influence on the temperature distribution.

The temperature field (Fig. 6.34) becomes less gradient with increasing distance a^*.

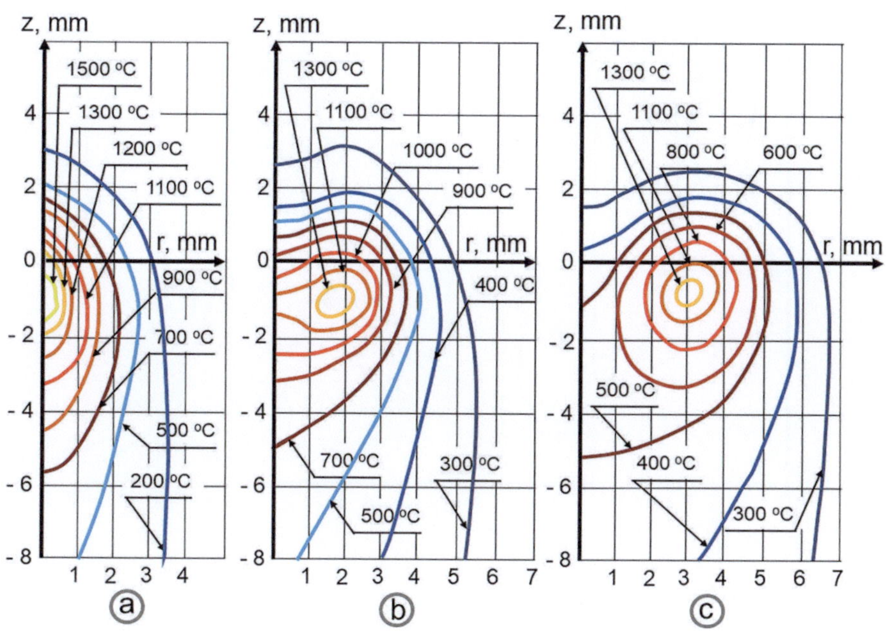

Fig. 6.34 Temperature isotherms on the wedge surface at different distances a* of the maximum laser intensity from the cutting edge ($\varphi_1 = \pi/3$; $r_0 = 2$ mm; V = 1 m/min): **a** at a* = 0; **b** at a* = 2 mm; **c** at a* = 3 mm

The **effect of the radius of the heating spot r_0** on the temperature regime of a wedge during laser hardening.

Changing the radius of the heating spot r_0 in the wedge changes both the nature of the temperature field and the temperature level (unlike a semi-infinite array, where the radius r_0 affects only the temperature level).

It is convenient to use the relative temperature θ when analyzing the peculiarities of the temperature field change. To determine the effect of the radius of the heating spot, the relative temperature is defined as the ratio $\theta_i^* = T_i/T^*_{max}$, where T^*_{max} is the maximum temperature of the wedge at $r_0 = 0.5$.

At small values of r_0, the temperature distribution in the cross section is extreme (curves 1, 2, 3 in Fig. 6.35).

With an increase in r_0 (curve 4 in Fig. 6.35), a monotonic decrease in the wedge temperature θ* occurs with distance from the edge.

In the zone of maximum heat input, the temperature decreases significantly as the radius of the heating spot r_0 increases. As you move away from the maximum heat input, the temperature levels off for different radii r_0.

The **effect of angle φ_1** of wedge sharpening on the temperature regime.

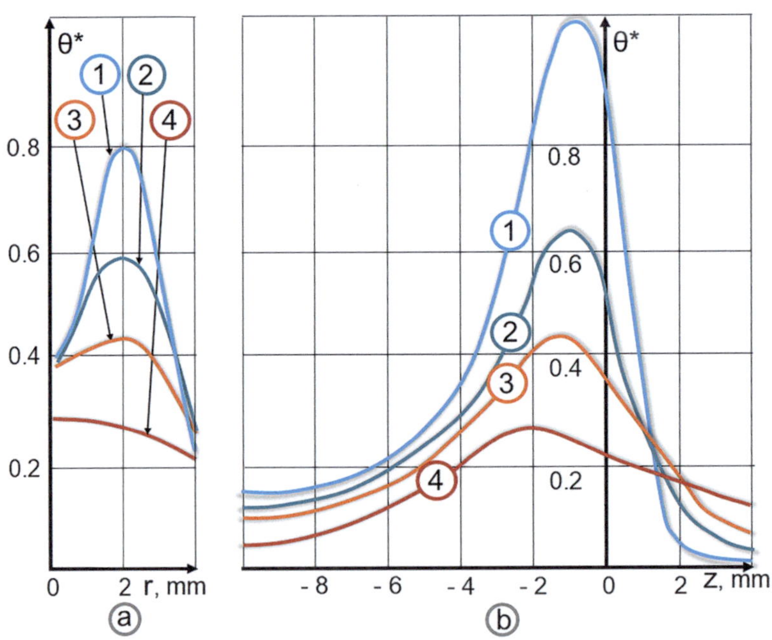

Fig. 6.35 Distribution of the relative temperature θ* on the wedge surface when the radius of the heating spot r_0 changes at $\varphi_1 = \pi/3$; $a^* = 2$ mm; V = 1 m/min (**a** in cross section at z = 1 mm; **b** in longitudinal section at r = 2 mm): 1—at $r_0 = 1$ mm; 2—at $r_0 = 1.5$ mm; 3—at $r_0 = 2$ mm; 4—at $r_0 = 3$ mm

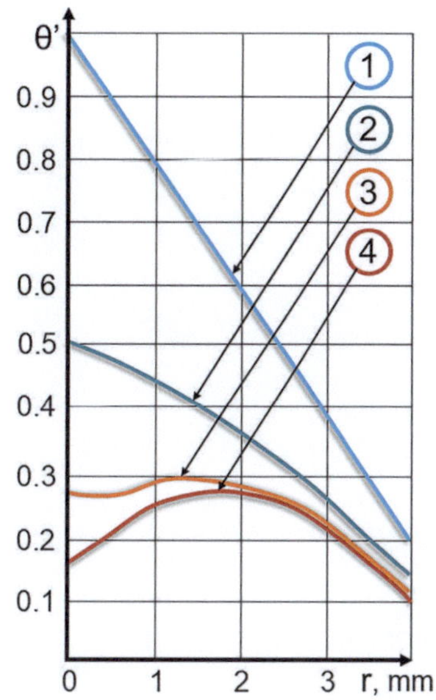

Fig. 6.36 Distribution of the relative temperature θ′ on the surface in cross-section when the angle φ_1 of the wedge sharpening changes (z = 1 mm; a* = 2 mm; V = 1 m/min): 1—at $\varphi_1 = \pi/12$; 2—at $\varphi_1 = \pi/6$; 3—at $\varphi_1 = \pi/3$; 4—at $\varphi_1 = \pi$

To determine the effect of the wedge sharpening angle, the relative temperature is defined as the ratio $\theta_i' = T_i/T'_{max}$, where T'_{max} is the maximum temperature of the wedge at $\varphi_1 = \pi/12$.

With an *increase* in the angle φ_1, ceteris paribus, the temperature level of the wedge generally *decreases* (Fig. 6.36).

The *largest* temperature changes with the change in angle φ_1 are observed *near the edge*. The changes decrease with distance from the edge in the radial direction.

As the angle φ_1 increases, the lower surface of the wedge that is not processed ceases to influence the formation of the wedge temperature field. Therefore, the influence of the angle φ_1 manifests itself differently depending on the range of its change:

at $\varphi_1 < \pi/6$, the change in angle significantly affects the temperature level in the high-temperature zone (curves 1, 2 in Fig. 6.36);
when $\varphi_1 > \pi/3$, the change in angle ceases to affect the temperature level (curves 3 and 4 in Fig. 6.36).

The **adequacy of the mathematical model** was verified by comparing the calculations with the experimental data (Fig. 6.37):

(a) the shape and size of the hardened zone were calculated from the shape and size of the isotherms that limit the zones of structural and phase transformations;
(b) experimentally, the shape and dimensions of the hardened zone were determined by conducting metallographic studies.

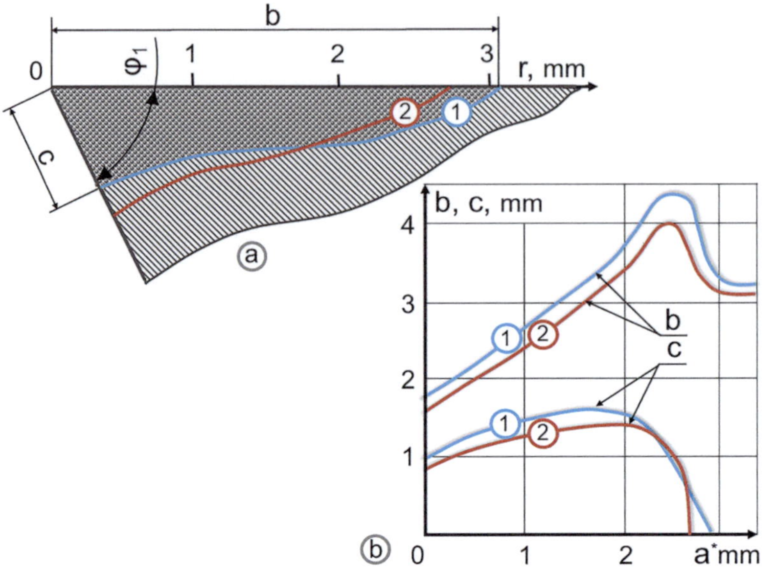

Fig. 6.37 Comparison of calculations (1) with experimental data (2): **a** shape of the hardened zone ($\varphi_1 = \pi/3$; $r_0 = 2$ mm; a* = 2 mm; V = 1 m/min); **b** change in the dimensions 'b' and 'c' of the hardened zone when changing the distance a* of the laser beam axis to the cutting edge ($\varphi_1 = \pi/3$; $r_0 = 2$ mm; V = 1 m/min)

6.4.2 Determination of Critical Values of the Laser Processing Parameters

Types of defects during laser hardening of blade tools with wedge-shaped working elements:

(a) *Melting* of the cutting edge.
(b) *Melting* the area adjacent to the cutting edge.
(c) *Displacement* of the hardened zone away from the cutting edge (the cutting edge remains unhardened).
(d) *Lack of hardening microstructure* on the cutting edge (due to insufficient heating temperature and/or cooling rate).

For each defect, a ***limit state*** can be set that determines the onset of defect formation and delimits the areas of acceptable and unacceptable values of the laser processing parameters (Fig. 6.38):

(a) for melting the cutting edge—the beginning of melting of the cutting edge;
(b) for melting the area adjacent to the cutting edge—the beginning of melting of the workpiece surface at a distance from the cutting edge;
(c) for displacement of the hardened zone away from the cutting edge—location of the boundary of the hardened zone on the cutting edge;

DEFECT TYPE	DEFECT FORMATION SCHEME	LIMIT STATE
Defect-free cutting edge	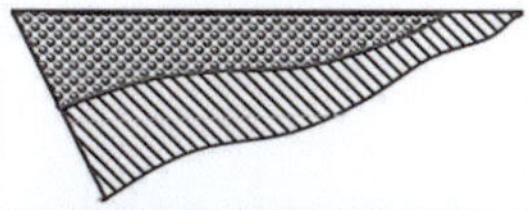	
Melting of the cutting edge	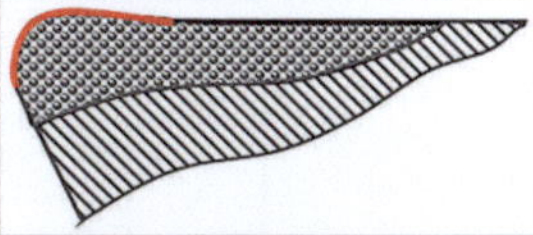	Beginning of melting of the cutting edge
Melting of the area adjacent to the cutting edge	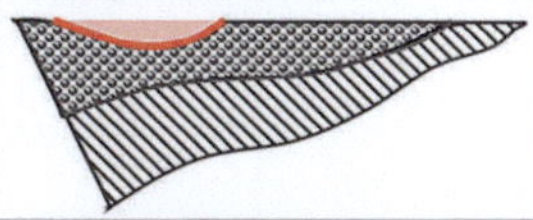	Beginning of melting of the surface being processed
Offset of the hardening zone from the cutting edge	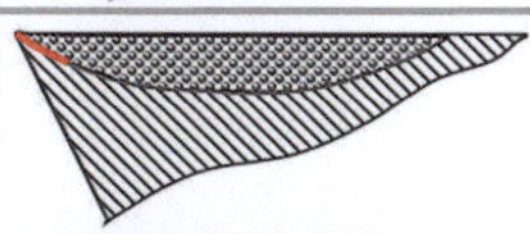	Touching the border of the hardening zone with the cutting edge
Lack of microstructure hardening on the cutting edge		The heating temperature of the cutting edge is less than the temperature of the structural phase transformation

Fig. 6.38 Limit states of defect formation in a blade tool

(d) for the absence of a hardened microstructure on the cutting edge—'the heating temperature of the cutting edge being less than the temperature of the structural-phase transformation.

The ***critical values of the parameters of*** the laser processing are the parameters corresponding to the *limit state.*

To determine the critical values of the parameters of the laser processing, a computational model of the heat transfer process in the limiting states of defect formation under the conditions of heat exchange between the wedge and the environment is considered. The temperature regime of the wedge is most effectively controlled by two parameters of the laser processing: the laser beam speed and the distance of the laser beam axis to the cutting edge.

The following combinations of parameters *are set*:

- the density of the supplied heat flux q_{max};
- the radius of the heating zone r_0;
- the ambient temperature T_e;
- heat transfer coefficient α ($q_{(max)} = 4.78 \times 10^7$ W/m^{2}K).

The critical values of the laser beam velocity *are determined* depending on the distance of the laser beam axis to the cutting edge:

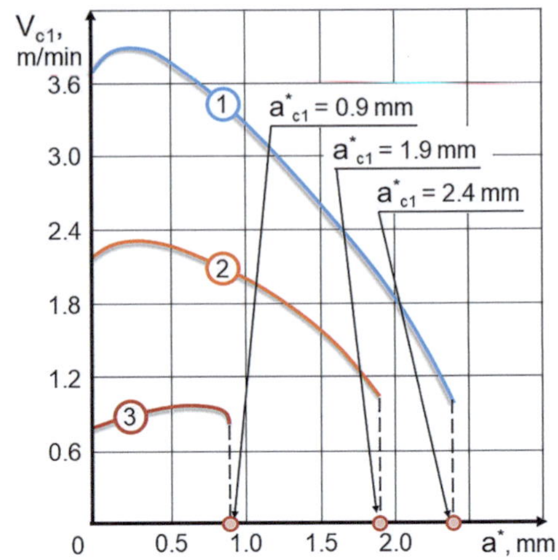

Fig. 6.39 Dependence of the first critical speed V_{c1} on the distance a* of the laser beam axis to the cutting edge ($r_0 = 2$ mm): 1—at $\varphi_1 = \pi/4$; 2—at $\varphi_1 = \pi/3$; 3—at $\varphi_1 = \pi/2$

(a) V_{c1} is the *first critical velocity* corresponding to the beginning of the cutting edge melting;
(b) V_{c2} is the *second critical velocity*, which corresponds to the beginning of the melting of the processed surface;
(c) V_{c3} is the *third critical velocity* corresponding to the contact with the boundary of the cutting-edge hardening zone.

The dependence of the **first critical velocity** V_{c1} *on the distance of the laser beam axis to the cutting edge* (Fig. 6.39) has an extremum. The presence of a maximum is due to the ratio of the effects of the two factors:

- the value of the total heat flux Q supplied to the wedge;
- heat transfer conditions in the part.

As the distance a^* increases, the total heat flux increases, which leads to an increase in the metal temperature and, consequently, an increase in the first critical velocity V_{c1}.

On the other hand, an increase in the distance a^* is accompanied by an improvement in the conditions of heat outflows from the heating zone. This leads to a decrease in the temperature of the high-temperature zone, and, accordingly, a drop in the first critical velocity V_{c1}.

The heat balance is manifested as follows:

- at small distances a^*, the heat flux supplied to the wedge has a dominant influence;
- at large distances a^*, heat dissipation has a dominant influence.

Starting from a certain critical distance a^*_{c1}, no penetration of the cutting edge occurs at any speed. For all $a^* \geq a^*_{c1}$, only the surface area located at a distance from the edge can be penetrated.

The first critical velocity V_{c1} depends significantly on the *wedge angle φ_1*.

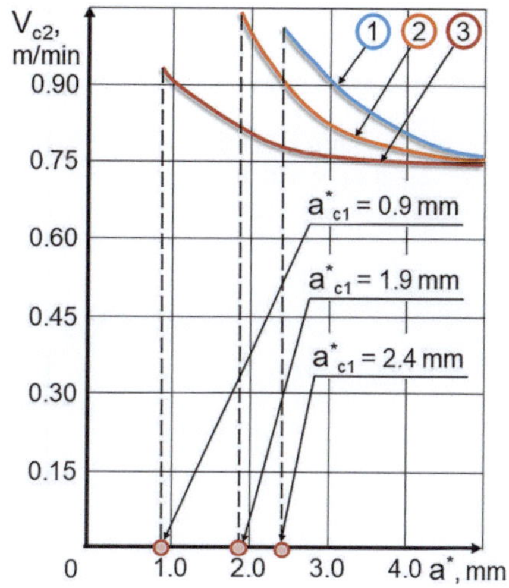

Fig. 6.40 Dependence of the second critical speed V_{c2} on the distance a* of the laser beam axis to the cutting edge ($r_0 = 2$ mm): 1—at $\varphi_1 = \pi/4$; 2—at $\varphi_1 = \pi/3$; 3—at $\varphi_1 = \pi/2$

As the wedge angle increases, the value of a^*_{c1} decreases significantly: at $\varphi_1 = \pi/4$, $a^*_{c1} = 2.4$, and at $\varphi_1 = \pi/2$, $a^*_{c1} = 0.96$.

The first critical velocities, which determine the critical distances a^*_{c1}, are very close in value for different angles φ_1 (0.9–1.0 m/min in Fig. 6.39).

Melting of the surface to be processed at a distance from the cutting edge begins at distances a^* that exceed a^*_{c1}. With the growth of a^* above a^*_{c1}, a decrease in the **second critical speed** V_{c2} is observed (Fig. 6.40).

As the distance a^* increases, the changes in the second critical velocity V_{c2} slow down.

When the distance a^* reaches certain values (different for different wedge sharpening angles), V_{c2} stabilizes, since with the growth of a^* the temperature field in the high-temperature zone approaches the heating conditions of a semi-infinite array.

The stabilization of the second critical velocity V_{c2} occurs later at small wedge angles φ_1.

The **third critical velocity** V_{c3} corresponds to the contact with subsequent separation of the hardening zone and cutting edge in the absence of melting.

The separation of the hardening zone from the cutting edge occurs at a certain critical value $a^* = a^*_{c2}$ (Fig. 6.41). The value of the critical distance a^*_{c2} depends on the wedge's sharpening angles φ_1—the smallest sharpening angles φ_1 correspond to large critical distances. For example, when $\varphi_1 = \pi/2$, the critical distance $a^*_{c2} = 0.74$, while when the angle decreases to $\varphi_1 = \pi/4$, the critical distance increases to $a^*_{c2} = 2.1$ (Fig. 6.41).

The third critical velocity V_{c3} monotonically decreases with increasing distance a^* at all values of angles φ_1 (Fig. 6.41).

When a certain critical value of $a^* = a^*_{c3}$ is reached, two types of defects are formed simultaneously—displacement of the hardening zone from the cutting edge and melting of the zone adjacent to the cutting edge. The value of the critical

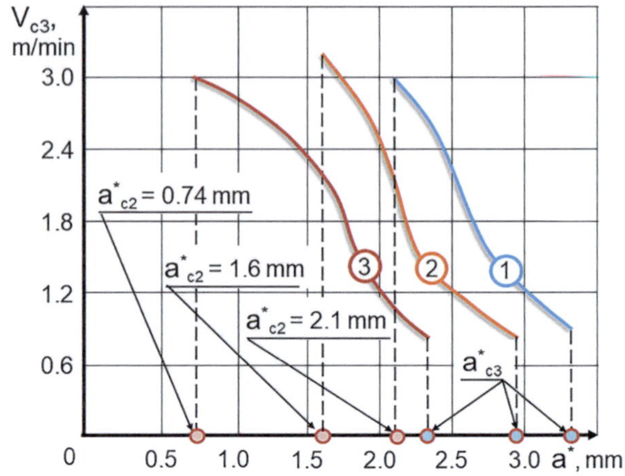

Fig. 6.41 Dependence of the third critical speed V_{c3} on the distance a* of the laser beam axis to the cutting edge ($r_0 = 2$ mm): 1—at $\varphi_1 = \pi/4$; 2—at $\varphi_1 = \pi/3$; 3—at $\varphi_1 = \pi/2$

distance a^*_{c3} significantly depends on the wedge sharpening angle φ_1—the larger the sharpening angle φ_1, the smaller, ceteris paribus, the value of a^*_{c3} (Fig. 6.41).

6.4.3 Experimental Verification of the Critical Values of the Laser Processing

The tests were performed on samples made of CT80 tool carbon steel in the delivery state and after quenching and low tempering.

Two types of samples were used:

- prismatic with dimensions of 40 × 40 × 5 mm;
- wedge-shaped with sharpening angles of 5°, 15°, 30°, 45°, 60° and 90°.

Prior to laser treatment, the samples were coated with an absorbent ZnO-based coating with a thickness of 20–30 μm.

The laser processing was performed with the following parameters:

- power P = 1000 W;
- the laser beam velocity V varied in the range of 0.2–4.5 m/min;
- focal length F = 300 mm;
- the diameter of the focusing spot d_0 varied in the range of 4–7 mm (regulated by the distance from the lens to the processing surface);
- the distance *a** of the laser beam axis to the cutting edge (on samples with wedge-shaped edges) varied in the range of 0–5 mm.

On wedge-shaped samples, the temperatures obtained by computer modeling were experimentally verified by chromium-aluminum thermocouples.

After the laser irradiation, the samples were cut by an electric spark method and metallographic analysis was performed.

The magnitude and nature of the distribution of residual stresses were determined by the X-ray method.

The wear resistance of the structures formed by laser irradiation was evaluated on a friction machine according to the "flat sample-roller" contact scheme. Rollers made of 1274 type tool steel (composition: C = 0.95%; Si = 0.25%; Mn = 0.30%; Cr = 1.25%) with a hardness of HRC55 served as the counter body.

During laser processing of CT80 *normalized steel* with a power density of 4×10^3 W/cm^2 and laser beam speeds of less than 0.6 m/min, which ensures *metal melting*, three layers are formed in the surface layer:

(a) The *surface layer* that crystallized from the melt consists of dispersed rounded grains of martensitic-austenitic structure with a microhardness of 9500–10,500 MPa. The presence of residual austenite is recorded on diffractograms taken from the laser heating zones. The amount of residual austenite increases with a decrease in the processing speed and reaches 35% at a speed of 0.2 m/min.
(b) The *middle layer* was formed in the zone where the heating temperature did not exceed the critical point A_1. The structure is characterized by increased dispersion and increased hardness of 11,000–11,500 MPa.
(c) The *transition layer* is located at the boundary of the heat affected zone with the base. The homogenization of austenite does not have time to occur, and martensite with a lower carbon content is present in the structure along with high-carbon martensite. The microhardness decreases at the edge of the transition layer to the original.

During laser processing of *normalized* CT80 steel with the initial structure of lamellar pearlite, *two layers* are formed if the heating temperature exceeds the A_1 point but *does not exceed the melting point*:

(a) The *surface layer*, where structural-phase transformations occur in the solid state, has a microstructure of highly dispersed martensite, residual austenite, and not pronounced troostite formations along the boundaries of primary austenitic grains (Fig. 6.42a). The hardness is 11000–12,000 MPa. The shape and size of the layer are determined by the irradiation parameters and vary in a wide range. Depending on the irradiation parameters, only the dispersion of the structure and its phase composition change. As the processing speed increases, the dispersion increases and the content of residual austenite increases.
(b) The *transition layer* is located at the boundary of the heat-affected zone with the base. The microstructure gradually changes from martensitic-troostitic to the original pearlite (Fig. 6.42b).

During laser processing of CT80 *hardened* steel with an initial martensitic-troostitic structure, *two layers* are formed when the surface *melting does not occur*:

(a) The *surface layer* has a microstructure of medium-needle martensite, which is poorly etched (Fig. 6.43a). A gradual increase of residual austenite and crushing of martensite needles is observed on deeper layers.

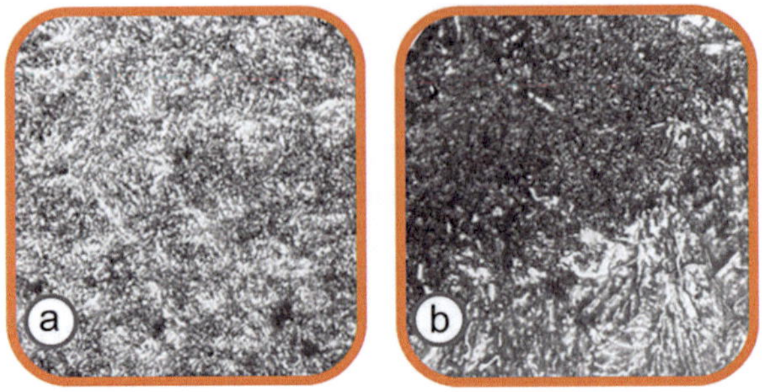

Fig. 6.42 Microstructure of surface (**a**) and transition (**b**) layers in normalized CT80 steel after laser treatment when heating temperatures exceeds the A_1 point, but does not exceed the melting point, ×500

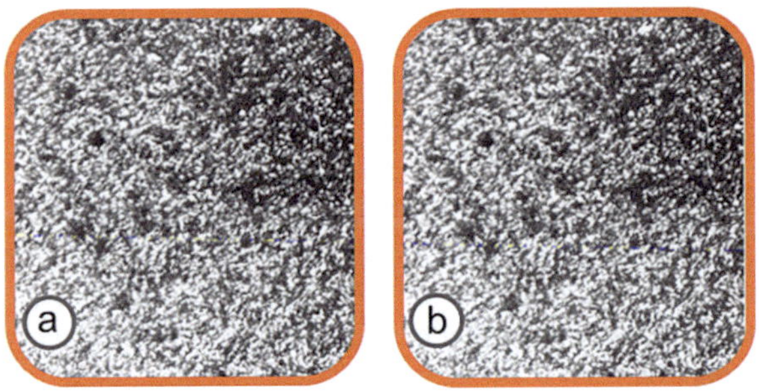

Fig. 6.43 Microstructure of the surface (**a**) and transition (**b**) layers in hardened CT80 steel when laser processing does not cause surface melting, ×500

(b) The *transition layer* is a tempering zone and has a troostite structure with cementitic carbides. The microstructure gradually changes from troosto-cementitic to martensitic-troostitic structure at the base boundary (Fig. 6.43b).

Thus, the *features of laser hardening* of CT80 tool carbon steel are:

- The heat affected zones, where structural and phase changes occur, have *clear boundaries* outlined by the corresponding isotherms (Fig. 6.44), which allows controlling the temperature distribution in the surface layer and assessing the accuracy of modeling thermal processes.
- The *microhardness* of the surface layers after laser irradiation with different parameters (both with and without surface melting) *weakly depends* on the *initial state* of the steel (Fig. 6.45), and is practically *independent of the irradiation parameters.*
- *In the hardened layers* obtained *with* surface *melting* at low processing speeds (V < 0.4 m/min), hot *microcracks* appear in the center of the zones on the surface, which develop during the crystallization and cooling of the molten metal due to high thermal stresses.

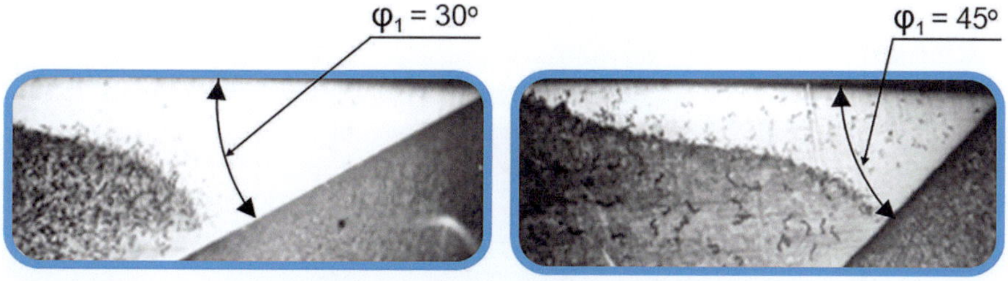

Fig. 6.44 Shape of the laser hardening zone in the cutting edges of a wedge made of CT80 tool carbon steel

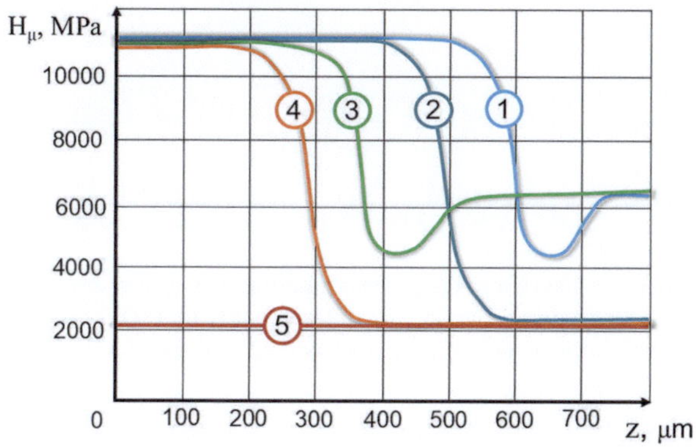

Fig. 6.45 Distribution of microhardness H_μ along the depth z depending on the initial state of steel and irradiation parameters: 1—Wp = $8 \cdot 10^4$ W/cm^2, V = 0.5 m/min (with surface melting), initial condition—bulk hardening; 2—Wp = $8 \cdot 10^4$ W/cm^2, V = 0.5 m/min (with surface melting) initial condition—normalization; 3—Wp = $2 \cdot 10^4$ W/cm^2, V = 1.5 m/min (without surface melting) initial condition—bulk hardening; 4—Wp = $2 \cdot 10^4$ W/cm^2, V = 1.5 m/min (without surface melting) initial condition—normalization; 5—without laser processing

- The size of the hardened layer under the same irradiation conditions in *pre-hardened* steel is 20–30% *larger* than in *normalized* steel (Fig. 6.45).

Based on the critical values of the laser processing mode, the ***areas of technological parameters*** of forming a hardened layer of ***guaranteed quality*** in a wedge can be determined (Fig. 6.46). Within the area, the hardened layers differ only in geometric dimensions, which are easy to control.

Laser hardening in the area of guaranteed quality technological parameters produces a favorable distribution of *residual stresses* in terms of strength and wear resistance:

- compressive residual stresses are formed in the hardened steel layer;
- in the tempered zone, the stresses change their sign and become tensile, gradually decreasing as the wedge moves away from the cutting edge.

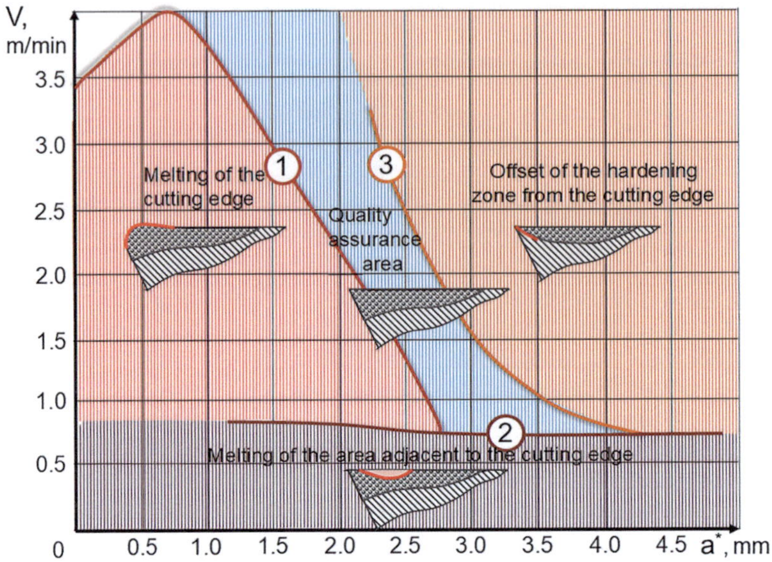

Fig. 6.46 An example of a diagram of the process mode area (V—laser beam speed; a*—distance of the laser beam axis to the cutting edge) of guaranteed quality for CT80 steel ($\varphi_1 = 45°$, P = 1000 W, d_0 = 5 mm): 1—first critical speed V_{c1}; 2—second critical speed V_{c2}; 3—third critical speed V_{c3}

Wear resistance due to high dispersion of the microstructure, metastability, high content of residual austenite and high hardness after laser hardening increases by 1.6–2 times. A significant (twofold) decrease in the friction coefficient is observed. The wear process is characterized by smoothness, absence of chips, absence of brittle fracture and adhesions.

6.5 Laser Thermal Cycling of Plasma Coatings

Laser thermal cycling is the action of a thermal cycle "heating–cooling" on the metal surface with supervelocities of 10^5–10^6 °C/s, which leads to a significant *redistribution of alloying elements* in the solid phase and associated recrystallization, structural-phase transformations, and changes in the elastic-strain state.

The spectral analysis shows that the redistribution of alloying elements away from the surface for a wide range of steels (C22, C45, X165CRMOV12, 34NiCrMoV14-5, 34CrNiMo6, X10CrNiTi18-10) has the following patterns [16]:

(a) the concentration of alloying elements changes at a considerable depth (200–350 μm), despite the short duration of the laser pulse;

(b) the concentration of various alloying elements (C—in all steels; Cr—in X165CRMOV12 steel; Mn, Ni, Ti—in X10CrNiTi18-10 steel) varies equally, despite the difference in diffusion coefficients by 5–6 orders of magnitude.

Thus, the *redistribution of alloying elements* during laser thermal cycling *is not* carried out *by the diffusion mechanism.*

The concentration of alloying elements increases from the depth and reaches a maximum at the surface. Moreover, *pulsed* laser irradiation increases the efficiency of the redistribution process.

Laser thermal cycling only moves the alloying elements to a depth. To deliver the right alloying elements to the surface in the right amount, plasma pre-spraying is effective.

Combined plasma-laser coating consists of two sequential processes: plasma *sputtering* for controlled surface alloying and subsequent laser *thermal cycling* to distribute the alloying elements over the depth and ensure favorable microstructure and mechanical properties of the surface layer.

Principles of combined coating application:

(1) during the coating process, it *is not allowed to melt* both the base and the coating, as this leads to a change in their chemical composition and properties, worsens the surface roughness;
(2) thermal processes during coating application must ensure a favorable distribution of residual stresses—residual *compressive stresses* must be present in the coating and in the area of the joint with the base of the part, which contribute to increased fatigue resistance;
(3) the coating in the transition layer should have a *metallurgical connection* with the base, which significantly increases the adhesion strength;
(4) the *duration and localization* of the coating process should be *minimal* so that its microstructure does not undergo undesirable changes.

The **mechanism of laser thermal cycling** consists in the occurrence of a ***thermal shock*** during ultrafast heating and excitation of an ***elastic wave*** with the subsequent development of three phenomena (Fig. 6.47):

(1) *tunneling* of alloying elements along with dislocations from the surface deep into the metal and uniform distribution over the depth in the laser heating zone;
(2) grain *grinding*;
(3) *recrystallization* in the deeper layers of the metal due to an increase in the concentration of alloying elements at high temperature.

In addition, localized residual deformations and stresses occur as a result of volumetric expansion under spatial constraints (Sect. 5.2).

The *highest efficiency* of the laser thermal cycling process is achieved at laser irradiation parameters that provide:

(1) maximum *surface temperature* up to 1250–1300 °C to prevent melting of the surface layer;

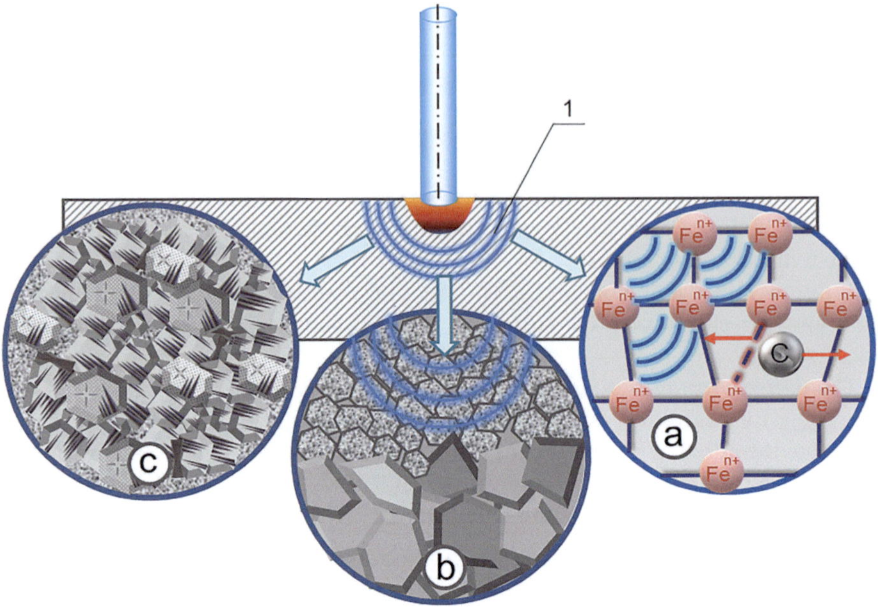

Fig. 6.47 Schematic of the laser thermal cycling mechanism (1—elastic wave): **a** tunneling of alloying elements; **b** grain refinement; **c** recrystallization

(2) heating *temperature at the interface* of the coating with the base metal at 1000 °C for uniform distribution of alloying elements and passing all temperature intervals of recrystallization;
(3) the *depth of the heat-affected zone* is higher than the thickness of the sprayed coating.

Laser thermal cycling of alloy steels and plasma-coated alloys initiates an effective mass transfer that can be used for localized microalloying. Microalloying gives the surface of products the desired properties and increases the adhesive strength of the coating to the base.

The task of determining the effective parameters of laser irradiation is solved by calculating temperature fields. In addition, temperature field calculations make it possible to determine the stress–strain state.

The **mathematical model** of combined plasma-laser coating is a calculation of the temperature field of a moving laser beam in a two-layer medium, provided that the top layer is a single-phase coating applied by plasma spraying (Fig. 6.48).

The temperature distribution is calculated based on the parameters of the laser irradiation, taking into account the properties of the coating and the base.

Parameters of the laser irradiation mode:

- the laser radiation power P;
- the diameter of the focusing zone d_0;
- the speed of the laser beam V;

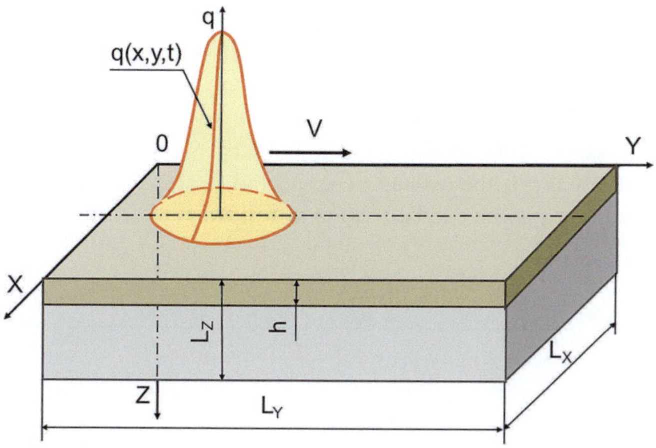

Fig. 6.48 Calculation diagram of the laser beam action on a part with a sprayed coating

- the heating time t_h;
- the pause time between heating pulses t_p;
- the number of heating pulses n.

Properties of the coating and the base:

- the metal density ρ;
- the metal heat capacity c;
- the metal thermal conductivity λ.

A part in the form of a rectangular parallelepiped with sides L_x, L_y, L_z is coated with a layer h and has an initial temperature T_s. The surface of the part is in the process of heat exchange with the environment with temperature T_e.

The upper sprayed surface is irradiated with a laser beam with an arbitrary power distribution across the cross-section. The laser beam moves along an arbitrary trajectory with a speed V.

The power distribution is described by the law W(x, y, t), and its action causes a heat flux on the sprayed surface with a power density distribution q(x, y, t).

After a certain period of time τ (depending on the thermal and physical properties of the metal), a quasi-stationary state is established, in which the metal heating zone is characterized by unchanged dimensions and moves along with the source of thermal influence.

In general, the nonstationary temperature distribution T(x, y, z, t) in a metal sample is described by a nonlinear heat conduction equation:

$$c\rho\frac{\partial T}{\partial t} = \frac{\partial}{\partial x}\left(\lambda\frac{\partial T}{\partial x}\right) + \frac{\partial}{\partial y}\left(\lambda\frac{\partial T}{\partial y}\right) + \frac{\partial}{\partial z}\left(\lambda\frac{\partial T}{\partial z}\right) \tag{6.17}$$

where c(z, T) is the effective value of the heat capacity of the material, which takes into account the presence of phase transitions and the structure of the coating;

ρ(z, T) is the density of the material;

λ(z, T) is the effective value of thermal conductivity.

For the bottom layer, the values of c, ρ, and λ are constant with respect to temperature, so the equation can be written as follows:

$$c\rho\frac{\partial T}{\partial t} = \lambda\left(\frac{\partial^2 T}{\partial x^2} + \frac{\partial^2 T}{\partial y^2} + \frac{\partial^2 T}{\partial z^2}\right)$$
$$x \in [0, L_x];\ y \in \left[0, L_y\right];\ z \in [0, L_z] \tag{6.18}$$

Starting condition: $T(x, y, z, 0) = T_S$.

Boundary condition on the machined surface in the laser exposure zone:

$$\lambda\frac{\partial T(x, y, 0, t)}{\partial z} + q(x, y, t) = 0 \tag{6.19}$$

Boundary condition outside the laser radiation zone on the surface being processed:

$$\lambda\frac{\partial T(x, y, 0, t)}{\partial z} + \alpha\left[T(x, y, 0, t) - T_e\right] = 0 \tag{6.20}$$

Boundary conditions on other surfaces of the computational domain:

$$\lambda\frac{\partial T(x, y, L_z, t)}{\partial z} + \alpha\left[T(x, y, L_z, t) - T_e\right] = 0 \tag{6.21}$$

$$\lambda\frac{\partial T(0, y, z, t)}{\partial z} + \alpha\left[T_e - T(0, y, z, t)\right] = 0 \tag{6.22}$$

$$\lambda\frac{\partial T(L_x, y, z, t)}{\partial x} + \alpha\left[T(L_x, y, z, t) - T_e\right] = 0 \tag{6.23}$$

$$\lambda\frac{\partial T(x, 0, z, t)}{\partial y} + \alpha[T_e - T(x, 0, z, t)] = 0 \tag{6.24}$$

$$\lambda\frac{\partial T\left(x, L_y, z, t\right)}{\partial y} + \alpha\left[T\left(x, L_y, z, t\right) - T_e\right] = 0 \tag{6.25}$$

$$T_c(x, y, h, t) = T_b(x, y, h, t) \tag{6.26}$$

where T_c is the temperature in the sputtered layer;

T_b is the temperature of the base metal.

In solving the problem, a coordinate difference scheme was used, having an adaptive grid that changed at each time step and condensed the nodes in areas of higher gradients of the desired function.

The optimal laser heating parameters, which ensures a maximum surface temperature of up to 1250–1300 °C and a temperature at the interface between the coating and the base of 1000 °C, is determined by varying the calculations, first of all, the diameter of the focusing zone and the laser radiation power (Fig. 6.49).

The slowness of the temperature change with depth is determined by the thermal properties of the composite layer of the coating and the base.

Laser thermal cycling of uncoated steels

Pulsed laser radiation provides a much *greater redistribution* of alloying elements compared to continuous laser radiation.

Laser irradiation parameters that cause heating with and without surface melting provide the same chemical composition of the surface layer. However, the depth distribution of alloying elements and non-metallic inclusions changes.

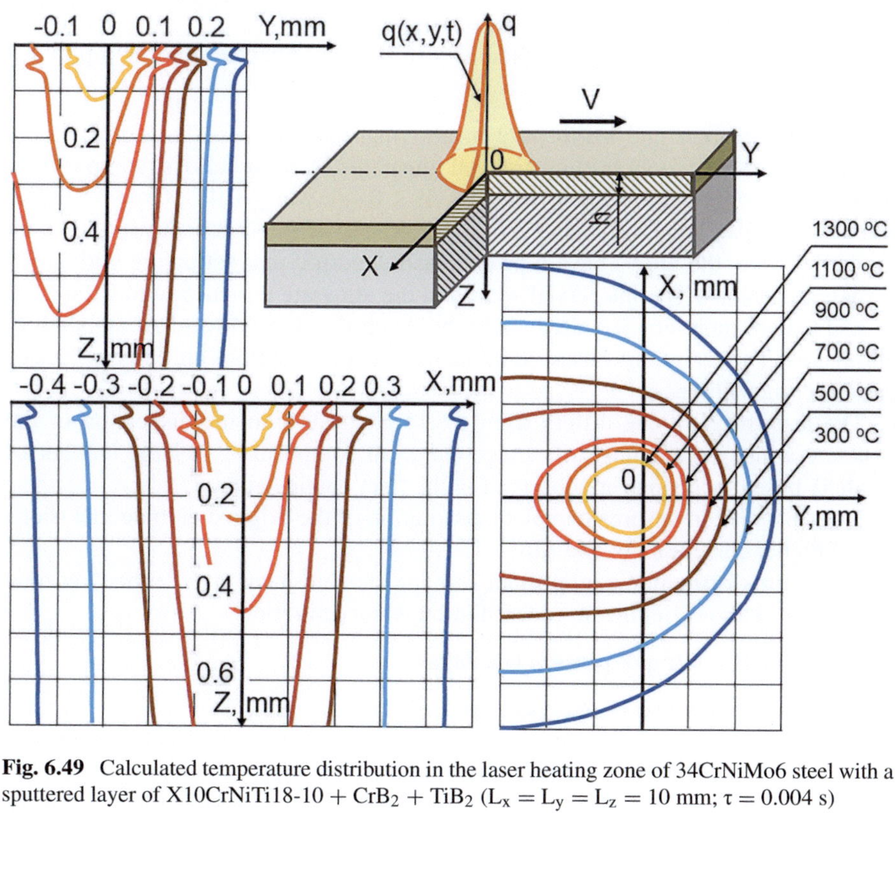

Fig. 6.49 Calculated temperature distribution in the laser heating zone of 34CrNiMo6 steel with a sputtered layer of X10CrNiTi18-10 + CrB_2 + TiB_2 ($L_x = L_y = L_z = 10$ mm; $\tau = 0.004$ s)

When HS6-5-2 steel is treated with pulsed laser radiation with surface melting (energy density $W_e = 15$ J/mm^2, pulse duration, $\tau = 6$ ms), a heat-affected zone with a depth of 300 microns is formed containing two layers:

(a) The molten layer ($h \leq 150$ μm), where structural and phase transformations occurred from the liquid phase, has a microhardness of 5000–7000 MPa. Moreover, the minimum microhardness of 5000–5500 MPa is in the near-surface layer.
(b) The transition layer (150 μm $< h \leq 300$ μm), where structural-phase transformations occurred in the solid phase, has a maximum microhardness of 10,500–11,500 MPa.

The layers have structures formed as a result of auto hardening and contain martensite, residual austenite and carbides. The carbides are concentrated in the transition layer.

The ***mobility of alloying elements*** under laser thermal cycling varies:

- Carbon is the most mobile and migrates in the solid phase to a distance of at least 150 μm in 6 ms, which is not explained by the diffusion mechanism. The carbon content as a result of migration can vary by 8–10 times.
- The content of Mo, W and V increases in the heat-affected zone by up to 1.5 times.

Grain ***refinement*** and ***uniformity*** of alloying elements ***distribution*** under the influence of laser thermal cycling are well illustrated by nickel alloys (composition: Ni = 60–68%; Fe $\leq$ 2%; C = 0.13–0.2%; Si $\leq$ 0.4%; Mn $\leq$ 0.4%; Cr = 9.5–12%; Mo = 3.5–4.5%; W = 4.5–5.5%; Co = 4–5.5%; Ti = 2.5–3.2%; Al = 5–6%). In the initial state, the alloys have a pronounced dendritic microstructure with small carbide inclusions. The inter-axial spaces in the alloy are enriched with Cr and Ti; W and Co are more concentrated along the dendritic axes; Mo and Al are slightly eliminated. The secretions observed in the interdendritic regions are Mn carbides of various compositions.

The laser-heated zone differs from the original alloy in the dispersion of the structure while maintaining an average chemical composition. A large number of small Ti-based inclusions are observed in the laser heating zone.

Laser irradiation ensures complete dissolution of the original carbide inclusions with a higher content of W and Mo.

A uniform distribution of almost all alloying elements over the laser heating zone is observed with no significant redistribution in concentration.

Laser thermal cycling of plasma coatings

Plasma coatings of the Cr-Ti-Ni system ***in the initial state*** provide functional purpose, productivity of the application process, layer thickness uniformity, and continuity at an <u>*optimum thickness*</u> of 50–100 μm. In sputtered coatings, the main volume (70–80%) is made up of white layers, which form due to rapid crystallization from the melt on a cold base. White layers have high microhardness (1250–1270 MPa) and

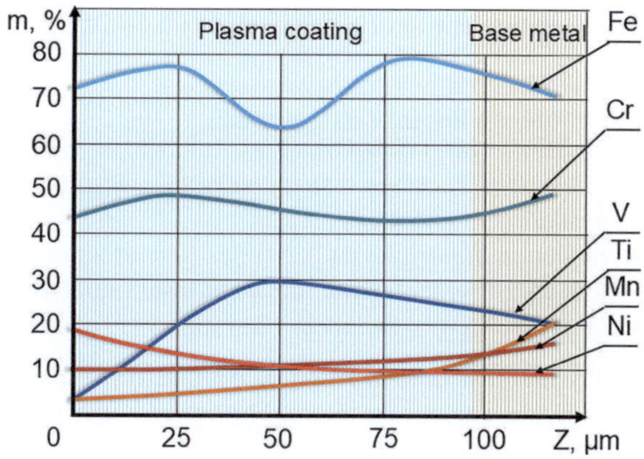

Fig. 6.50 Distribution of elements (m, %—mass percentage) over the thickness z of the V-Ti-Ni plasma coating and the base metal X10CrNiTi18-10 after four cycles of laser thermal cycling

low porosity (2–10%). The second component is areas with a gray eutectic structure and microhardness of 830–850 MPa.

Laser thermal cycling of plasma coatings causes mutual ***mass transfer of alloying elements*** between the coating and the base with the formation of transition layers.

The direction of movement depends on the composition and concentration of alloying elements in the coating and base:

- for V-Ti-Ni coating and X10CrNiTi18-10 base vanadium, boron, titanium and manganese in the base, nickel in the coating (Fig. 6.50);
- for Cr-Ti-Ni coating and 34CrNiMo6 base, chromium is observed to move into the base, titanium and nickel—into the coating.

The ***microhardness*** of all structural components decreases slightly after four thermal cycles (Table 6.2), which is associated with the decomposition of the supersaturated solid solution of the base. White layers are characterized by a constant decrease in microhardness with an increase in the number of thermal cycles. At the same time, after six cycles, the hardness of the white layers and eutectic areas become approximately equal. A further increase in the number of cycles leads to a significant decrease in the microhardness of the white layers compared to the eutectic components.

Table 6.2 Changes in microhardness of structural components of plasma coating made of eutectic V-Ti-Ni alloy powder under laser thermal cycling

Structural components of plasma coating	Volumetric content of the structural components, %	Micro-hardness, MPa (before thermal cycling)	Micro-hardness, MPa (after thermal cycling with the number of thermal cycles N)				
			N = 2	N = 4	N = 6	N = 8	N = 10
Eutectics	12–14	9750	9100	8740	7800	7150	7175
White layers	70–80	13,400	10,250	9100	7750	6800	6750

The redistribution of alloying elements and the presence of a transition layer leads to:

- increase in the *adhesive strength* of the plasma coating by 5 times (from 14–18 MPa to 80–110 MPa);
- relaxation processes that *increase the plasticity* of the coating and *prevent the formation of cracks*;
- *reduction of porosity* by 1.5 times (the effect of porosity reduction is manifested to a greater extent in the modes of laser radiation with melting);
- *increased wear resistance*.

The ***increase in wear resistance*** is due to an increase in the number of areas with a more plastic eutectic structure. The eutectic structure is capable of absorbing a large proportion of energy and relaxing frictional stresses. In addition, the increase in the elastic modulus during the decay of white layers increases the micro-plasticity and the stress of the beginning of the plastic flow of the coating (Fig. 6.51).

Technological recommendations

Laser thermal cycling of plasma coatings, both by pulsed and continuous radiation, is carried out at the temperature of the upper limit of the laser heating cycle of $\boldsymbol{0.75{\cdot}T_m}$ (or ***800–1000*** °C). This temperature ensures the absence of morphological changes in the eutectic crystals of the introduction phases.

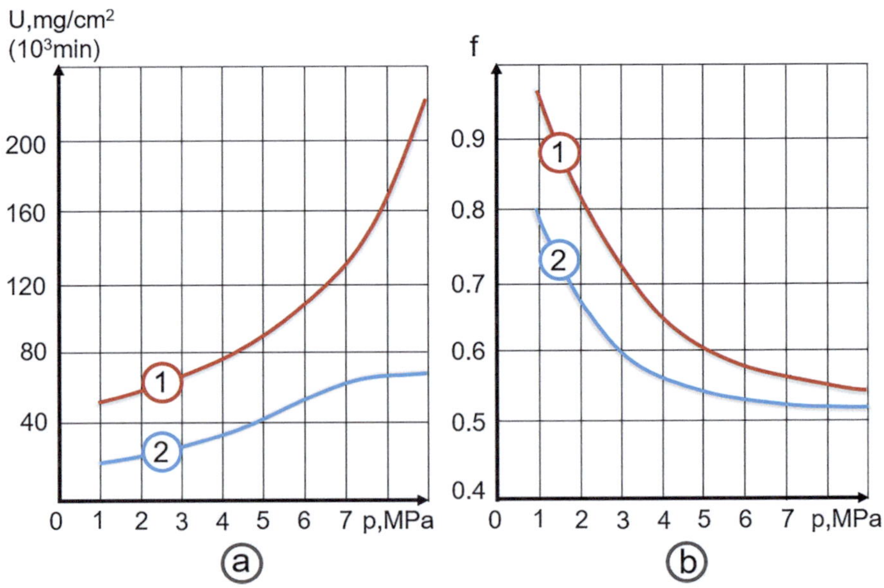

Fig. 6.51 Tribotechnical properties of V-Ti-Ni plasma coating (1—without thermal cycling; 2—after four cycles of laser thermal cycling): **a** reduced wear value U; **b** friction coefficient f

The largest *sizes of the heat-affected zone* and *redistribution of alloying elements* between the coating and the substrate are observed at ***four thermal cycles*** of laser radiation and a plasma coating ***thickness of 50–60 μm***.

A further *increase* in the number of cycles has virtually no effect on the dimensional characteristics of the heat-affected zone, but *reduces the microhardness.* Continued coagulation of the introduction phases in the white layers with increasing cycles leads to a decrease in strength.

For plasma coatings on products with a *small area* of functional surfaces (50–80 cm^2), it is effective to use laser thermal cycling with *pulsed* high-frequency radiation of low power (300–400 W).

On *large-sized* products, it is advisable to use laser thermal cycling with high-power (1000–2500 W) *continuous* laser radiation. The thickness of the plasma coating can be increased to 100 μm.

Examples of combined plasma-laser coating in industry

Laser thermal cycling is used to improve the quality of plasma coatings on parts of oil and gas production equipment, food processing machinery, tools and parts of medical equipment.

The ***auger and sleeves*** for supplying aggressive mixture under a pressure of up to 100 atm at a temperature of 650 °C are made of X10CrNiTi18-10 stainless steel. A wear-resistant V-Ti-Ni coating was applied to the working surfaces by plasma spraying. After plasma spraying, 25% of the surface area was treated with 4 cycles of continuous laser thermal cycling at a power of P = 1000 W, a focusing zone diameter d_0 = 5 mm, and a laser beam speed V = 0.8 m/min. To increase the absorption capacity, a ZnO-based coating was applied to the irradiated surfaces.

As a result, the adhesion strength of the plasma coating to the substrate increased by 4–5 times, and the service life—by 2–3 times.

Piston ring of an internal combustion engine made of gray cast iron. The Fe-Ni-Cr alloy was applied to the working surfaces by plasma spraying. After plasma spraying, 25% of the surface area was treated with 4 cycles of continuous laser thermal cycling at a power of P = 1000 W, a focusing zone diameter d_0 = 5 mm, and a laser beam speed V = 0.8 m/min.

As a result, the layer with the changed structure exceeded the coating thickness. The adhesion strength of the plasma coating to the substrate increased from 25 to 90 MPa, and the wear resistance increased by 3–4 times.

6.6 Laser Thermal Deformation Synthesis of Diamond-Containing Composite Coatings

6.6.1 Laser Thermal Deformation Synthesis Technology

The technology for manufacturing diamond tools using laser radiation, which is very attractive to the global scientific community [17–21], developed by specialists from the Institute of Superhard Materials of the National Academy of Sciences of Ukraine and the Department of Laser Systems and Advanced Technologies of the Igor Sikorsky Kyiv Polytechnic Institute. The developments made by the authors of this section include original methods for synthesizing diamond-containing composites by laser axial and radial sintering of working layers of disc tools and replaceable inserts of large-sized drum-type tools.

The main *tasks* of the technology for manufacturing tools from superhard materials, including *diamond*, are:

- obtaining the required shape and dimensions;
- ensuring the strong fixation of abrasive grains on working surfaces;
- maintaining cutting ability during operation.

Laser thermal deformation synthesis of diamond-containing composite coatings is carried out in two stages (Fig. 6.52):

(1) The *creation of a composite pool* 1 involves laser melting 2 of the product surface (base 3) with the simultaneous addition of metal powder 4 and artificial diamonds 5. The diamond content is about 20% of the binder mass. The travel speed V of the laser beam must ensure that the diamonds remain in the range from the melting point of the base to the temperature of oxidation and effect on the hardness of diamonds (750–800 °C) ***for no more than 0.4 s***(the optimum is 0.3 s). During the crystallization of the composite pool, two types of bonds are formed: a *metal bond* with the base and a *mechanical lock connection* 6 between diamond grains and the bond.

2) *Plastic deformation* of the composite layer during cooling to 650–500 °C by roller 7, to which force F is applied at a distance $\boldsymbol{l_F = 16\text{–}20\ mm}$ from the laser beam axis. High-temperature deformation leads to the following: it presses the diamond grains deep into the binder (in the molten binder, the diamond grains are immersed only 2/3 of the grain height) and strengthens the mechanical interlocking of the grains; it forms structures of plastic strengthening of the surface layer.

The *advantages* of laser thermal deformation synthesis of diamond-containing composite coatings are due to the *high speed* of non-contact energy input into the material and its *precise dosing*, and consist in the possibility of:

- minimizing the time of high-temperature exposure to diamonds and avoiding deterioration of physical properties;

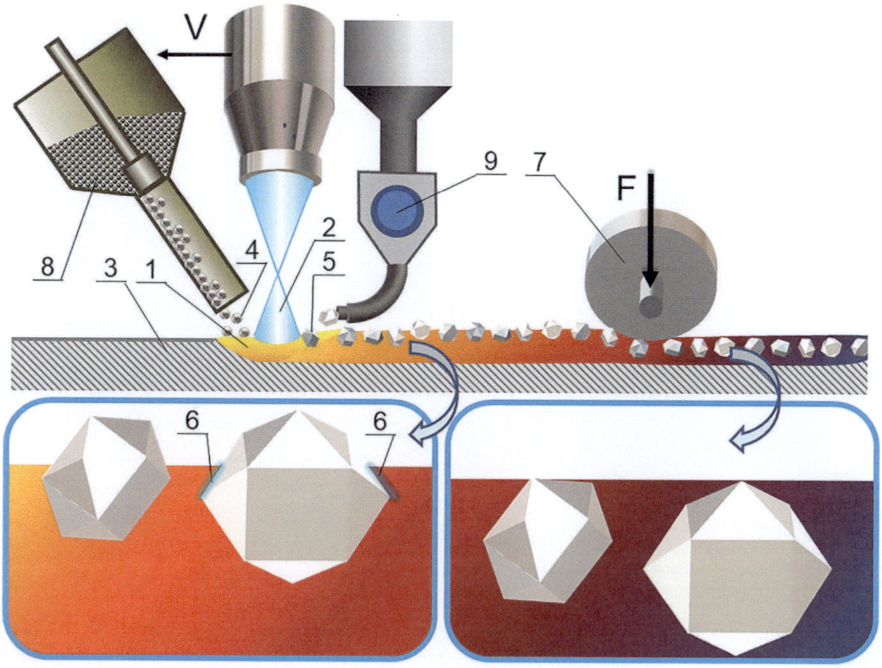

Fig. 6.52 Diagram of laser thermal deformation synthesis of diamond-containing composite coatings: 1—composite pool; 2—laser beam; 3—base; 4—metal binder powder; 5—artificial diamonds; 6—mechanical interlocking of diamond grains with the binder; 7—deforming roller; 8—metal powder dispenser; 9—diamond dispenser

- regulating the temperature over a wide range;
- regulating the temperature exposure zone;
- a wide range of binding components with different functional properties, easily controlled composition and concentration of composites, and obtaining of a highly dispersed structure;
- ensuring a metal bond between the composite layer and the steel body of the tool (base);
- significant increase in the productivity of diamond-coated tool manufacturing.
- layer-by-layer formation of single- or multi-row diamond-coated working surfaces and, thus, obtaining tools of any configuration.

Research into the effect of heating temperatures and laser irradiation conditions on the mechanical properties of synthetic and natural diamonds was carried out according to two schemes (Fig. 6.53):

(a) *direct irradiation* of diamonds placed on a graphite substrate (Fig. 6.53a);
(b) by *indirect heating* of diamonds *in melts* of various powder bonds (Fig. 6.53b).

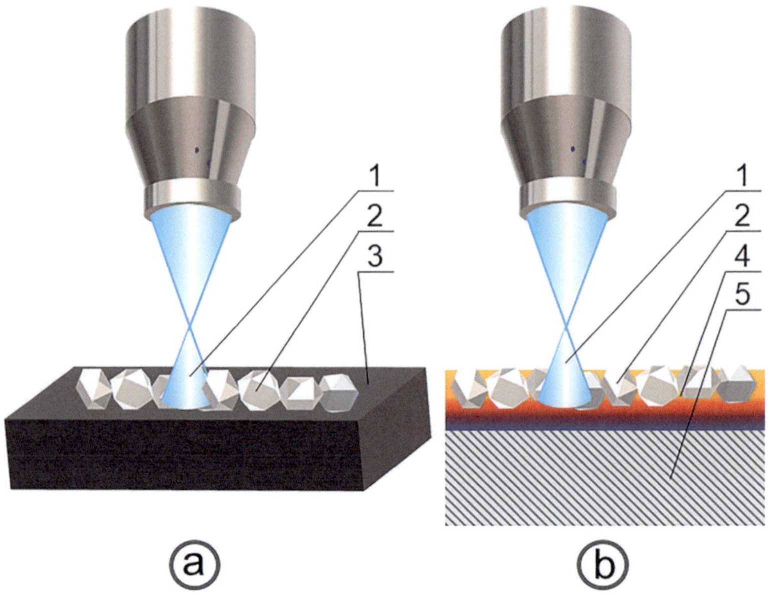

Fig. 6.53 Schemes for studying laser heating temperatures and irradiation conditions (1—laser beam; 2—diamonds; 3—graphite substrate; 4—powder bond melt; 5—base): **a** direct irradiation of diamonds; **b** indirect heating of diamonds in a melt

With *direct irradiation* of synthetic diamonds, it is practically impossible to determine the amount of absorbed energy, since diamonds are partially transparent to laser radiation.

When diamonds are *indirectly heated* in powder bond melts, the following parameters can be clearly determined:

(a) the maximum heating temperature of the diamond grain—is equal to the melting temperature of the bond;
(b) the duration of high-temperature exposure—is regulated by the speed of the laser beam.

Indirect heating by laser radiation allows for liquid-phase sintering of diamond-containing composites based on a wide range of bonds, both with low-melting and high-melting components.

Quality assurance requirements for high-temperature processing of diamonds:

(1) *limiting the time* of diamonds' exposure to temperatures above the oxidation temperature (750–800 °C), $t < 0.4$ s;
(2) high *purity* of synthetic diamonds;
(3) ensuring that the diamond grains are *wetted* by the binder.

Limiting the time diamonds are exposed to temperatures above the oxidation temperature (750–800 °C), $\boldsymbol{t < 0.4}$ ***s***. This is achieved in two ways:

(a) By *reducing the intensity* of laser radiation W_p and (or) increasing the travel speed of laser beam V. There are areas of optimal laser irradiation modes (W_p, t) in which, firstly, the bond melts and a mechanical lock connection of diamond grains with the bond is formed, and secondly, no cracks are formed and diamonds are not oxidized (Fig. 6.54). When the intensity of laser radiation increases ($W_p \geq 6 \times 10^7$ W/m^2), the bond temperature rises and cracks form in the diamonds along the bonding plane (Fig. 6.55a). As the temperature rises, additional cracks appear across the crystal edges (Fig. 6.55b). With a further increase in temperature, the cracks spread. The release of the metal catalyst used in diamond synthesis (green dots in Fig. 6.55c) through the cracks on the diamond surface is observed, indicating the direction of crack propagation from the grain center to the surface.

(b) *Limiting* the heating *temperature* of diamonds by using a low-melting-point bond, such as bronze. When using low-melting-point bonds, diamonds do not oxidize even at $t > 0.4$ s (regions 3 and 4 in Fig. 6.54).

High purity and initial strength of synthetic diamonds. Impurities and associated microdefects in diamonds lead to the formation of cracks on the surface of the grains and their propagation along the facets. High-purity diamonds exposed to laser irradiation for up to 0.3 s practically do not lose their strength (Fig. 6.56, c,d).

Wetting of diamond grains with the binder is achieved by:

(a) adding *cobalt* to the binder composition;

(b) the formation of chromium, cobalt, and tungsten *carbides* between the diamond and the binder.

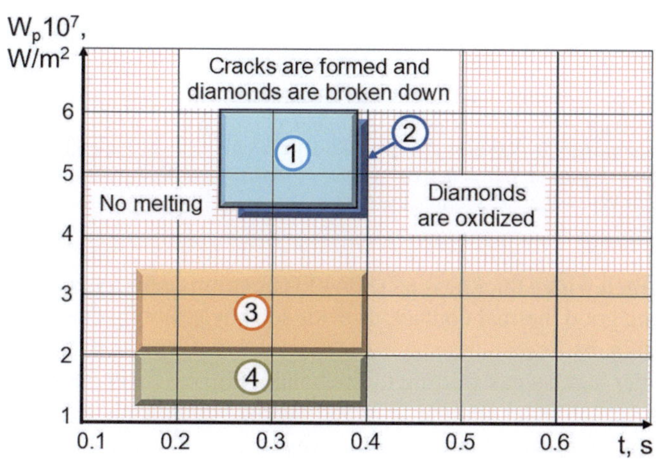

Fig. 6.54 Areas of optimal laser irradiation parameters (W_p, t) for indirect heating of diamonds in melts of various powder bonds: 1—powder Ni (base)-Cr(18%)-B(3%)-Si (4.3%)-C(0.8%)-Fe(3.5%) + WC(35%) + Co; 2—powder X10CrNiTi18-10 + TiB_2 + CrB_2 + Co; 3—powder Cu + Sn + Co; 4—powder Cu + Sn

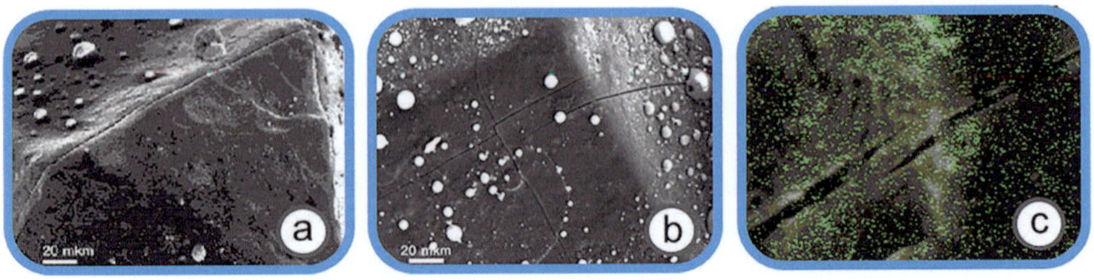

Fig. 6.55 Cracks in diamond grains at elevated heating temperatures: **a** cracks along the bonding plane; **b** cracks across the crystal edges; **c** crack propagation and metal catalyst release. *Note* Cracks in diamonds can form during the cooling stage under the action of shrinkage force. Shrinkage force arises due to the significant difference in the linear expansion coefficients of diamond ($\alpha = 1 \times 10^{-6}$ 1/deg.) and the bond (for Ni-Cr-Co alloy $\alpha = 10 \times 10^{-6}$ 1/deg.). However, for small artificial diamonds used in cutting tool coatings (up to 500 μm), the shrinkage force is insignificant (up to 50 N)

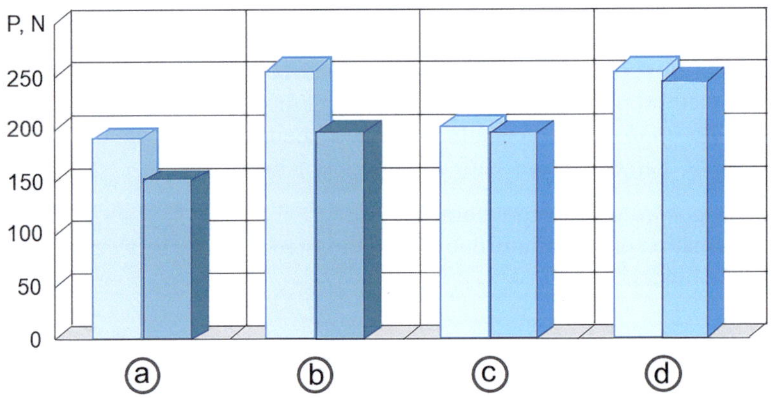

Fig. 6.56 Average destructive load P for initial (light color) and irradiated (dark color) diamonds of different purity: **a** ACC 125 diamonds (425–300 μm) at Wp = $1.7{\cdot}10^3$ W/cm^2, t = 0.09 s; **b** ACC125 diamonds (500–400 μm) at Wp = $0.6{\cdot}10^3$ W/cm^2, t = 0.15 s; **c** high-purity ACC 160 diamonds (400–315 μm) at Wp = $0.4{\cdot}10^3$ W/cm^2, t = 0.18 s; **d** high-purity ACC 160 diamonds (600–400 μm) at Wp = $0.2{\cdot}10^3$ W/cm^2, t = 0.24 s

A ***cobalt*** melt with a thickness of 180–400 angstroms densely covers all diamond faces, creating good thermal contact, as evidenced by scanning electron microscopy data (Fig. 6.57a, b).

On the other hand, weak wetting of the bonds (Fig. 6.57c) creates uneven contact between the molten bond and the diamond facets, which leads to a decrease in the strength of the diamond's bond with the base and uneven heating of the diamond with the possibility of crack formation.

The *risk* of cobalt's negative impact on the mechanical properties of diamonds is related to the fact that cobalt melts at a temperature of 1450 °C and dissolves carbon well in its liquid state. However, with a short duration of the liquid phase (10^{-2}–10^{-3} s), diffusion processes are unlikely to occur. Nevertheless, to eliminate

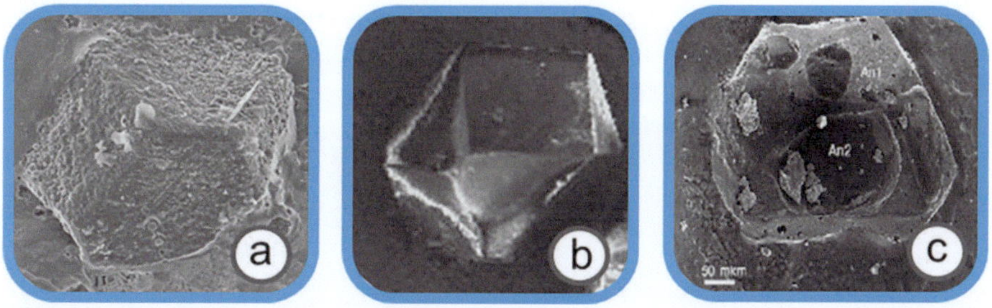

Fig. 6.57 Wetting of diamond grains with a binder: **a** diamond grain densely covered with a layer of cobalt (×1000); **b** diamond grain after treatment in a cobalt-containing binder at the optimal laser irradiation mode; **c** diamond grain after treatment in a binder without cobalt (spots of non-wetting)

the possibility of carbon dissolution, it is advisable to add carbide-forming metals or carbides to the cobalt bond.

During the laser synthesis of diamond-bonded composite coatings, a uniform distribution of bonding elements is observed due to the intensification of <u>*concentration redistribution*</u> processes, as evidenced by the chemical composition maps in the transition layer at the boundary between diamond and Ni-Cr-Co bonding (Fig. 6.58).

Wetting of diamond surfaces also enhances the formation of <u>*chromium, cobalt, and tungsten*</u> ***carbide*** compounds between the diamond and the bond.

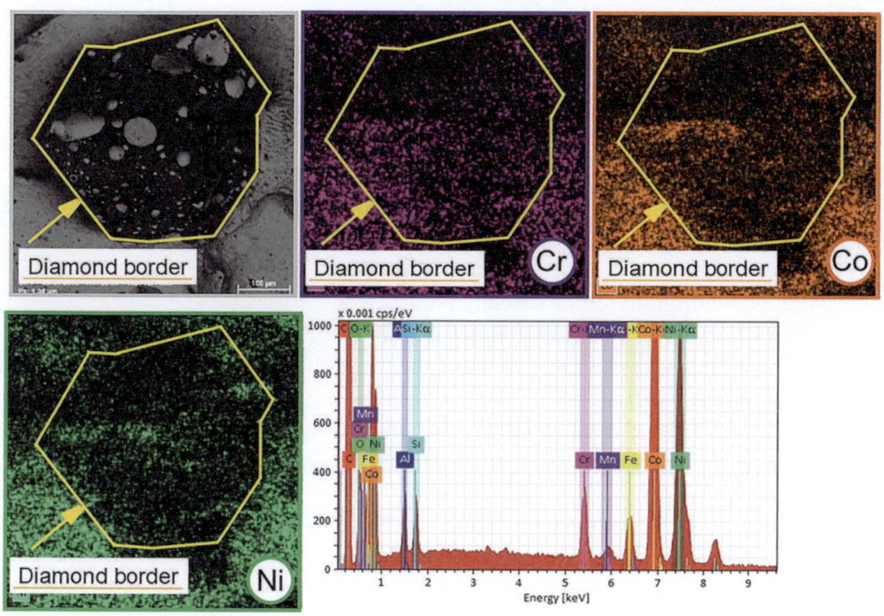

Fig. 6.58 Maps of chemical composition distribution in the 'diamond—Ni-Cr-Co binder' system

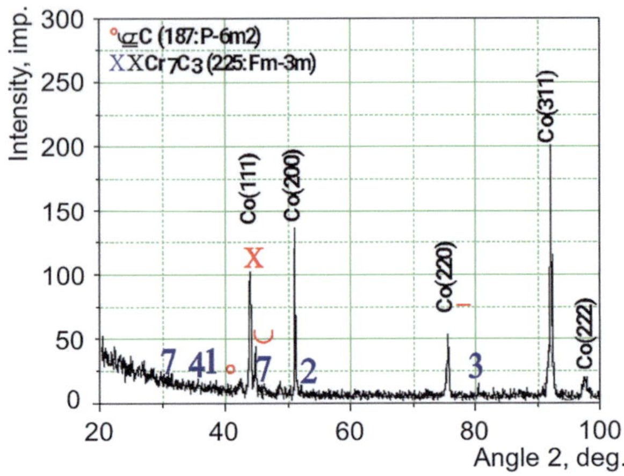

Fig. 6.59 Phase composition of a diamond-containing composite

Phase analysis (Fig. 6.59) shows the formation of Cr carbides in the bond during laser synthesis. Diffraction maxima from a solid solution based on cobalt, carbon, and chromium carbide Cr_7C_3 were detected. The content of chromium carbide Cr_7C_3 in the diamond-implanted composite is ~ 7 wt%, and in artificial diamonds ~ 13 wt% (the basis is a face-centered solid solution based on cobalt ~ 80 wt%). During laser sintering, carbon on the diamond surfaces interacts with chromium, which is a component of the Ni-Cr-Co bond. The formed Cr_7C_3 carbide nuclei subsequently grow epitaxially. The carbide formation process is accompanied by a decrease in the interfacial tension between the diamond and the bond, which improves the wetting of the diamond.

The chromium carbide layer has high adhesion strength to diamond and high thermal conductivity.

Metallographic studies have shown that after laser sintering, the bonds of composites have highly dispersed structures. With increasing distance from the surface, the dispersion of structures decreases and the size of the crystals of the strengthening phases increases.

The microhardness of structures varies in depth within the range of 7000–9000 MPa, depending on the processing parameters.

The maximum hardness is in the upper highly dispersed part of the molten layer. The hardness increases with the growth of the proportion of the harder eutectic and the hardness of the matrix.

The **tribotechnical characteristics** (amount and nature of wear, friction coefficient) of the composites were determined under dry friction conditions at a sliding speed of 0.1 m/s at specific loads of 1–9 MPa.

Iron-based powder composites (such as X10CrNiTi18-10 + TiB_2 + CrB_2) after laser sintering have greater wear resistance compared to nickel-based powder composites (such as Ni(base)-Cr(18%)-B(3%)-Si(4.3%)-C(0.8%)-Fe(3.5%) + WC(35%)). Moreover, at high specific loads (up to 9 MPa), the difference is very significant. This is due to the influence of oxide processes on the friction process and the formation of secondary structures capable of self-organization.

Profilograms of friction surfaces after wear tests show that the deposited layers wear evenly, without brittle destruction. This is due to the high dispersibility of the structure and the uniform distribution of hard strengthening and plastic phases.

6.6.2 *Mathematical Modeling of Laser Thermal Deformation Synthesis of Diamond-Containing Composite Coatings [19, 22]*

Structural and phase changes in composite coatings are determined by the temperature regime of laser heating and subsequent cooling. The characteristics of the temperature field (temperature level and distribution in volume, heating and cooling rates) are determined by the heat source, the characteristics of the materials being processed, and the conditions of interaction.

Problem statement. To apply a diamond layer, the steel body of the cutting wheel is placed between two matrix discs (Fig. 6.60).

The lower matrix disc has a recess into which a mixture of metal and diamond powders is poured during the tool manufacturing process. The mixture is exposed to a laser beam moving at a speed V. High-temperature heating causes the low-melting

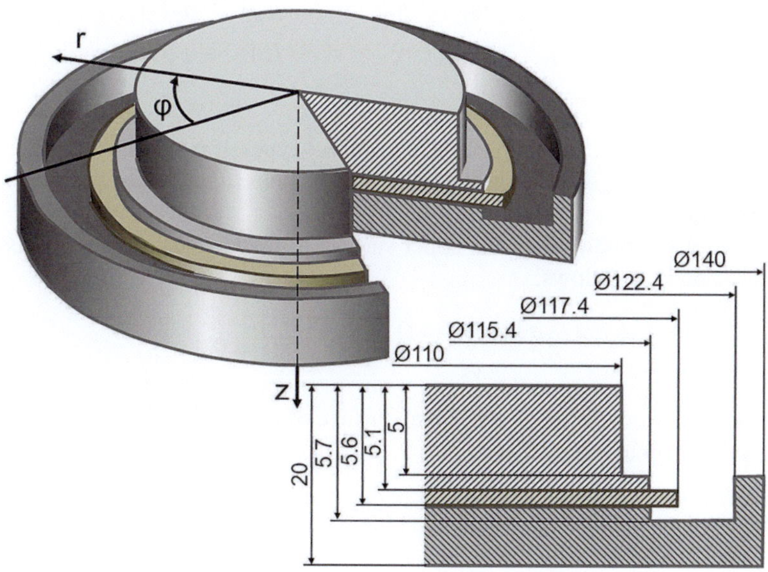

Fig. 6.60 Schematic diagram of a cutting wheel with a diamond coating

component of the composite mixture to melt and the edge of the disc tool body to melt. Objective is to calculate the temperature field.

Mathematical model

The following designations are used in the mathematical model:

r, φ, z—coordinates in a cylindrical system (physical domain);
x, y, z—coordinates in the Cartesian system (calculation domain);
t—time;
t_e—the simulation time.
z_i—depth coordinate of the laser radiation influence zone;
r_s, r_e, φ_e—dimensions of the computational domain;
c—heat capacity coefficient;
ρ—material density;
λ—effective thermal conductivity;
q—radiation power density;
α—heat transfer coefficient;
T_e—ambient temperature;
T_s—initial temperature of the object;
L_z—layer thickness.

The temperature field T(r, φ, z, t) is described by a three-dimensional heat conduction equation in cylindrical coordinates r, φ, z:

$$c\rho\frac{\partial T}{\partial t} = \lambda\left(\frac{1}{r}\cdot\frac{\partial T}{\partial r} + \frac{\partial^2 T}{\partial r^2} + \frac{1}{r^2}\cdot\frac{\partial^2 T}{\partial \varphi^2} + \frac{\partial^2 T}{\partial z^2}\right)$$
$$r \in [r_s, r_e]; \quad \varphi \in [0, \varphi_e]; \quad z \in [0, L_z]; \quad t \in [0, t_e] \tag{6.27}$$

Initial condition: $T(r, \varphi, z, 0) = T_s$.

Since the object of calculation consists of three types of materials (matrices, disc, and powder mixture), each of the parameters c, ρ, λ has three values and depends on the coordinates.

The boundary condition on the upper surface of the physical domain in the laser radiation zone is as follows:

$$\lambda\frac{\partial T(r, \varphi, z_i, t)}{\partial z} + q(r, \varphi, t) = 0$$

The boundary condition on the upper surface of the physical domain outside the laser radiation zone is as follows:

$$\lambda\frac{\partial T(r, \varphi, z_i, t)}{\partial z} + \alpha(T(r, \varphi, z_i, t) - T_e) = 0$$

The boundary conditions on the lower and side surfaces of the physical domain model heat exchange with the environment:

$$\lambda \frac{\partial T(r, \varphi, L_z, t)}{\partial z} + \alpha(T(r, \varphi, L_z, t) - T_e) = 0$$
$$\lambda \frac{\partial T(r_e, \varphi, z, t)}{\partial z} + \alpha(T(r_e, \varphi, z, t) - T_e) = 0$$

The boundary conditions at the edges of the calculation domain at $\varphi = 0$; $\varphi = \varphi_e$; $r = r_s$ are as follows:

$$\partial T/\partial n = 0$$

where n is the direction of the normal to the surface.

Example solution for the dimensions specified in the diagram (Fig. 6.60).

We assume a uniform distribution of the power density of the laser beam q over a rectangular area, the length of which coincides with the width of the recess of the lower matrix (7 mm), and the width—with the depth of the recess 0.7 mm (0.006 rad).

The physical domain (cylinder system) can be represented in the Cartesian coordinate system with axes 0x, 0y, 0z as a computational domain in the form of rectangular parallelepipeds (Fig. 6.61). The coordinates r, φ, z correspond to the coordinates x, y, z.

The high power of laser radiation and the small size of the light spot lead to large temperature field gradients (Fig. 6.62).

In the calculation domain, the initial Eq. (6.27) has the following form:

$$\frac{\partial T}{\partial t} = a\left(\frac{1}{x}\frac{\partial T}{\partial x} + \frac{\partial^2 T}{\partial x^2} + \frac{1}{r^2}\frac{\partial^2 T}{\partial y^2} + \frac{\partial^2 T}{\partial z^2}\right) \quad (6.28)$$

where $a = \lambda/(c\rho)$ is the thermal conductivity coefficient.

Each splitting step for a given problem on a non-uniform grid implements the transition from the (k − 1)th time layer to the (k + 1)th and consists of the following six steps:

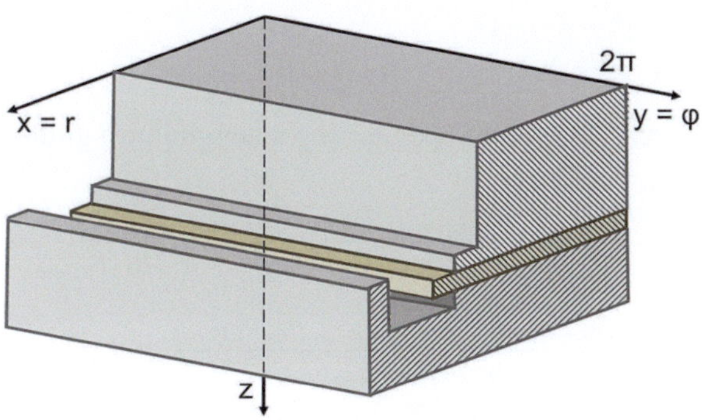

Fig. 6.61 Calculation domain of a diamond-coated cutting wheel

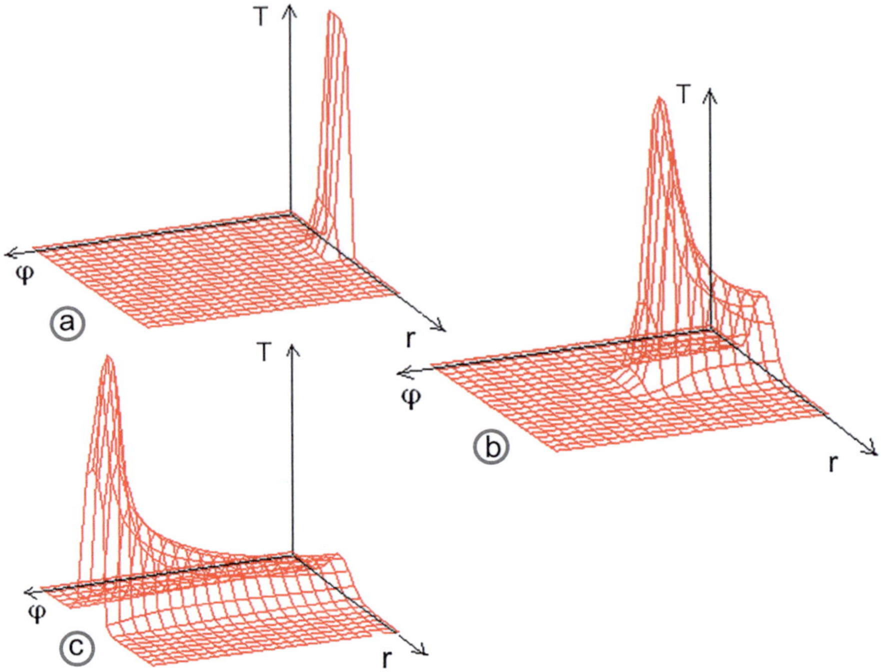

Fig. 6.62 Temperature distribution T at different times t: **a** at t = 0.05 s; **b** at t = 0.4 s; **c** at t = 0.95 s

(1) $\frac{u^{k-\frac{2}{3}}-u^{k-1}}{\tau}=\Lambda_3\frac{u^{k-\frac{2}{3}}+u^{k-1}}{2};$

(2) $\frac{u^{k-\frac{1}{3}}-u^{k-\frac{2}{3}}}{\tau}=\Lambda_2\frac{u^{k-\frac{1}{3}}+u^{k-\frac{2}{3}}}{2};$

(3) $\frac{u^{k}-u^{k-\frac{1}{3}}}{\tau}=\Lambda_1\frac{u^{k}+u^{k-\frac{1}{3}}}{2};$

(4) $\frac{u^{k+\frac{1}{3}}-u^{k}}{\tau}=\Lambda_1\frac{u^{k+\frac{1}{3}}+u^{k}}{2};$

(5) $\frac{u^{k+\frac{2}{3}}-u^{k+\frac{1}{3}}}{\tau}=\Lambda_2\frac{u^{k+\frac{2}{3}}+u^{k+\frac{1}{3}}}{2};$

(6) $\frac{u^{k+1}-u^{k+\frac{2}{3}}}{\tau}=\Lambda_3\frac{u^{k+1}+u^{k+\frac{2}{3}}}{2}.$

where Λ_1, Λ_2, Λ_3—difference operators on a non-uniform grid:

$$\Lambda_1 u^k_{i,j,m}=a\left(\frac{1}{x_i}\cdot\frac{u^k_{i+1,j,m}-u^k_{i-1,j,m}}{2h_{1,c}}+\frac{h_{1,i+1}u^k_{i-1,j,m}-2h_{1,c}u^k_{i,j,m}+h_{1,i}u^k_{i+1,j,m}}{h_{1,c}h_{1,i}h_{1,i+1}}\right);$$

$$\Lambda_2 u^k_{i,j,m}=\frac{a}{x_i^2}\cdot\frac{h_{2,j+1}u^k_{i,j-1,m}-2h_{2,c}u^k_{i,j,m}+h_{2,j}u^k_{i,j+1,m}}{h_{2,c}h_{2,j}h_{2,j+1}};$$

$$\Lambda_3 u^k_{i,j,m}=a\frac{h_{3,m+1}u^k_{i,j,m-1}-2h_{3,c}u^k_{i,j,m}+h_{3,m}u^k_{i,j,m+1}}{h_{3,c}h_{3,m}h_{3,m+1}}.$$

At each of the six stages, systems of linear equations with a three-diagonal matrix are solved.

6.6.3 Diamond Cutting Wheels

Difficulties in sintering diamond-containing composites from a *free state* mixture:

- it is difficult to control the size and shape of the beads (Fig. 6.63a), since the parameters of laser irradiation are aimed at preserving the mechanical properties of diamonds;
- there is an accumulation of diamonds on the surface of the melt bond with the formation of characteristic regular groups (Fig. 6.63b) due to differences in specific gravity (the specific gravity of diamonds is approximately half that of the bond).

The *purpose* of laser thermal deformation synthesis of diamond-impregnated composite coatings is to ensure the specified shape and dimensions of the coating with guaranteed preservation of the mechanical properties of diamonds.

The shape and dimensions of the diamond-containing coating are ensured by the use of *forming matrices* in which:

(a) a cutting wheel disc is *fixed*;
(b) a *composite pool* is created and formed from the molten edge of the disc, molten powder, and diamonds;

Fig. 6.63 Sintering of diamond-reinforced composites from a free state mixture: general view of the bead; characteristic regular groups of diamonds in the mixture

(c) the coating formed as a result of crystallization of the composite pool at a certain distancc from the laser heating spot is *mechanically compacted* by a deforming roller, *pressing* the diamond grains deep into the bond.

Mathematical modeling of thermal processes, presented in Sect. 6.6.2, establishes:

- the *laser* irradiation *parameters* that ensure the melting of the edge surface of the cutting disc and the melting of the powder mixture while simultaneously *eliminating* the loss of mechanical properties of diamonds and welding the composite to the surfaces of the forming matrix;
- the temperature of the composite and, accordingly, the *magnitude and coordinates of the point of application of the deforming force* relative to the axis of the laser beam.

Based on the analysis, two *technological schemes* for laser thermodeformation synthesis of diamond-containing composite coatings on thin cutting discs are proposed:

(1) axial laser thermodeformation synthesis;
(2) edge laser thermodeformation synthesis.

For both schemes, the optimal *plastic deformation parameters* are:

- deformation temperature T = 600–500 °C;
- deformation force F = 400–600 N;
- distance between the point of application of the deforming force and the laser beam axis l_F = 12–16 mm;
- metal powder consumption of the bond—0.3–0.5 g/s.

The technological scheme of axial laser thermal deformation synthesis of diamond-containing composite coatings (Fig. 6.64) is implemented by feeding a powder mixture of the bond and diamonds from dispenser 1 into the mold 2 with a cutting wheel 3 body located in it. The mold is made of metal with high thermal conductivity (e.g., copper) and is cooled with water. The powder is filled to the level of the mold clamp protrusion 4. At the same time, a laser beam 5 scans in the radial direction across the width Δr of the diamond-bonded layer and melts the edge of the cutting wheel body and the metal bond. The scanning frequency of the beam ensures a quasi-continuous heat flow with sufficient intensity to melt the metal. The crystallized but still plastic bond with diamonds is compacted by a deforming roller 6 when the mold is moved. As a result of deformation, the diamond grains are pressed deep into the bond and the composite is shaped into the required dimensions and profile.

The technological scheme of edge laser thermal deformation synthesis of diamond-containing composite coatings (Fig. 6.65) is implemented by creating a composite pool 2 at the edge of the cutting wheel body 3 using a laser beam 1. Metal powder 4 is fed into the composite pool from the metal powder dispenser 5, and artificial diamonds 6—from the diamond dispenser 7. The body of the cutting wheel is fixed in a movable matrix 8. The movable matrix rotates and provides a

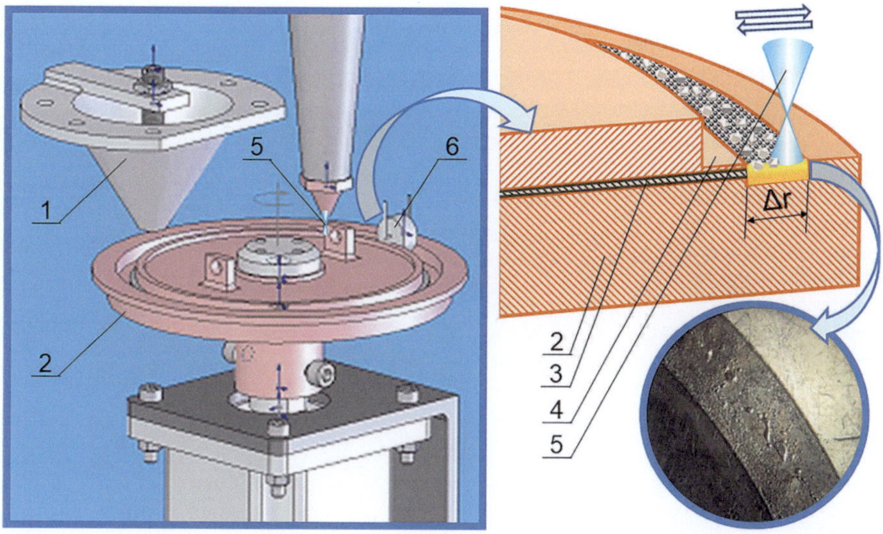

Fig. 6.64 Technological scheme of axial laser thermal deformation synthesis of diamond-containing composite coatings: 1—dispenser of powder mixture of the bond and diamonds; 2—mold; 3—body of the cutting wheel; 4—mold clamp protrusion; 5—laser beam; 6—deforming roller

linear speed V for the movement of the laser beam. The shape and dimensions of the synthesized coating are provided by a fixed forming matrix 9. The forming matrix consists of two half-matrices. The movable and fixed matrices are made of metal with high thermal conductivity (e.g., copper). The movable matrix is cooled with water. The molten and crystallized diamond-containing binder is compacted by a deforming roller 10.

The technological scheme of end-face edge laser thermal deformation synthesis allows the layer-by-layer creation of multilayer diamond-containing composite coatings.

6.6.4 Instrumental Drum with Replaceable Cutting Inserts with Diamond-Containing Coating

Large-sized parts made of titanium alloys, high-alloy corrosion-resistant steels and alloys, composite materials (carbon, carbon–carbon, organic and glass plastics) are widely used in mechanical engineering, aircraft and shipbuilding. Diamond abrasive tools are used in the manufacturing process. The most productive method is grinding with diamond drums. However, existing grinding drums have low adhesive strength of the diamond-containing composite layer (no more than 200–250 MPa), do not allow quick replacement of worn cutting elements, and do not allow the creation

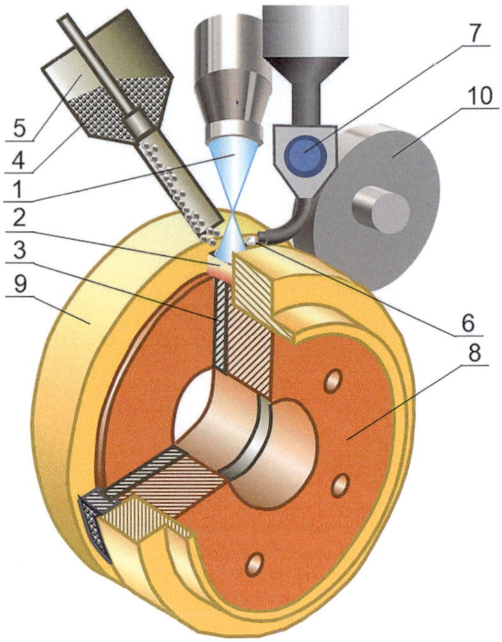

Fig. 6.65 Technological scheme of edge laser thermal deformation synthesis of diamond-containing composite coatings: 1—laser beam; 2—composite pool; 3—cutting wheel housing; 4—metal powder binder; 5—metal powder dispenser; 6—artificial diamonds; 7—diamond dispenser; 8—movable matrix; 9—fixed forming matrix; 10—deforming roller

of working surfaces with different configurations of cutting edges and chip removal gaps.

The drum with replaceable cutting inserts (Fig. 6.66) has a metal body 1 in the form of a hollow cylinder with T-shaped longitudinal grooves (dovetail type). The grooves are located at an angle of 5° to the axis of rotation of the drum body. Replaceable cutting inserts 2 are fixed in the grooves. The lower shelf of the insert fits into the T-shaped groove of the cylinder. The upper shelf of the insert is profiled to the working diameter of the cylinder. A diamond-imitation composite coating 3 with a thickness of 2–3 mm is synthesized on the surface of the upper shelf of the insert. Flanges 4 are fixed on both ends of the drum body. The inner surfaces of the flanges have recesses that correspond to the profile of the metal inserts and provide additional fixation of the inserts. The end flanges are attached to the drum body with screws 5.

The *use of replaceable cutting inserts* improves the quality of the coating, increases the manufacturability of the drums, reduces the cost and simplifies maintenance during operation due to the following:

- small inserts simplify the technology and equipment of laser thermal deformation synthesis compared to a large-sized drum;
- the range of bond powders and base metals is significantly expanded;
- it becomes possible to manufacture tools with diamond-containing coatings of any spatial configuration;

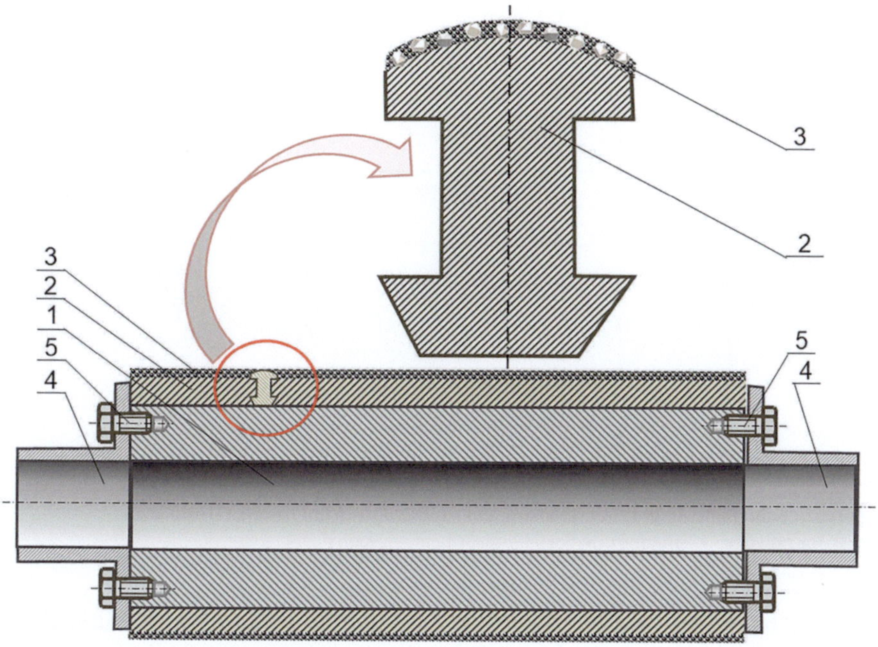

Fig. 6.66 Tool drum with replaceable cutting inserts made of diamond-containing composites: 1—cylindrical drum body; 2—replaceable cutting inserts; 3—diamond-containing composite coating; 4—end flanges; 5—screws

- when the coating wears out, there is no need to replace the drum (only the inserts are replaced);
- the inserts are easily repaired by building up layers of coating.

Mathematical modeling of laser sintering of a rectangular variable cutting insert of a tool drum [8, 9]

Objective is to calculate the characteristics of the temperature field (temperature values and distribution by volume, heating and cooling rates) during laser melting of powder material in a rectangular metal insert depending on the heat source, the characteristics of the materials being processed, and the conditions of interaction.

In the calculation scheme (Fig. 6.67), the working surface of the drum is intermittent and consists of a set of metal replaceable inserts 1 in the form of a rectangular parallelepiped with a length of 100 to 800 mm and a width of 20 to 60 mm, arranged at a certain pitch along the entire length of the drum. The inserts have rectangular recesses 2 for the composite powder mixture 3.

The initial temperature of the replaceable insert is equal to the ambient temperature T_e.

The laser heat source moves at a speed V. At the focus on the treated surface, the laser beam has an elliptical shape. The intensity distribution of the heat source is

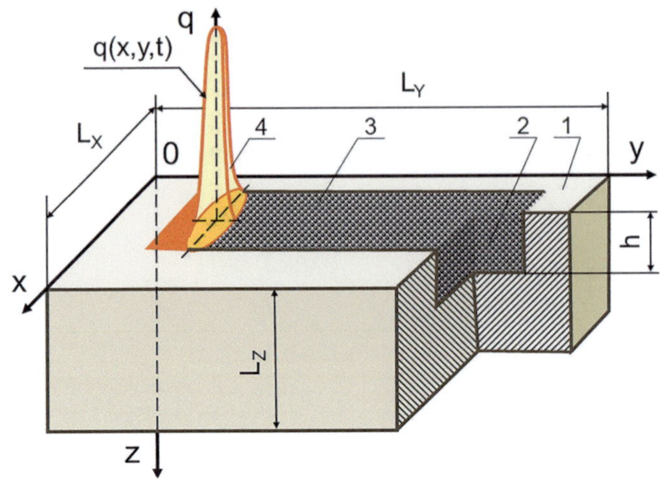

Fig. 6.67 Schematic diagram of a rectangular replaceable drum insert: 1—replaceable insert; 2—recess for powder; 3—binder powder; 4—laser heat source

described by the law W(x, y, t). Under the action of laser radiation, a heat flux with power distribution q(x, y, t) arises on the surface.

The powder material is located in the channel of the replaceable insert. We assume that the powder is a solid layer of homogeneous material that melts under the action of laser radiation.

The temperature field in the replaceable insert under the action of laser radiation is described by a three-dimensional differential equation with given initial and boundary conditions:

$$c\rho\frac{\partial T}{\partial t} = \lambda\left(\frac{\partial^2 T}{\partial x^2} + \frac{\partial^2 T}{\partial y^2} + \frac{\partial^2 T}{\partial z^2}\right)$$
$$x \in [0, L_x]; \quad y \in \left[0, L_y\right]; \quad z \in [0, L_z]; \quad t \in [0, t_e] \tag{6.29}$$

where c is the heat capacity coefficient;

ρ is the density of the material;

λ is the effective thermal conductivity;

L_x, L_y, L_z are the width, length, and depth of the replaceable insert, respectively;

t_e is the simulation time.

Initial condition: $T(x, y, z, 0) = T_s$.

where T_s is the initial temperature of the object.

Boundary condition on the processing surface in the laser radiation zone:

$$\lambda\frac{\partial T(x, y, 0, t)}{\partial z} + q(x, y, t) = 0 \tag{6.30}$$

Boundary condition on the treatment surface outside the radiation zone:

$$\lambda \frac{\partial T(x, y, 0, t)}{\partial z} + \alpha\big[T(x, y, 0, t) - T_e\big] = 0 \tag{6.31}$$

where T_e is the ambient temperature.

Boundary conditions on other faces where heat loss occurs:

$$\lambda \frac{\partial T(x, y, L_z, t)}{\partial z} + \alpha\big[T(x, y, L_z, t) - T_e\big] = 0 \tag{6.32}$$

$$\lambda \frac{\partial T(0, y, z, t)}{\partial x} + \alpha\big[T_e - T(0, y, z, t)\big] = 0 \tag{6.33}$$

$$\lambda \frac{\partial T(L_x, y, z, t)}{\partial x} + \alpha\big[T(L_x, y, z, t) - T_e\big] = 0 \tag{6.34}$$

$$\lambda \frac{\partial T(x, 0, z, t)}{\partial y} + \alpha[T_e - T(x, 0, z, t)] = 0 \tag{6.35}$$

$$\lambda \frac{\partial T\big(x, L_y, z, t\big)}{\partial y} + \alpha\big[T\big(x, L_y, z, t\big) - T_e\big] = 0 \tag{6.36}$$

To solve a three-dimensional non-stationary partial differential equation, the finite difference method is used, namely, a coordinate splitting scheme in an adaptive grid.

Let the approximate solution be found on the (k–1)th layer on an uneven grid at points (x_i, y_i, z_m), where $i = 0,1,\ldots,n_1$; $j = 0,1,\ldots,n_2$; $m = 0,1,..,n_3$.

Let us denote the value of the grid steps in the direction: 0X by $h_{1,i} = x_i - x_{i-1}$; 0Y – by $h_{2,j} = y_j - y_{j-1}$; 0Z – by $h_{3,} = z_m - z_{m-1}$.

Let us also denote by $h_{1,c}$, $h_{2,c}$, $h_{3,c}$ the mean value of two adjacent steps for the node (x_i, y_i, z_m):

$$h_{1,c} = \frac{h_{1,i} + h_{1,i+1}}{2} = \frac{x_{i+1} - x_{i-1}}{2};$$
$$h_{2,c} = \frac{h_{2,j} + h_{2,j+1}}{2} = \frac{y_{j+1} - y_{j-1}}{2};$$
$$h_{3,c} = \frac{h_{3,m} + h_{3,m+1}}{2} = \frac{z_{m+1} - z_{m-1}}{2}.$$

The time step is $\tau = \tau_k = t_k - t_{k-1}$.

For a non-uniform grid, we will approximate the derivatives of the function $y = y(x)$ using the following formulas:

$$y''(x_i) \approx \frac{1}{h_c} \cdot \left(\frac{y_{i+1} - y_i}{h_i} - \frac{y_i - y_{i-1}}{h_i}\right) = \frac{h_{i+1}y_{i-1} - 2h_c y_i + h_i y_{i+1}}{h_{i+1}h_c h_i}$$
$$h_c = (h_i + h_{i+1})/2 \tag{6.37}$$

Let $u^k_{i,j,m}$ denote the value of the approximate solution at the point (x_i, y_j, z_m, t_k), and $T = T^k_{i,j,m}$ denote the value of the exact solution.

The coordinate splitting scheme for Eq. (6.37) is implemented in six steps and has second-order accuracy in time and spatial variables:

(1) $\frac{u^{k-\frac{2}{3}}-u^{k-1}}{\tau} = \Lambda_3 \frac{u^{k-\frac{2}{3}}+u^{k-1}}{2}$;

(2) $\frac{u^{k-\frac{1}{3}}-u^{k-\frac{2}{3}}}{\tau} = \Lambda_2 \frac{u^{k-\frac{1}{3}}+u^{k-\frac{2}{3}}}{2}$;

(3) $\frac{u^{k}-u^{k-\frac{1}{3}}}{\tau} = \Lambda_1 \frac{u^{k}+u^{k-\frac{1}{3}}}{2}$;

(4) $\frac{u^{k+\frac{1}{3}}-u^{k}}{\tau} = \Lambda_1 \frac{u^{k+\frac{1}{3}}+u^{k}}{2}$;

(5) $\frac{u^{k+\frac{2}{3}}-u^{k+\frac{1}{3}}}{\tau} = \Lambda_2 \frac{u^{k+\frac{2}{3}}+u^{k+\frac{1}{3}}}{2}$;

(6) $\frac{u^{k+1}-u^{k+\frac{2}{3}}}{\tau} = \Lambda_3 \frac{u^{k+1}+u^{k+\frac{2}{3}}}{2}$.

where Λ_1, Λ_2, Λ_3 are difference operators, taking into account (5.2):

$$\Lambda_1 u^k_{i,j,m} = \frac{h_{1,i+1}u^k_{i-1,j,m} - 2h_{1,c}u^k_{i,j,m} + h_{1,i}u^k_{i+1,j,m}}{h_{1,c}h_{1,i}h_{1,i+1}};$$

$$\Lambda_2 u^k_{i,j,m} = \frac{h_{2,j+1}u^k_{i,j-1,m} - 2h_{2,c}u^k_{i,j,m} + h_{2,j}u^k_{i,j+1,m}}{h_{2,c}h_{2,j}h_{2,j+1}}; \; ;$$

$$\Lambda_3 u^k_{i,j,m} = \frac{h_{3,m+1}u^k_{i,j,m-1} - 2h_{3,c}u^k_{i,j,m} + h_{3,m}u^k_{i,j,m+1}}{h_{3,c}h_{3,m}h_{3,m+1}}.$$

At each of the six stages, a series of nonlinear equations arises. For example, the equations of the first stage system in index notation are as follows:

$$\frac{u^{k-\frac{2}{3}}_{i,j,m} - u^{k-1}_{i,j,m}}{\tau} = \frac{a}{2} \times \left(\frac{h_{3,m+1}u^{k-\frac{2}{3}}_{i,j,m-1} - 2h_{3,c}u^{k-\frac{2}{3}}_{i,j,m} + h_{3,m}u^{k-\frac{2}{3}}_{i,j,m+1}}{h_{3,c}h_{3,m}h_{3,m+1}} + \frac{h_{3,m+1}u^{k-1}_{i,j,m-1} - 2h_{3,c}u^{k-1}_{i,j,m} + h_{3,m}u^{k-1}_{i,j,m+1}}{h_{3,c}h_{3,m}h_{3,m+1}} \right) \tag{6.38}$$

Let's denote: $\sigma_i = \frac{a\tau}{2h_{1,c}h_{1,i}h_{1,i+1}}$; $\sigma_j = \frac{a\tau}{2h_{2,c}h_{2,i}h_{2,i+1}}$; $\sigma_m = \frac{a\tau}{2h_{3,c}h_{3,i}h_{3,i+1}}$.
Then Eq. (6.38) will be formulated as follows:

$$\begin{aligned} &- h_{3,m+1}\sigma_m u^{k-\frac{2}{3}}_{i,j,m-1} + \left(1 + 2h_{3,c}\sigma_m\right)u^{k-\frac{2}{3}}_{i,j,m} - h_{3,m}\sigma_m u^{k-\frac{2}{3}}_{i,j,m+1} \\ &= \left(1 - 2h_{3,c}\sigma_m\right)u^{k-1}_{i,j,m} + \sigma_m\left(h_{3,m}u^{k-1}_{i,j,m-1} + h_{3,m}u^{k-1}_{i,j,m+1}\right) \end{aligned} \tag{6.39}$$

Since the calculation scheme for the replaceable insert includes two types of materials, the thermophysical properties are selected depending on the location of the calculation node. The ratio of thermophysical properties forms a system of linear algebraic equations of size (n_1 + 1)(n_2 + 1)(n_3 + 1) with respect to the unknown values of the grid function on the time layer with number (k – 2/3).

When calculating temperatures, the following parameters are specified:

- the dimensions of the replaceable insert;
- the dimensions of the recess for the powder;
- the thermophysical properties (heat capacity, thermal conductivity, density, heat transfer coefficient) of the replaceable insert;
- the thermophysical properties of the powder;
- laser radiation parameters (shape and dimensions of the laser beam, total power, power distribution in the focus zone, laser beam travel speed);
- calculation parameters (acceptable error, size of the difference grid steps, and time steps).

The result of the calculations is the temperature distribution in the variable insert at any given moment.

Example of calculating the parameters of the laser sintering process for a rectangular replaceable cutting insert of a tool drum [20]

Initial data:

- material of the replaceable insert—41Cr4 steel;
- powder—X10CrNiTi18-10 + TiB_2 + CrB_2;
- thickness of the diamond-containing layer—0.3–0.6 mm;
- laser beam diameter—3, 4, 5 mm;
- bond melting temperature—1250 °C;
- laser radiation power—1500 W;
- laser radiation wavelength—1.06 μm;
- thermal conductivity—29 W/(m·K);
- thermal conductivity—3.4×10^{-6} m^2/s;
- absorption capacity of the bond—0.7.

The dependence of the laser beam speed on the thickness of the diamond-containing layer at different focusing zone diameters is shown in Fig. 6.68.

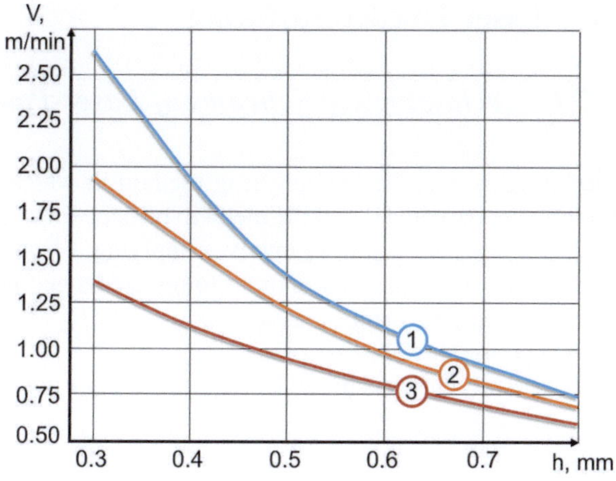

Fig. 6.68 Dependence of the laser beam travel speed V on the thickness of the diamond-containing layer h: **a** with a focusing zone diameter $d_0 = 3$ mm; **b** with a focusing zone diameter $d_0 = 4$ mm; **c** with a focusing zone diameter $d_0 = 5$ mm

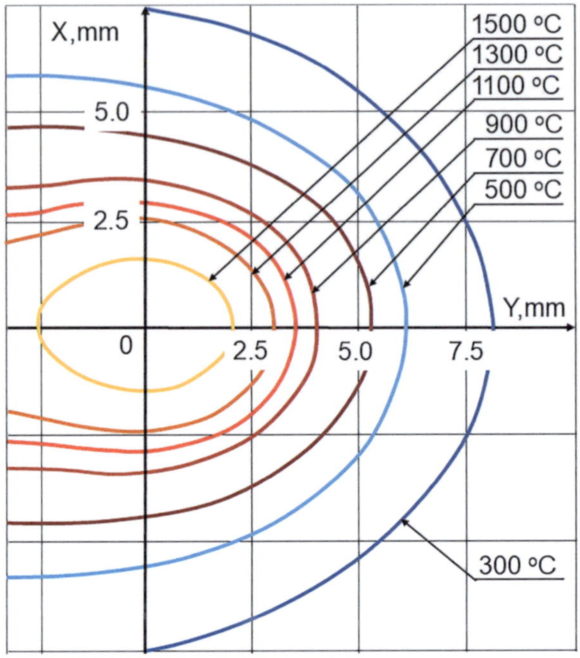

Fig. 6.69 Temperature distribution on the surface of the replaceable insert (d_0 = 5 mm, V = 0.6 m/min, h = 0.5 mm)

Based on the analysis of the temperature distribution in the replaceable insert (Fig. 6.69), taking into account that the plastic deformation of the composite layer should be carried out during cooling to 650–500 °C, the ***optimal distance*** for applying the deforming force $\boldsymbol{l_F}$ ***= 16–20 mm*** from the laser beam axis. The deforming force F = 300–500 N.

6.7 Laser-Powder Surfacing

6.7.1 Technological Schemes of Laser-Powder Surfacing

Surfacing is a surface engineering method in which a surface layer is formed by melting the base metal and the metal to be deposited to form a *single surfacing pool.*

Laser powder surfacing involves applying a layer of functional material to the surface of a metal product in order to restore its geometric dimensions and shape or improve the performance characteristics of the layer, which has a metallurgical bond with the base [11, 23].

Laser surfacing has the following *advantages* over other surfacing methods [23]:

- shallow penetration depth, which *minimizes the proportion of base metal in the deposited* metal and allows obtaining the desired composition and properties of the deposited metal in the first layer;
- high energy concentration, which allows for *precise restoration* of the dimensions of worn parts with a minimum level of residual stresses and deformations;
- high temperature, which allows surfacing *refractory* composites.

Laser surfacing is performed in one or two stages:

(1) In ***one-step*** surfacing, the filler material is fed directly into the surfacing zone in the form of powders, solid or flux-cored wires, and rods. The use of powders provides a wide range of options for shaping the properties of the filler material.
(2) In a ***two-step*** surfacing process, the filler material is applied to the surface of the product by one of the following methods before surfacing:

   - plasma or detonation spraying, which allows for thick deposited layers;
   - in bulk;
   - powder pastes;
   - powder with a binding component (glue);
   - electrolytic deposition.

At the *second stage*, the deposited filler material is melted with a laser beam by sequentially scanning the surface to be deposited. In multilayer surfacing, a layer of new filler material is applied after each pass. The main disadvantages of two-stage laser surfacing are low productivity due to the high labor intensity of the first stage of the process and uneven coating caused by the surface tension forces of the liquid metal. When laser melting coatings with a thickness of more than 0.5 mm, there is a tendency for shrinkage cracks to form due to a large temperature difference.

Thus, the most promising is a one-step laser surfacing process using a filler material in the form of a powder—***laser-powder surfacing***.

In the process of laser-powder surfacing (Fig. 6.70), a laser beam 1 with a focus spot diameter d_0 melts the surface of the workpiece 2 and moves at a speed V. The filler powder 3 is fed at an angle α in the transport gas flow 4, melted by the laser beam and enters the surfacing pool 5 in a liquid state. The transport gas is also a shielding gas. The filler powder can also be supplied by a colloidal solution jet, which eliminates spraying and significantly reduces the mass consumption of powder G.

Depending on the functional purpose of the coating and, consequently, the composition and properties of the powder, the laser power distribution in the focus spot $W_p(t)$ is determined;

(a) uneven with increasing power (Fig. 6.70a);
(b) uneven with decreasing power (Fig. 6.70b);
(c) conditionally uniform multi-mode (Fig. 6.70c).

The molten powder is mixed with the molten base metal in the surfacing pool. During the crystallization of the surfacing pool, the deposited layer 6 is formed. The quality criteria for the surfacing process are:

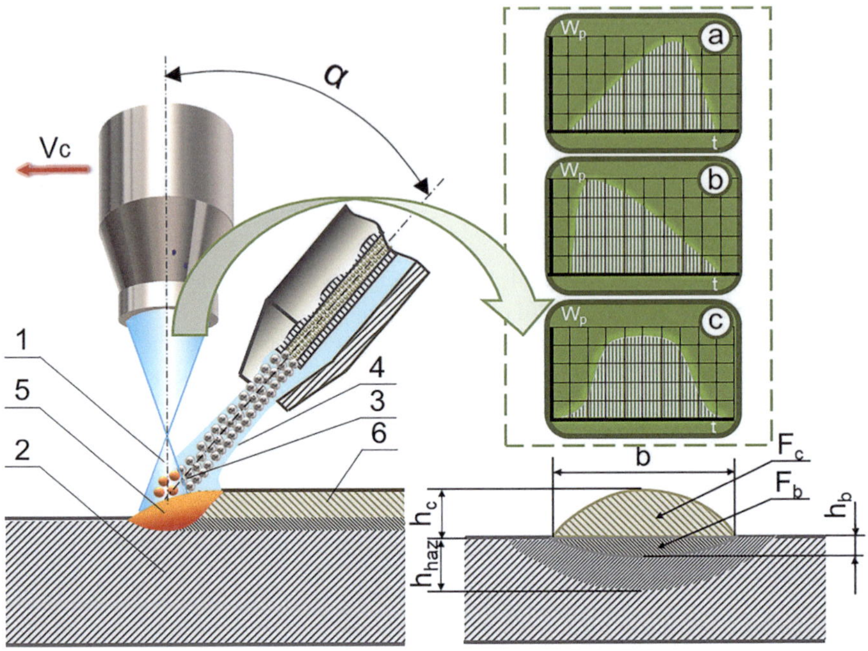

Fig. 6.70 Scheme of the laser-powder surfacing process: 1—laser beam; 2—workpiece; 3—filler powder; 4—transport gas flow; 5—surfacing pool; 6—deposited layer (**a** uneven distribution of laser radiation power in the focus spot $W_p(t)$ with increase; **b** uneven distribution of laser radiation power in the focus spot $W_p(t)$ with decrease; **c** conditionally uniform multimode distribution of laser radiation power in the focus spot $W_p(t)$)

(a) *Depth of penetration*—the distance h_b (Fig. 6.70), at which a single surfacing pool of the metal to be deposited and the base metal is formed.

(b) The *proportion of base metal in the deposited metal* γ—the ratio of the penetration area F_b (Fig. 6.70), which is proportional to the rate of base metal to the total cross-sectional area of the deposited bead $(F_b + F_c)$, which is proportional to the total amount of molten metal:

$$\gamma = \frac{F_b}{F_b + F_c} \tag{6.40}$$

The quality of the surfacing process can be ensured by ***minimizing the penetration depth and the proportion of the base metal in the deposited metal***.

When designing the technological process of laser-powder surfacing, it is necessary to imagine the scheme of connection between physical processes and the formation of coating properties (Fig. 6.71).

The *purpose* of laser-powder surfacing is to provide the surface layer of a part with a specified set of *performance characteristics*: wear resistance, contact strength, fatigue strength, corrosion resistance, and heat resistance.

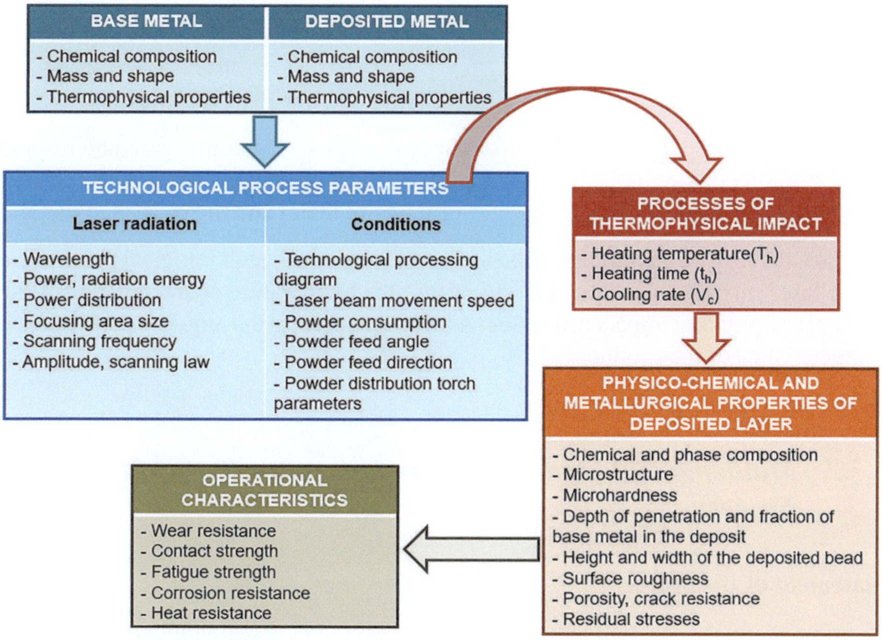

Fig. 6.71 Diagram of the relationship between technological parameters and product performance during laser-powder surfacing

The performance characteristics of a part are determined *by the physical, chemical and metallurgical properties of the deposited layer* (chemical and phase composition, microstructure, microhardness, penetration depth, proportion of base metal in the deposited metal, height and width of the deposited bead, surface roughness, porosity, crack resistance, residual stresses). The physical, chemical and metallurgical properties of the deposited layer are formed as a result of interrelated *processes of thermal and physical effects* of laser radiation on the workpiece: absorption of laser radiation, heating, melting, diffusion, hydrodynamic processes in the melt and its crystallization, structural and phase transformations in the solid state, and thermal conductivity. When the powder is melted, the surface layer of the base material is partially melted and subsequently cooled at ultra-high speed, a coating is formed of three layers: a surface layer dominated by structures formed from the powder melt; a transition layer formed as a result of mechanical mixing, diffusion, or similar mass transfer processes between the deposited and base metal in liquid and solid state; and a heat affected zone of the base metal that does not melt but undergoes structural changes. The transition layer determines the adhesive strength (adhesion of the deposited layer to the base).

The processes of thermal and physical impact [11] are largely determined by three *temperature characteristics*:

- heating temperature (T_h);

- heating time (t_h);
- cooling rate (V_c).

The temperature characteristics of the thermal impact are determined by:

- *technological parameters of the process* (laser radiation parameters and processing conditions);
- *composition and physical properties* of the base metal and the deposited metal.

Thus, the task of technological design is to assign such technological parameters of the laser surfacing process that will ensure the temperature characteristics of the thermal and physical impact processes and, as a result, form the required performance characteristics of the product.

6.7.2 Mathematical Modeling of the Laser-Powder Surfacing Process

Statement of the problem

In order to relate the temperature characteristics and technological parameters, thermal fields in two-layer coating-substrate systems under the action of laser radiation with melting of the coating were calculated. In this case, the coating can be either single-phase or multiphase (composite type).

According to the technology, the powder material is continuously fed into the melting zone by a flow of transport gas. For simplicity, the coating material is assumed to be a continuous medium and is pre-applied to the substrate, after which it is melted under the influence of laser radiation.

The design scheme (Fig. 6.72) is a rectangular parallelepiped with sides L_x, L_y, L_z (the base) with a coating of thickness h. The surface of the coating is exposed to a laser beam moving at a constant speed V along the 0x axis. All the surfaces of the substrate are in the process of heat exchange with the environment, whose temperature is T_e.

A change in the laser radiation power over time $W_p(t)$ causes a heat flux on the surface with a density distribution q(x, y, t). Different functions of the heat flux density distribution in the focusing zone are considered:

- Gaussian;
- uniform;
- non-uniform with a shifted maximum to the edge of the focusing spot in the direction of the laser beam movement;
- uneven with a shifted maximum to the edge of the focus spot in the opposite direction to the laser beam movement.

It is necessary to determine the *temperature distribution* in the coating and the transition layer of the substrate under the action of a moving laser beam.

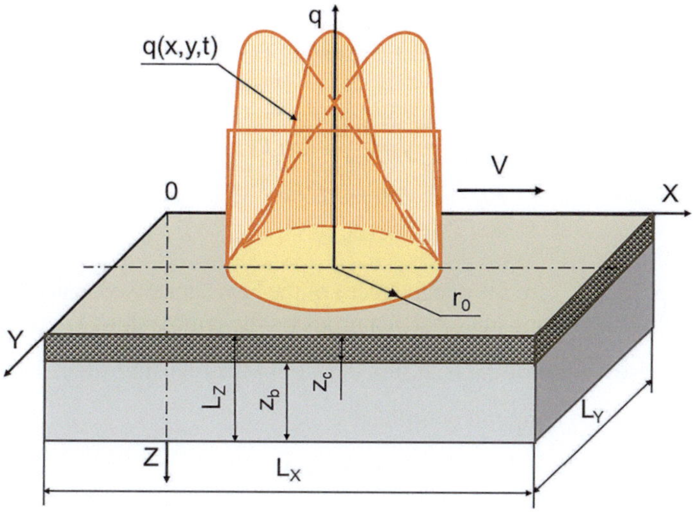

Fig. 6.72 Calculation scheme for modeling thermal processes during laser-powder surfacing

A two-dimensional (plane) problem is solved in the coordinate system associated with the heat source.

When solving the thermal problem, the *following assumptions* are made:

- the material to be deposited is a continuous medium;
- there is perfect thermal contact between the coating and the substrate;
- there are no hydrodynamic, aerodynamic, or thermo-chemical phenomena in the laser impact zone.

The thermal process is described by a nonlinear unsteady two-dimensional heat conduction equation with boundary conditions of the second kind, an initial condition, and an implicit scheme for selecting phase transition boundaries: (6.41)–(6.44).

$$c(T)\rho(T)\frac{\partial T}{\partial t} = \frac{\partial}{\partial x}\left(K(T)\frac{\partial T}{\partial x}\right) + \frac{\partial}{\partial z}\left(K(T)\frac{\partial T}{\partial z}\right) \tag{6.41}$$

$$c_v(T) = c(T) + \delta Q_f\left(T - T_f\right) \tag{6.42}$$

$$\mathrm{T_s} = 25\ ^{\circ}\mathrm{C} \tag{6.43}$$

$$\left.\frac{\partial T}{\partial x}\right|_{x=0;x=L_x} = 0;\ -K(T)\left.\frac{\partial T}{\partial z}\right|_{z=0;z=L_z} = (1 - R(T))q(x, t);\ \left.\frac{\partial T}{\partial z}\right|_{z=0;z=L_z} = 0 \tag{6.44}$$

Taking into account the above assumptions is achieved by the fact that the thermophysical characteristics of the base depend not only on the temperature, but also on the spatial coordinates:

$$\begin{aligned} c &= c_1(T, x, y)|_{z=z_b} + c_2(T, x, y)|_{z=z_c} \\ \rho &= \rho_1(T, x, y)|_{z=z_b} + \rho_2(T, x, y)|_{z=z_c} \\ K &= K_1(T, x, y)|_{z=z_b} + K_2(T, x, y)|_{z=z_c} \end{aligned} \quad (6.45)$$

where c(T), ρ(T), K(T) are the dependences of the heat capacity, density and thermal conductivity of the coating and base materials on the temperature T;

q(x, t) is the distribution of the heat flux density in the focusing zone;

x, y, z, t—spatial and temporal coordinates;

$c_v(T) = c(T)\rho(T)$ is the volumetric heat capacity;

δ is the Dirac function;

Q_f is the latent heat of phase transformation;

T_s is the initial temperature of the workpiece;

R(T) is the dependence of the laser reflection coefficient on the temperature T;

z_b is the thickness of the base layer;

z_c is the thickness of the coating layer.

The system of Eqs. (6.41)–(6.45) is solved by the method of finite differences on a moving (associated with the focusing zone) grid, with an implicit scheme for approximating derivatives.

In this case, the dependences of thermal conductivity $\lambda_f(T)$, density $\rho_f(T)$, heat capacity $C_f(T)$ on temperature, which are included in (6.41), are interpolated by cubic splines:

$$\begin{aligned} &S(x) = yi + bi(x - xi) + ci(x - xi)2 + di(x - xi)3 \\ &xi \le x \le xi + 1 \end{aligned} \quad (6.46)$$

where: i is the number of points x and y to be interpolated;

S(x)—the value of the function and argument x;

b_i, c_i, d_i—spline coefficients.

The thermophysical properties of a composite coating containing i = 1…n components are determined according to the conditions:

$$c\Sigma = \Sigma ciki \quad (6.47)$$

$$K\Sigma = \Sigma Kiki \quad (6.48)$$

$$\rho\Sigma = \Sigma\rho iki \tag{6.49}$$

where c_i, K_i, ρ_i are the dependences of the heat capacity, thermal conductivity, and density of the i-th component of the composite material;

k_i is the amount of the i-th component.

Calculations of changes in time of temperatures and heating (cooling) rates in the surface layer at different distributions of the heat flux density in the focusing zone of continuous laser radiation are shown in Figs. 6.73, 6.74, 6.75, and 6.76.

Initial data for numerical calculation:

- laser radiation power P = 1500 W;
- diameter of the focusing zone $d_0 = 2r_0 = 2.5$ mm;
- laser beam travel speed V = 0.2; 0.5 m/min;
- thickness of the coating layer $z_c = 0.5$ mm;
- base—steel 2C45;
- powder composition: 1.2–1.8% B, 0.2–0.6% C, 2.3–2.8% Si, 0.1% Fe, 94.9% Ni;
- powder melting point $T_m = 1400$ °C;
- dimensions of the part: $L_x = 500$ mm; $L_y = 20$ mm; $L_z = 6$ mm.

The *maximum heating temperature* (graphs a in Figs. 6.73, 6.74, 6.75 and 6.76) is the same for Gaussian and non-uniform distributions of the heat flux density (T_{max} = 1670 °C). With a uniform distribution of the heat flux density, the maximum heating temperature decreases (T_{max} = 1500 °C). The heating temperature reaches its maximum most rapidly when the heat flux density is unevenly distributed with a

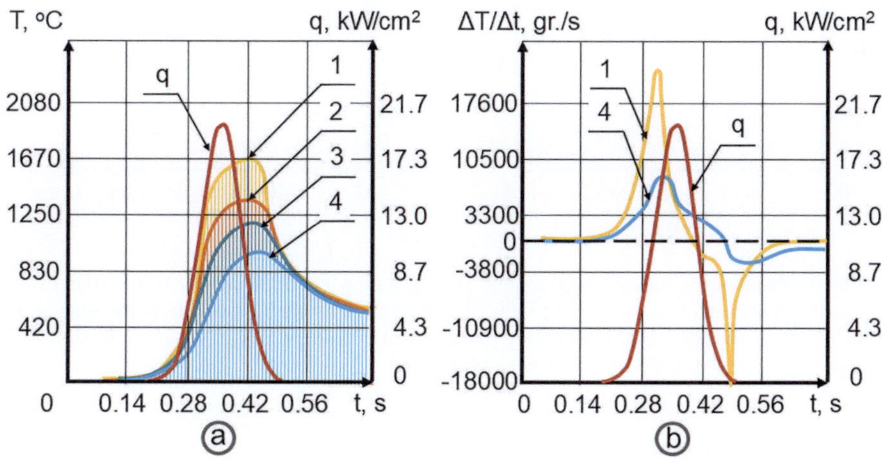

Fig. 6.73 Variation in time t of the temperature T (**a**) and the heating (cooling) rate ΔT/Δt (**b**) in the surface layer with a Gaussian distribution of the heat flux density q at different depths z: 1—at z = 0 mm; 2—at z = 0.25 mm; 3–at z = 0.45 mm; 4–at z = 0.85 mm

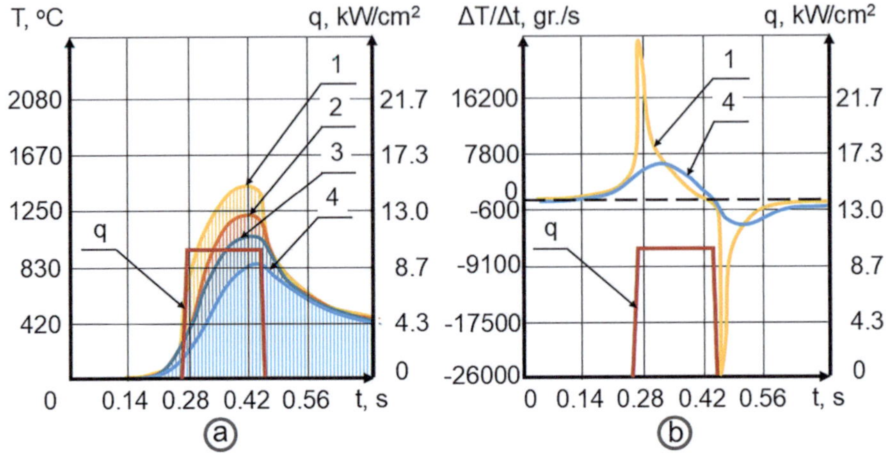

Fig. 6.74 Variation in time t of temperature T (**a**) and heating (cooling) rate ΔT/Δt (**b**) in the surface layer with a uniform distribution of heat flux density q at different depths z: 1—at z = 0 mm; 2—at z = 0.25 mm; 3—at z = 0.45 mm; 4—at z = 0.85 mm

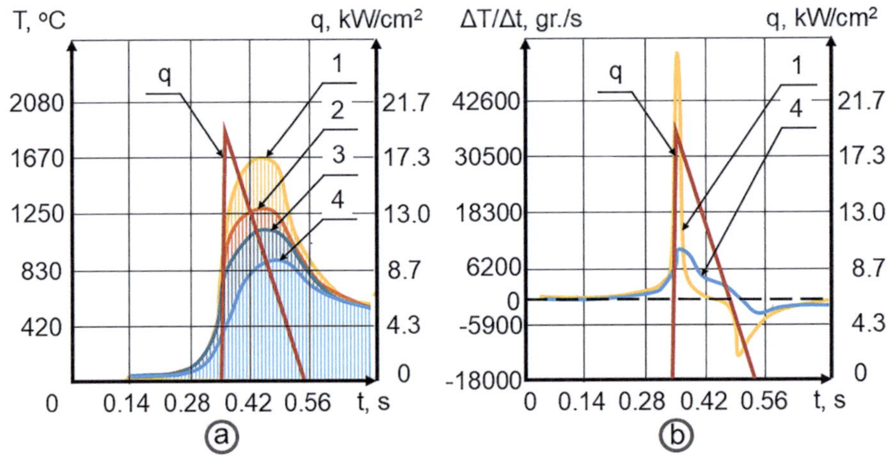

Fig. 6.75 Variation in time t of temperature T (**a**) and heating (cooling) rate ΔT/Δt (**b**) in the surface layer with an uneven distribution of heat flux density q with a shifted maximum to the edge of the focusing spot in the direction of laser beam movement at different depths z: 1—at z = 0 mm; 2—at z = 0.25 mm; 3—at z = 0.45 mm; 4—at z = 0.85 mm

shifted maximum to the edge of the focusing zone in the direction of the laser beam movement and when the Gaussian distribution is sharp.

The surface <u>*heating rate*</u> (graphs b in Figs. 6.73, 6.74, 6.75 and 6.76) is maximum ($\Delta T/\Delta t = 52{,}000$ deg/s) at an uneven distribution of the heat flux density with a shifted maximum to the edge of the focusing zone in the direction of laser beam

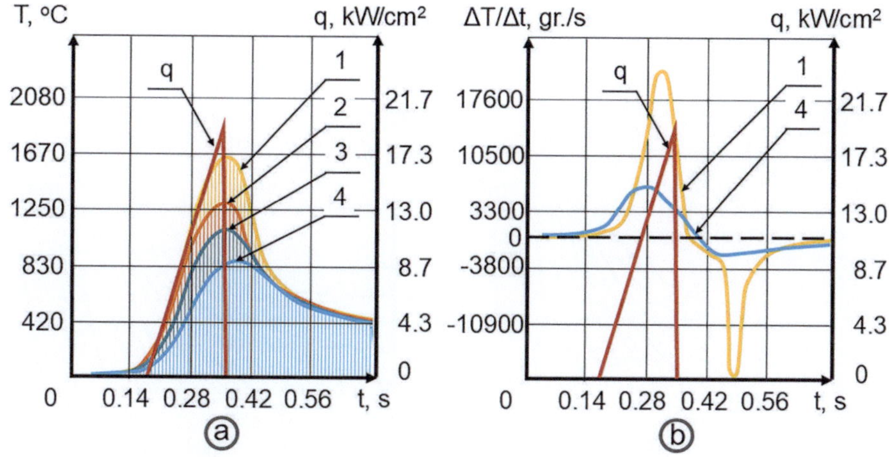

Fig. 6.76 Variation in time t of temperature T (**a**) and heating (cooling) rate ΔT/Δt (**b**) in the surface layer with an uneven distribution of heat flux density q with a shifted maximum to the edge of the focusing spot in the opposite direction to laser beam movement at different depths z: 1—at $z = 0$ mm; 2—at $z = 0.25$ mm; 3—at $z = 0.45$ mm; 4—at $z = 0.85$ mm

movement, which exceeds the heating rate by 2.5 times at all other distributions of the heat flux density.

The <u>*existence time of the liquid phase*</u> (the time interval at which the surface heating temperature exceeds the melting temperature—curves 1 in Figs. 6.73, 6.74, 6.75 and 6.76) is the longest (0.14 s) with an uneven distribution of the heat flux density with a shifted maximum to the edge of the focusing zone in the direction of the laser beam movement (Fig. 6.75b). The liquid phase lifetime decreases in the sequence of heat flux density distributions: Gaussian distribution → non-uniform distribution with a shifted maximum to the edge of the focusing zone in the opposite direction to the laser beam movement → uniform distribution (minimum liquid phase lifetime).

The <u>*cooling rate*</u> of the surface decreases with the increase of the liquid phase lifetime and is minimal ($\Delta T/\Delta t = -8000$ deg/s) at an uneven distribution of the heat flux density with a shifted maximum to the edge of the focusing zone in the direction of the laser beam movement (Fig. 6.75b). The cooling rate increases by 2.5 times with the Gaussian distribution of the heat flux density and with an uneven distribution with a shifted maximum to the edge of the focusing zone in the opposite direction to the laser beam movement. The cooling rate increases by 3.5 times with a uniform distribution of the heat flux density.

The **shape of the melting isotherm** determines the depth of propagation and the shape of the <u>*surfacing pool*</u> and affects the formation of stress and crack formation. The shape of the melting isotherms (in the example under consideration, $T_m = 1400$ °C) is determined from the calculations of temperature fields at different distributions of the heat flux density (Figs. 6.77, 6.78, 6.79 and 6.80).

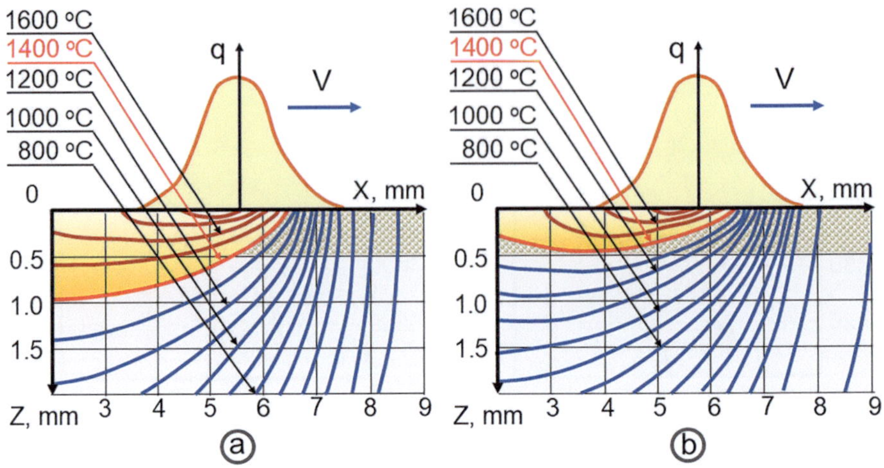

Fig. 6.77 Temperature distribution in the longitudinal section for the Gaussian distribution of the heat flux density in the focusing spot at different laser beam displacement speeds V: **a** at V = 0.2 m/min; **b** at V = 0.5 m/min

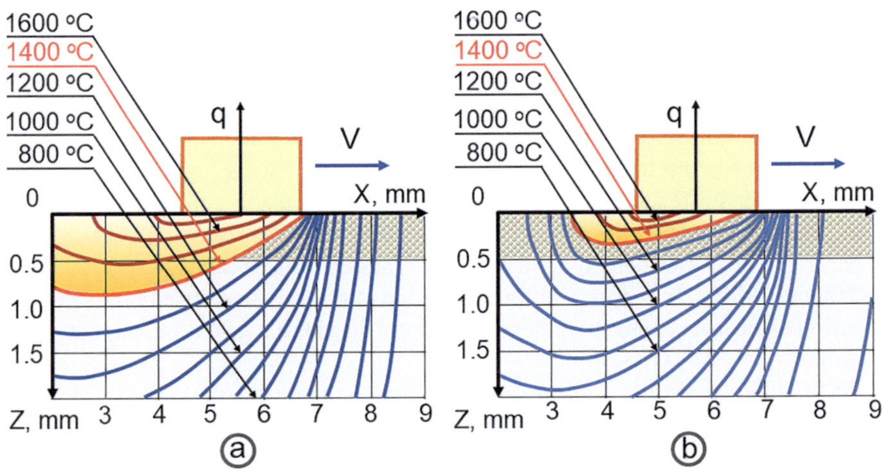

Fig. 6.78 Temperature distribution in the longitudinal section for the uniform distribution of the heat flux density in the focusing spot at different laser beam displacement speeds V: **a** at V = 0.2 m/min; **b** at V = 0.5 m/min

The ***depth*** of propagation of the melting isotherm is effectively controlled by the irradiation parameters, primarily by the speed of the laser beam, which makes it possible to ensure:

- minimum depth of penetration of the base;
- minimum proportion of the base metal in the deposited metal;

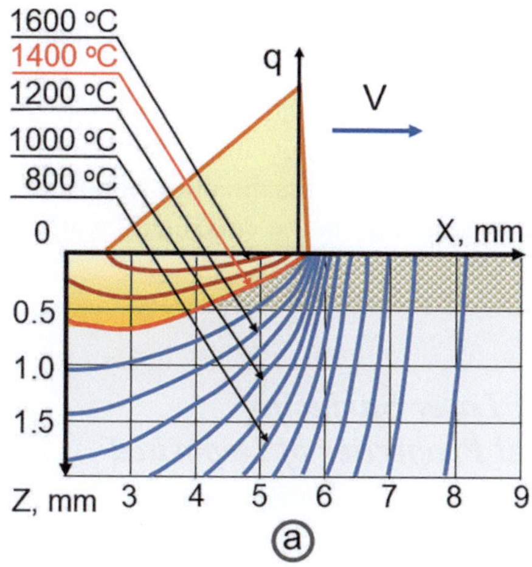

Fig. 6.79 Temperature distribution in the longitudinal section for the uneven distribution of the heat flux density with a shifted maximum to the edge of the focus spot in the direction of laser beam movement at different laser beam displacement speeds V: **a** at V = 0.2 m/min; **b** at V = 0.5 m/min

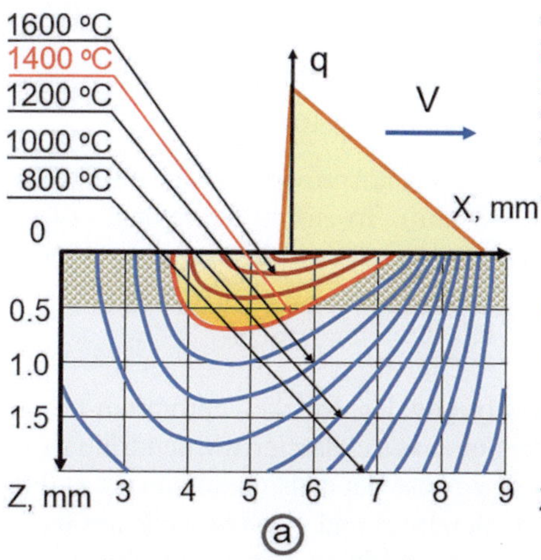

Fig. 6.80 Temperature distribution in the longitudinal section for the uneven distribution of the heat flux density with a shifted maximum to the edge of the focus spot in the opposite direction to laser beam movement at different laser beam displacement speeds V: **a** at V = 0.2 m/min; **b** at V = 0.5 m/min

- controlled properties of the deposited layer;
- metallurgical bonding and high adhesion strength of the coating to the substrate.

The optimum value of the laser beam travel speed (in the example under consideration V = 0.2 m/min) ensures a melting isotherm propagation depth greater than the powder layer thickness for all heat flux density distributions.

The ***shape*** of the melting isotherm largely depends on the *law* of heat flux density *distribution*.

When the heat flux density is unevenly distributed with a shifted maximum to the edge of the focus spot in the direction of the laser beam movement, the resulting melting isotherms are extended along the deposited layer in the direction opposite to the laser beam movement. This reduces the cooling rate in the deposited layer and helps to reduce the risk of cracking.

6.7.3 Technological Parameters of Laser Surfacing and Physical and Mechanical Properties of Deposited Layers

Two *schemes* of powder *feeding* into the laser beam zone are used:

(a) following the laser beam;
(b) towards the laser beam.

The scheme of powder feeding *following* the laser beam provides:

- stable metallurgical bond and higher adhesive strength;
- greater efficiency in terms of laser energy consumption (by 25–30%);
- greater efficiency in terms of powder consumption (by 10–15%).

The scheme of powder feeding *towards* the laser beam provides a greater thickness of the deposited layer in one pass (up to 2–3 mm). To ensure the strength of the coating's adhesion to the substrate, it is necessary to increase the laser power (up to 3–5 kW).

Nickel-based surfacing powders

The ***microstructure of the deposited layers*** of powdered nickel–chromium-boron-silicon alloys is characterized by high dispersion, has a characteristic dendritic structure with cells of the γ-phase. The amount of γ-phase depends mainly on the carbon and boron content of the alloy. Between the dendrites and γ--phase cells are inclusions of the eutectic γ- + Ni_3B and accumulations of highly dispersed carbides such as $Cr_{23}C_6$.

The ***microhardness*** of the deposited layers is determined by the structural and phase composition of the initial powder material and is weakly dependent on the laser beam speed (Fig. 6.81).

The microhardness gradually decreases *with depth* from 7000 to 14,000 MPa in the near-surface layer to 4000 MPa in the fusion zone (Fig. 6.82). The maximum microhardness $H_\mu = 14{,}000$ MPa is achieved by the deposited layer of TiC powder. However, this alloy is susceptible to the formation of cracks, pores, micro- and macro-irregularities during surfacing. To avoid defects, TiC powder should be supplemented with substances that increase the fluidity and wettability of the melt, as well as plasticizers to relax the residual stresses that cause cracking.

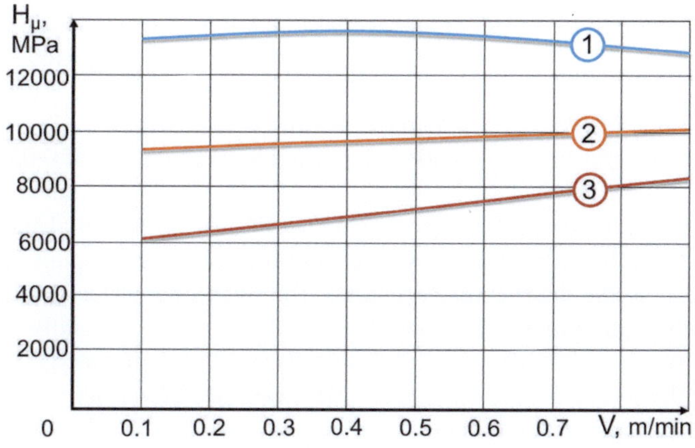

Fig. 6.81 Influence of the laser beam speed V on the surface microhardness H_μ of the deposited layer for different structural and phase composition of the initial powder material (P = 1.5 kW; d_0 = 2.0 mm; G = 0.4 g/s): 1—TiC powder; 2—mixture: 65% powder (Ni (base) + 0.8% C + 3.8% B + 4% Si + 17% Cr + 5% Fe) + 35% WC powder; 3—powder (Ni (base) + 0.8% C + 3.8% B + 4% Si + 17% Cr + 5% Fe). The base is C45 steel

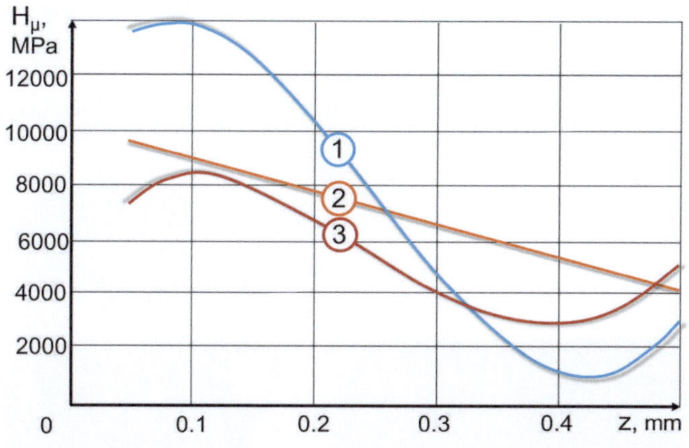

Fig. 6.82 Distribution of microhardness H_μ by depth z of the deposited layer (P = 1.5 kW; d_0 = 2.0 mm; G = 0.4 g/s; V = 0.4 m/min): 1—TiC powder; 2—mixture: 65% of powder (Ni (base) + 0.8% C + 3.8% B + 4% Si + 17% Cr + 5% Fe) + 35% of WC powder; 3—powder (Ni (base) + 0.8% C + 3.8% B + 4% Si + 17% Cr + 5% Fe). The base is C45 steel

The ***formation of cracks*** decreases with a decrease in the laser power density and with a decrease in the laser beam speed. At low surfacing speeds, cracks appear mainly in the surface layer, and at higher speeds—in the fusion zone.

Iron-based surfacing powders (X10CrNiTi18-10 + TiB_2 + VC and X10CrNiTi18-10 + TiB_2 + CrB_2) are used to increase the wear resistance of parts operating at high temperatures and in aggressive environments. The powders are surfaced on a base of C45 steel or a heat-resistant alloy such as GC6K (60.1–68.4% Ni; 9.5–12.0% Cr; 5–6% Al; 4.5–5.5% W; 4–5.5% Co; 3.5–4.5% Mo; 2.5–3.2% Ti; $\leq 2\%$ Fe; $\leq 0.4\%$ Mn; $\leq 0.4\%$ Si; 0.1–0.2% C; $\leq 0.04\%$ Zr; 0.02% B) with the laser radiation parameters ($P = 1.5$ kW; $d_0 = 2$ mm; $V = 0.2$–0.8 m/min; powder flow rate $G = 0.1$–0.5 g/s), which provide a deposited layer thickness of 0.05–0.2 mm in one pass.

The ***microstructure*** of the heat-resistant GC6K alloy in the initial state is heterogeneous with the liquation of components and the presence of small inclusions of carbide phases in the interdendritic regions. The interstitial spaces in the metal are enriched with chromium and titanium. Tungsten and cobalt are mainly concentrated in the dendrite axes. Molybdenum and aluminum are slightly liquated. The emissions observed in the interdendritic regions are MeC carbides of various compositions.

When surfacing powdered material on the heat-resistant GC6K alloy, the base of which is X10CrNiTi18-10 stainless steel, and the strengthening phases are TiB_2 and CrB_2, a cast dendritic microstructure with a highly dispersed strengthening phase and eutectic local areas is formed (Fig. 6.83a). The fineness of the excess crystals of the hardening phases and eutectics is much higher than that of plasma coatings. The strengthening phases and the γ-phase are oriented in the direction of heat dissipation. The cooling rate in the near-surface layers is $(5–10)\cdot 10^4$ deg/s. As the distance from the surface increases, the dispersion of the structure decreases and the crystal size of the hardening phases increases.

The white particles in the image in the absorbed electrons have a small average atomic radius and are lighter than the matrix. The black inclusions contain a large

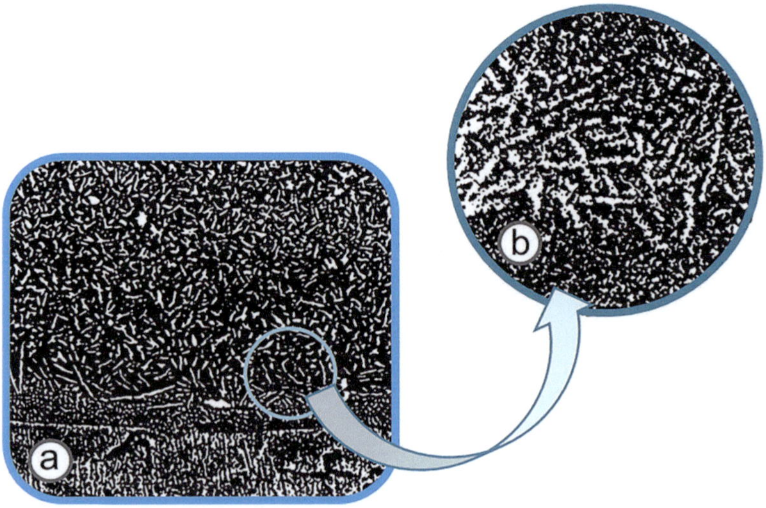

Fig. 6.83 Microstructure of the deposited layer (X10CrNiTi18-10+TiB_2+CrB_2) and the fusion zone with the GC6K base obtained by laser gas-powder deposition (×300) (**a**) and a fragment of the microstructure of the deposited layer in chromium radiation (×1000) (**b**)

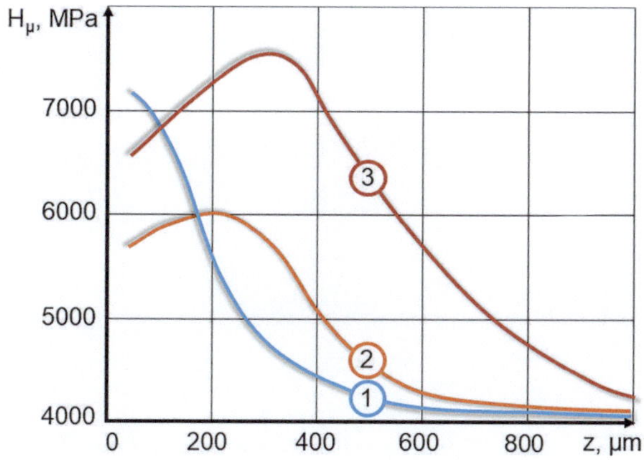

Fig. 6.84 Distribution of microhardness H_μ along the depth of the deposited layer (X10CrNiTi18-10 + TiB_2 + CrB_2) at different laser power density W_p values (V = 0.2 m/min): 1—at $W_p = 6$ kW/cm^2; 2—at $W_p = 11$ kW/cm^2, 3—at $W_p = 25$ kW/cm^2

amount of heavy elements, such as tungsten and molybdenum. The difference in the composition of the inclusions indicates the absence of areas of their mutual solubility.

In the fusion layer, the matrix material dissolves in the liquid phase of the coating. At the same time, the borides are retained, and the matrix melts and mixes with the GC6K melt. The needles of the hardening phases are predominantly chromium borides (Fig. 6.83b).

The ***microhardness*** along the depth of the deposited layer changes smoothly and depends on the laser power density (Fig. 6.84).

The microhardness of the deposited layers increases with an increase in the degree of dispersion of the structural components and an increase in the proportion of harder eutectic.

Surfacing of B_4C boron carbide powder cladded with Ni + Al is accompanied by strong absorption of laser radiation. The deposited layer has a uniform width and height and low surface roughness.

The ***microstructure*** of the layer has a highly dispersed block-cellular structure (Fig. 6.85). The grid of the hardening phase consists of intergranular boundaries that are much more dispersed than in the base material. The alloying elements in the deposited layer contain an increased amount of boron and are distributed evenly.

In the fusion zone, small allocations of carbides of the MeC type and segregation of W along the grain boundaries are formed. Outside the grains, ultra-dispersed chromium boride particles are observed. The concentration of alloying elements is equalized, which has a positive effect on the mechanical properties of the deposited layer.

The ***microhardness*** of the deposited layer reaches 6000–6500 MPa at the surface and decreases uniformly with depth (Fig. 6.86).

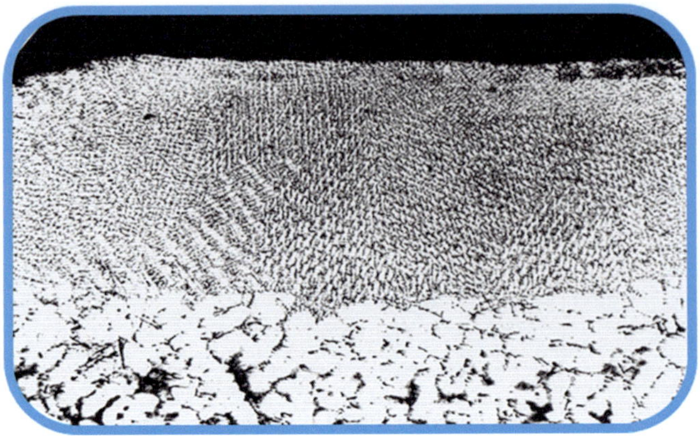

Fig. 6.85 Microstructure of the deposited layer of B_4C cladded with Ni and Al on GC6K base

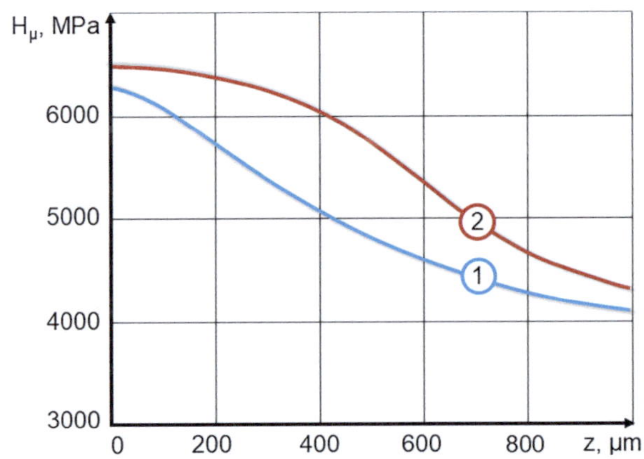

Fig. 6.86 Distribution of microhardness H_μ along the depth of the deposited layer B_4C at different values of power density W_p of laser radiation and speed V of the laser beam: 1—at Wp = 25 kW/cm^2, V = 2 m/min; 2—at Wp = 6 kW/cm^2, V = 0.8 m/min

Structures formed in the deposited layer during B_4C surfacing (as well as during X10CrNiTi18-10 + TiB_2 + CrB_2 surfacing) reduce the rate of negative recrystallization processes that usually occur in the heat-resistant GC6K alloy at temperatures of 750–900 °C. This leads to an increase in surface wear resistance at high temperatures.

The *technological conditions* for the formation of the required properties of the deposited layer and ensuring fusion are: $P \geq 0.9$ kW; $d_0 \leq 4.5$ mm; $V \leq 0.8$ m/min; $G = 0.1$–0.6 g/s.

The **level and depth distribution of residual stresses** are determined by two factors:

(a) *composition* of the deposited powder;
(b) level and nature of the distribution of the *heat flux density* of laser radiation.

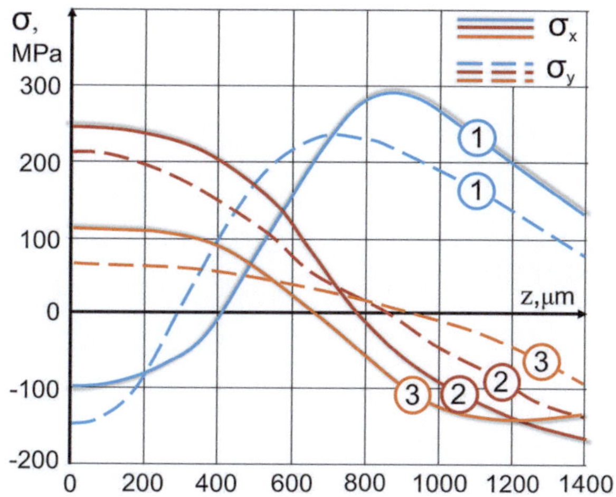

Fig. 6.87 Distribution of longitudinal (σ_x) and transverse (σ_y) residual stresses along the depth z of the deposited layer depending on the composition of the deposited powder: 1—X10CrNiTi18-10 + TiB_2 + CrB_2 powder; 2—Ni (base) + 0.8% C + 3.8% B + 4% Si + 17% Cr + 5% Fe powder; 3—mixture: 65% powder (Ni (base) + 0.8% C + 3.8% B + 4% Si + 1 7% Cr + 5% Fe) + 35% WC powder

It is advisable to estimate the residual stresses during laser surfacing by the X-ray $\sin^2\psi$ method and/or by measuring the change in deformation during layer-by-layer etching.

The ***composition of the deposited powder*** (Fig. 6.87) determines the fraction of phase stresses (Sect. 5.2).

In the deposited layer of iron-based powders of the X10CrNiTi18-10 + TiB_2 + CrB_2 type, residual compressive stresses of – 100 to – 160 MPa are formed due to structural and phase transformations with a volume effect (curves 1 in Fig. 6.87). With depth, the stresses decrease and turn into tensile stresses. The maximum tensile stresses occur in the fusion zone.

In the deposited layer of self-fluxing nickel alloys of the type Ni (base) + 0.8% C + 3.8% B + 4% Si + 17% Cr + 5% Fe, the thermal component of residual stresses is determinative (curves 2 and 3 in Fig. 6.87). Tensile stresses are formed on the surface at the level of 70–240 MPa, which decrease at a depth of 2/3 of the thickness of the deposited layer. Further, balancing effect of compressive stresses is observed.

The ***heat flux density*** of the laser radiation (Fig. 6.88) determines the rate of thermal stresses (Sect. 5.2).

The use of an uneven distribution of the heat flux density with a shifted maximum to the edge of the focusing spot in the direction of the laser beam movement will reduce the level of residual stresses, both tensile (2.2 times) and compressive (30%), compared to the Gaussian distribution.

The reduction in residual stresses is due to:

- increased duration of the liquid phase existence;
- elongation of the molten pool along the surfacing axis;
- reduction of the cooling rate and the possibility of relaxation of residual stresses due to plastic deformation of adjacent layers.

Cracks in laser surfacing occur at the stage of crystallization in two zones:

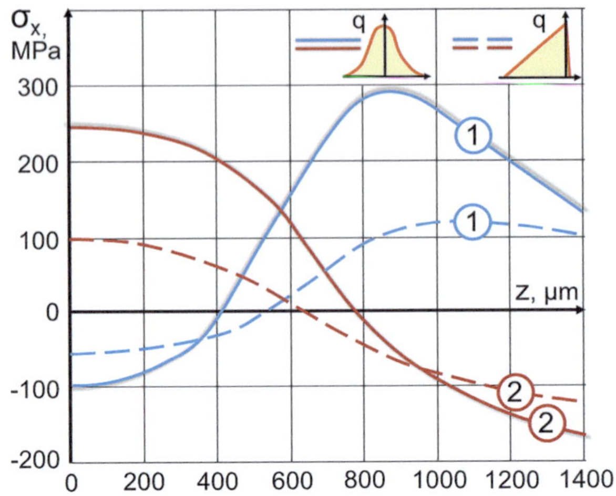

Fig. 6.88 Influence of the heat flux density distribution q in the focusing zone on the magnitude and nature of the distribution of longitudinal residual stresses σ_X: 1—X10CrNiTi18-10 + TiB_2 + CrB_2 powder; 2—Ni (base) + 0.8% C + 3.8% B + 4% Si + 17% Cr + 5% Fe powder

(a) *on the surface* of the crystallizing metal (crack nucleation) with further propagation deep into the deposited layer (predominant formation zone);

(b) *along the fusion line* between the coating and the base (crack initiation) with further propagation to the surface to almost the entire thickness of the deposited layer.

The ***mechanisms*** of crack formation are related to the formation of a stress–strain state (Sect. 5.2) and the ability of the metal to perceive thermal deformations in certain temperature intervals. Laser surfacing produces three types of *hot cracks*:

(1) *Solidification cracks*—formed in the deposited bead in the solid–liquid state of the metal in the temperature interval between liquidus and solidus. At temperatures close to liquidus, due to the significant amount of liquid phase, thermal deformations are compensated for by relative movement of solid phase areas and circulation of liquid phase between them. With further cooling, the volume of the liquid phase decreases, which leads to collision of crystallites during deformation and limits the circulation of the liquid phase. Thermal deformation is perceived only by the areas of crystallites that have formed contacts. Due to the limited total contact area, high stresses arise, which creates conditions for contact fracture and microcrack formation. During further cooling, the volume of fully solidified metal is involved in compensating for the thermal deformation. As a result, the deformation capacity of the metal increases sharply, reaches its maximum, and the conditions for the formation of crystallization cracks disappear.

(2) *Liquation cracks*—are formed, as a rule, in the fusion area and propagate into a deposited bead. The cause of liquation crack formation is low deformation capacity due to segregation of impurities at the boundaries of metal grains as a result of diffusion processes during the heating stage, followed by melting of the boundary as a result of laser irradiation. Crack formation occurs at a

temperature below the solidus temperature (solid state of the metal) after the primary crystallization is completed.

(3) *Precipitation induced cracks*—are formed in the deposited layer in the solid state as a result of intense diffusion of atoms during secondary crystallization and migration of grain boundaries. The precipitation of finely dispersed intermetallic and carbide-nitride phases from the solid solution causes grain embrittlement and hardening. Dispersion hardening of grains leads to localization of plastic deformations along the boundary areas with relative grain slippage and microcracks nucleation at the boundaries.

Factors contributing to crack formation during laser surfacing of powdered materials:

(a) High content of *alloying elements*, which leads to embrittlement of the deposited metal and structural transformations with volumetric changes and the formation of additional stresses of the second kind (at the grain level) and the third kind (at the lattice level). The introduction of components into the surfacing material that contribute to the formation of an increased amount of plastic components in the microstructure, namely, the γ-phase, significantly reduces the formation of cracks.

(b) High *cooling rate* of the deposited bead on the base metal that does not deform. This leads to large deformations during the crystallization process in the brittle temperature range. In addition, there is a high level of residual tensile stresses in the deposited bead.

(c) High amount of *solid inclusions* in the molten powder (impurities, refractory powders).

d) Significant *difference in the chemical composition* of the deposited powder and the base, as well as possible poor mutual solubility of the components.

Since crack formation is determined by the maximum temperature deformations and associated temporary stresses, the level of residual stresses is an indicator of crack formation. Technological methods for reducing residual stresses reduce thermal deformations and temporary stresses and thus prevent crack formation.

Parameters of the laser radiation affecting crack formation:

(a) Increasing the laser ***power density*** increases the temperature gradient, thermal deformation, temporary and residual stresses, and thus the number of cracks increases. Moreover, with an increase in power density, the maximum number of cracks shifts towards higher laser beam speeds. The *optimum* level of power density $W_p = 20$–$30\ \text{kW/cm}^2$ (at V = 0.5–1 m/min), at which the number of cracks decreases by 2–3 times (Fig. 6.89).

(b) The optimal ***distribution*** of the heat flux ***density***, ceteris paribus, can significantly reduce the number or completely eliminate the formation of surface cracks. The use of an *uneven distribution* of the heat flux density with a shifted maximum to the edge of the focus spot in the *direction* of the laser beam *movement*, along with a decrease in residual stresses, leads to the avoidance of cracks in comparison with the Gaussian distribution (Fig. 6.90).

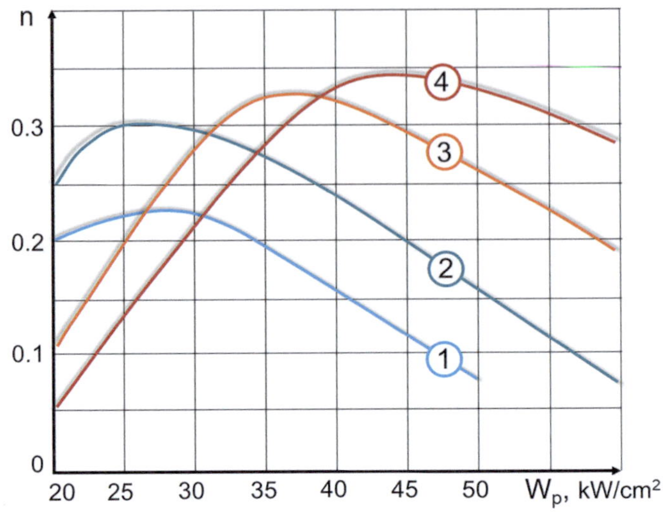

Fig. 6.89 Effect of laser radiation power density W_p and laser beam speed V on the specific number of cracks n in the deposited layer: 1—at V = 0.5 m/min; 2—at V = 1 m/min; 3—at V = 1.5 m/min; 4—at V = 2 m/min

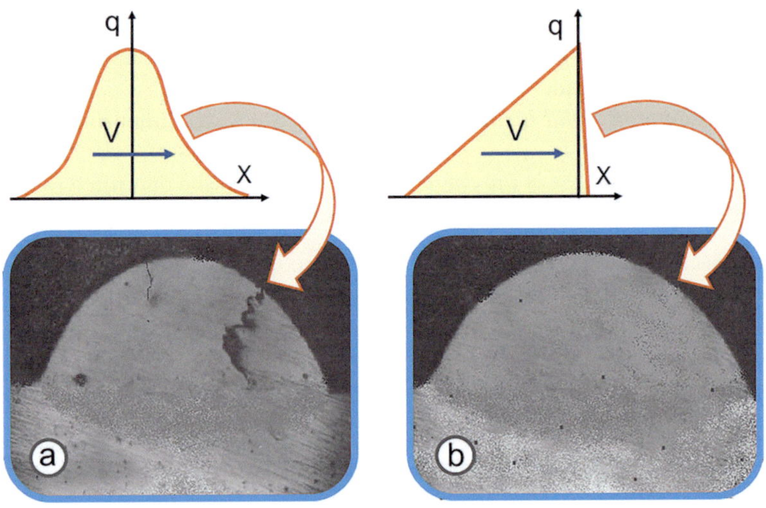

Fig. 6.90 Effect of heat flux density distribution q in the focusing zone on crack formation (W_p = 30 kW/cm^2): **a** Gaussian distribution of the heat flux density; **b** uneven distribution of the heat flux density with a shifted maximum to the edge of the focusing spot in the direction of the laser beam movement

Table 6.3 Wear intensity of deposited layers from X10CrNiTi18-10 + TiB_2 + CrB_2 powder at 800 °C

The method of surfacing	Wear intensity, mg/(cm^2 km)
Plasma spraying	35
Laser-powder surfacing	12

(c) Reducing the ***speed*** of the laser beam increases the mixing of filler materials and the base, reduces the cooling rate and, as a result, reduces the number of cracks. The *optimum* level of laser beam velocity is $V = 0.5–1$ m/min (at $W_p = 20–30$ kW/cm^2), at which the number of cracks is reduced by 2–3 times (Fig. 6.89).

The **tribotechnical properties** (wear resistance) of surfaces during laser surfacing with hypereutectic alloys containing excess diboride crystals are much better compared to plasma spraying. Laser surfacing reduces the size of excess crystals due to high crystallization rates and the formation of highly supersaturated solutions. The microstructure of the coating shifts towards eutectic. At the same time, the hardness of plasma coatings is higher than that of coatings obtained by laser deposition.

Analysis of friction surface profiles shows that laser deposited layers, unlike plasma coatings, wear evenly, without brittle fracture (Table 6.3). This is due to the high dispersion of the structure, uniform distribution of hardening phases, and the presence of a plastic phase. In addition, the layers obtained by laser surfacing have higher high-temperature wear resistance Increased heat resistance is associated with a decrease in its heterogeneity due to the formation of supersaturated solutions.

The wear resistance of the deposited layers depends on the type of powder material. The wear resistance of coatings made of steel-based powders during laser surfacing is superior to self-fluxing nickel-based alloys (Fig. 6.91), which is explained by the effect of oxidation and the formation of secondary structures capable of self-organization on the friction process.

The **adhesion strength** of the deposited coatings to the base is determined by the lower strength of the constituent materials of the joint and is 250–350 MPa, depending on the type of coating (Table 6.4). The adhesion strength of the deposited coatings with the substrate during laser surfacing is higher than the adhesion strength of plasma coatings.

6.7.4 Increasing the Service Life of Gas Turbine Blades of Aircraft Engines

The service life of aircraft engines is largely determined by the amount of wear on the contacting surfaces of the gas turbine blades and nozzle blades. Nickel alloys of the austenitic structure of the Ni-Co-Cr system are materials with high heat resistance and corrosion resistance. The formation of intermetallic compounds such as Ni_3(Ti,

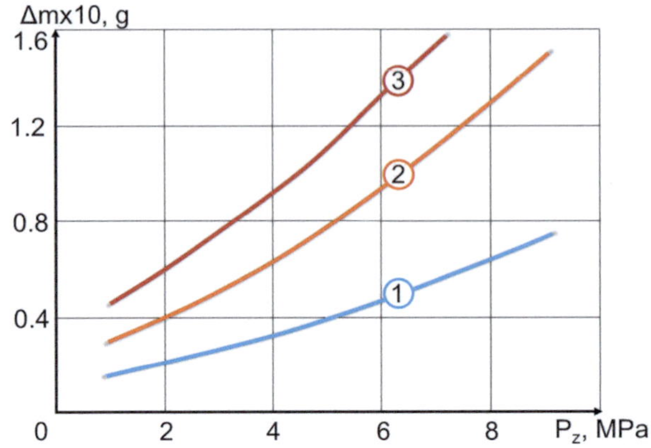

Fig. 6.91 Dependence of the wear value Δm of the deposited layers on the specific load P_z and the composition of the deposited material (under dry friction conditions with quenched C45 steel): 1—X10CrNiTi18-10 + TiB_2 + CrB_2 powder; 2—mixture: 65% powder (Ni (base) + 0.8% C + 3.8% B + 4% Si + 17% Cr + 5% Fe) + 35% WC powder; 3—powder Ni (base) + 0.8% C + 3.8% B + 4% Si + 17% Cr + 5% Fe

Table 6.4 Adhesion strength of deposited layers of different powders on the substrate (C45 steel)

Filler powders	Microhardness, MPa	Bond strength, MPa
Ni (base) + 0.8% C + 3.8% B + 4% Si + 17% Cr + 5% Fe	9200	252
Mixture: 65% of (Ni (base) + 0.8% C + 3.8% B + 4% Si + 17% Cr + 5% Fe) powder + 35% of WC powder	9600	275
X10CrNiTi18-10 + TiB_2 + CrB_2	7300	347
X10CrNiTi18-10 + TiB_2 + CrB_2 (Plasma spraying)	11,000	16–20

Note Laser surfacing parameters: laser radiation power (uneven distribution of heat flux density with a shifted maximum to the edge of the focus spot in the direction of the laser beam movement) P = 1200 W; diameter of the focus zone d_0 = 4 mm, laser beam movement speed V = 0.2 m/min; powder consumption G = 0.2 g/s. These parameters ensure the formation of a deposited layer with a height of 0.5–0.8 mm, a width of 3.1–3.5 mm, and a thickness of the fusion layer of 0.05–0.08 mm

Al, Nb) along the grain boundaries gives the alloy high strength at high temperatures and under cyclic alternating loads.

Gas turbine blades for aircraft engines (Fig. 6.92a) have Z-shaped elements with a contact shelf at the ends (Fig. 6.92b). In an assembled turbine, adjacent blades are in close contact with each other by contact shelves (Fig. 6.92c).

Gas turbine blades are operated under dynamic and fatigue loading in an aggressive environment at high temperatures (Table 6.5). When the turbine rotor rotates, the contact shelves of adjacent blades are subjected to friction with a force F_f, high-frequency vibration with a force F_v, and high-temperature oxidation. As a result, fretting corrosion develops on the contact surfaces. Depressions from 0.5 to 2.5 mm

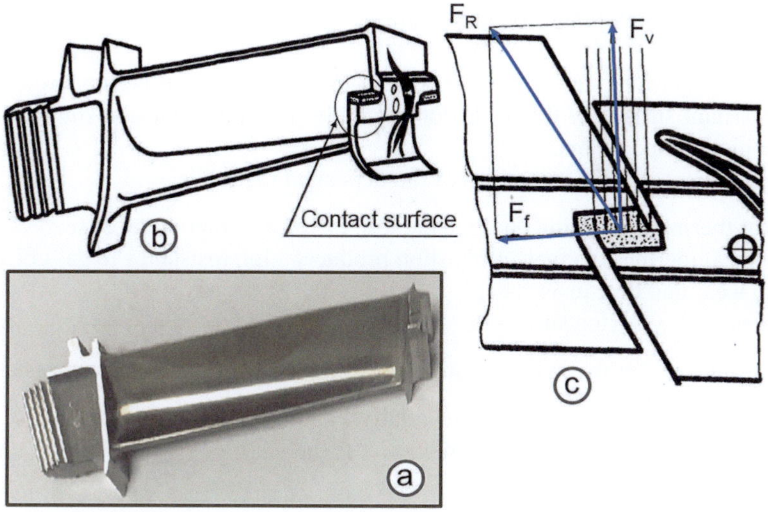

Fig. 6.92 Blade of a gas turbine of an aircraft engine: **a** general view; **b** diagram of a Z-shaped element with a contact shelf; **c** diagram of contact of the shelves of adjacent blades

are formed on the surfaces of the contact shelves. Unacceptable gaps appear between the blade shelves. The turbine rotor becomes unusable.

Increasing the blade service life and providing the possibility of its repair is possible by strengthening the contact shelves by applying wear-resistant material to the surface.

Possible methods of applying wear-resistant coatings to the blade contact shelves include electric-arc surfacing, plasma and detonation spraying, electric spark alloying, and laser surfacing.

Electric-arc surfacing causes the surface layer of the contact shelf to crack.

Plasma and detonation spraying form porous coatings with low adhesion strength to the substrate (15–20 MPa).

Table 6.5 Operating conditions of gas turbine blades of aircraft engines

Characteristics of operating conditions	Value
Temperature	900–1050 °C
Load	800–1200 N
Cyclic load	1600–2400 N
Contact surface area	20 mm^2
Turbine speed	9500 rpm
Environment	Aggressive
Speed of the environment	700 m/s
Maximum allowable wear	0.1 mm

Electric spark alloying produces coatings of low thickness (20–30 μm) and low productivity.

Laser gas-powder surfacing with powder injection makes it possible to obtain a wear-resistant surface of the blade contact shelf with high adhesion strength to the base at high temperatures.

Requirements for the properties of the deposited materials:

- linear thermal expansion coefficient close in value to that of the base;
- chemical inertness to fuel combustion products, high resistance to gas corrosion;
- high wear and heat resistance;
- processability during surfacing.

Requirements for the microstructure of materials to ensure high wear resistance:

- high-strength phase dispersed in a relatively soft and ductile matrix
- the largest possible volume fraction of the carbide phase during eutectic crystallization.

The **surfacing powder** from the alloy (stainless steel X10CrNiTi18-10 (base) + 16% TiB_2 + 16% CrB_2) in the form of a granular powder with a dispersion of 60–80 microns meet all the requirements. The composition of the alloy is close to eutectic.

Hypereutectic alloys containing excessive diboride crystals (such as TiB_2 and CrB_2) are brittle, prone to staining, and have increased wear, especially at increased specific loads. Melting *by laser irradiation* brings such alloys closer to eutectic alloys due to:

- formation of highly supersaturated solutions;
- partial dissolution of the strengthening phases
- grinding of the strengthening phases.

Accordingly, the transfer of eutectic alloys to the cast state in the process of laser surfacing shifts them towards hypoeutectic alloys.

The alloy (X10CrNiTi18-10 + TiB_2 + CrB_2) has good wettability, does not require heat treatment, and does not lose its initial strength (350–400 MPa) when heated to 900 °C.

During laser surfacing, a microstructure with crystals oriented in the direction of heat removal is formed due to directional crystallization, which significantly increases the heat resistance of alloys.

During laser surfacing of powder (X10CrNiTi18-10 + TiB_2 + CrB_2) on a base of heat-resistant alloy of the GC6K type (60.1–68.4% Ni; 9.5–12.0% Cr; 5–6% Al; 4.5–5.5% W; 4–5.5% Co; 3.5–4.5% Mo; 2.5–3.2% Ti; $\leq$ 2% Fe; $\leq$ 0.4% Mn; $\leq$ 0.4% Si; 0.1–0.2% C; $\leq$ 0.04% Zr; 0.02% B) cracks in the fusion zone do not form.

Parameters of the laser surfacing of the blade contact shelves:

- laser radiation power $P = 2.7$ kW;
- focal length $F = 300$ mm;
- diameter of the focusing zone $d_0 = 4$ mm;
- laser beam travel speed $V = 0.5–0.8$ m/min;

- distribution of laser radiation power in the focusing zone is uneven with a shifted maximum to the edge of the focusing spot in the direction of laser beam movement (provided by the position of the focusing plane and the angle of incidence of the laser beam on the sphere);
- powder consumption G = 0.3 g/s;
- powder injection angle $\alpha = 45^o$.

The ***size*** of the deposited layer is controlled by setting the parameters of the laser surfacing:

(a) ensuring the required *width* of the deposited layer—laser radiation power and diameter of the focusing zone;
(b) ensuring the required *height* of the deposited layer—powder consumption;
(c) ensuring the required *thickness of the fusion layer*—by laser radiation power and powder consumption.

The **microstructure** of the deposited layers is uniform and highly dispersed (Fig. 6.93a). The strengthening phases (mainly TiB_2) are evenly distributed over the volume. The crystal sizes of the hardening phases increase from the surface to the periphery of the deposited layer.

The microhardness of the deposited layer is 6800–7300 MPa, evenly distributed over the depth of the deposited layer, and gradually decreases in the fusion layer to the hardness of the base material (Fig. 6.93b).

The quality criteria for the laser surfacing process are the *alloy layer* thickness of 50–100 μm and the absence of cracks.

Tests have shown a significant increase in wear resistance at a temperature of 800 °C (by 2–3 times) of coatings obtained by laser surfacing compared to plasma spraying. At the same time, the microhardness of the surface layers of the plasma-sprayed coating is 10000–11,000 MPa.

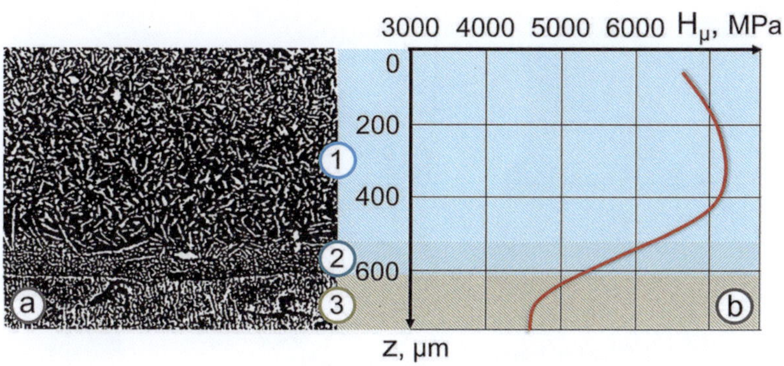

Fig. 6.93 Microstructure of the deposited layer with powder (X10CrNiTi18-10 + TiB_2 + CrB_2) on the heat-resistant alloy GC6K (a) and distribution of microhardness H_μ along the thickness z (b): 1—deposited layer; 2—fusion layer; 3—base

6.7.5 Increasing the Service Life of Drill Bits

The ***drill bit*** is the main tool for drilling wells, which destroys rocks and forms the wellbore (Fig. 6.94).

The drill bit (Fig. 6.94) is a body consisting of three shoulders 1. The shoulders are connected by welding. The shoulders have interchangeable bores 2. The cutters 3—steel disks with carbide inserts 4–are mounted on the interchangeable bores. The interchangeable bore with the cutter forms a sliding bearing.

The high energy intensity of rock fracturing requires more than 50 kW of power to be supplied to the bit through small contact surfaces, which causes high loads on drill bits:

- contact stresses in rolling bearings—4000–5000 MPa, in sliding bearings—30–40 MPa;
- axial loads—2–40 kN;
- coefficient of dynamism—1.3–1.5
- torque—up to 2500 N·m;
- rotational speed—0.7–20 s^{-1}.

The drill bit is washed with drilling fluid at a speed of 50–100 m/s. At large well depths and severe operating conditions, it is impossible to use rolling bearings in the design. Interchangeable bores (Fig. 6.94c) are used as sliding bearings. The

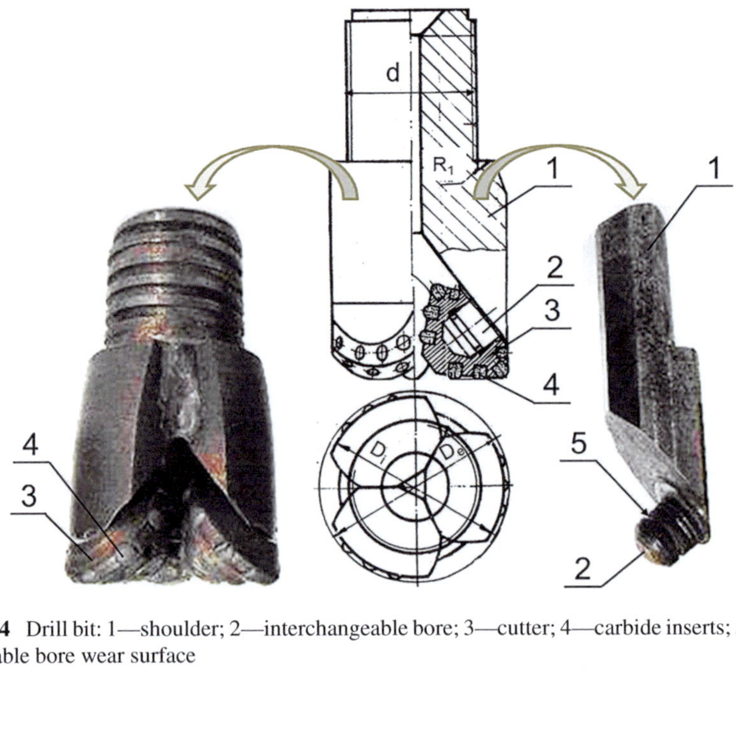

Fig. 6.94 Drill bit: 1—shoulder; 2—interchangeable bore; 3—cutter; 4—carbide inserts; 5—interchangeable bore wear surface

service life of the interchangeable bore is determined by the intensity of two related processes—abrasive wear and fatigue failure.

The intensity of abrasive wear of the interchangeable bore in a drilling fluid with abrasive particles is determined by the relative displacement rate; size and hardness of the abrasive particles; size, hardness and condition of the interchangeable bore surfaces; and type of lubricant.

Fatigue failure is caused by the accumulation of damage under cyclic contact loads. Fatigue failure occurs after a certain number of load cycles. Fatigue deformations cause metal embrittlement and increase the wear rate of friction surfaces.

The right choice of bearing materials and their chemical and thermal treatment technology can regulate the wear of the bearing elements and significantly increase the durability of the bits. The surface of the interchangeable bore should be much harder than its body.

Ways to increase the service life of drill bits:

(1) Increase the hardness of the interchangeable bore surface.
(2) Solid lubrication of the interchangeable bore—cutter friction unit.

Increasing the hardness of the interchangeable bore surface

Interchangeable bores are made by stamping from 9310H type chrome-nickel-molybdenum steel. To increase wear resistance, it is carburized and hardened by high frequency current heating.

In its ***original state*** after normalization, the steel structure consists of irregularly shaped ferrite grains of different orientations and unevenly distributed carbides. Small areas of pearlite are located between the grains (Fig. 6.95a).

Laser irradiation of steel in the delivery state after normalization and carburizing with quenching results in the formation of *two hardened layers*:

(a) *"white layer"* of hardened solid solution up to 1.3 mm deep has a microstructure of amorphized martensite with a small number of spheroidized carbide particles. In the lower part, there is a hypereutectic hardened layer (Fig. 6.95b) of coarse-needle martensite, residual austenite in the form of triangles located between

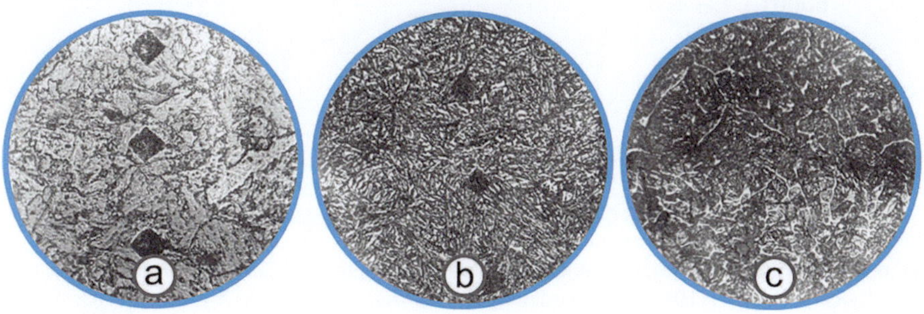

Fig. 6.95 Microstructure of 9310H type steel (×500): **a** in the initial state; **b** hypereutectic hardened layer; **c** transition layer

the martensite plates and a fine grid of free cementitic carbides oriented in the direction of heat removal.

(b) _transition layer_ from the hypereutectic hardened layer to the heat affected zone (Fig. 6.95c) also has a martensite with austenite structure. However, the martensite needles become smaller. At the boundary with the original metal, the martensite branching is almost invisible. The formation of troostite follows the pattern of the original austenitic grains.

The ***microhardness*** is unevenly distributed over the depth. When laser processing steel after carburizing at the initial parameters P = 1200 W, $d_0 = 5$ mm, V = 0.2–0.6 m/min, the microhardness is 4200–4500 MPa. With an increase in the laser processing speed, the dispersion of the structure increases, which leads to an increase in microhardness.

The _optimum_ laser processing _speed_ is 0.8–1.2 m/min, at which the hardened layer has a high and uniform microhardness of 7500–9000 MPa (curve 1 in Fig. 6.96).

Further increase in the laser processing speed leads to a decrease in the level of microhardness.

The technology of increasing the hardness of the interchangeable bore surface, consisting of carburizing, high-frequency current treatment and laser hardening, increases the service life of drill bits by 30–60%.

Solid lubrication of the Interchangeable bore—cutter friction unit

Solid lubrication involves creating a layer of antifriction material, such as a copper-based alloy, between the friction surfaces. To retain the antifriction material, evenly

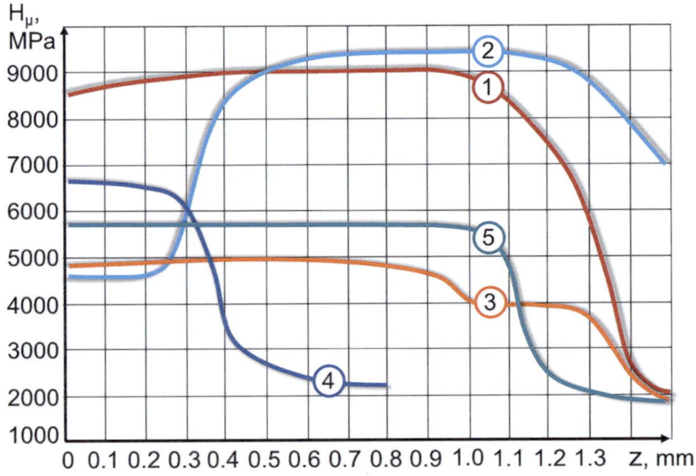

Fig. 6.96 Distribution of microhardness H_μ along the depth z of laser heating in 9310H type steel: 1—carburizing + HF current hardening + laser hardening ($d_0 = 5$ mm; V = 0.8 m/min; 2—carburizing + HF current hardening + laser hardening ($d_0 = 5$ mm; V = 0.4 m/min); 3—carburizing + laser hardening ($d_0 = 5$ mm; V = 0.8 m/min); 4—carburizing + laser hardening ($d_0 = 5$ mm; V = 1.6 m/min); 5—normalization + laser hardening ($d_0 = 5$ mm; V = 0.6 m/min)

spaced grooves are made on the interchangeable bore surface. The grooves are filled by laser-powder surfacing.

In the operating process, under the influence of the force P from the side of the cutter, the walls of the grooves are deformed along the depth ε(z)—Fig. 6.97. The antifriction material is squeezed out of the grooves into the contact zone and fills the micro-irregularities and gaps. This provides forced lubrication.

The shape of the groove cross section can be different (Fig. 6.98):

- rectangular
- arched
- trapezoidal;
- inverted trapezoidal.

Modeling of the stress–strain state of the surface of a heavily loaded friction unit determined that the best removal of the antifriction material above the interchangeable bore surface under external loading is provided by two shapes of the groove cross-section: inverted trapezoidal and rectangular. The *rectangular shape* is more technologically advanced.

Calculations for the rectangular cross-sectional shape of the change in ε(z) depending on the load P at different ratios of the width b and depth h of the groove show that the *optimal* dimensions are b = 2.0 mm; h = 1.2 mm.

The ***optimal composition*** of the filler powder according to the criterion of minimal wear is a mixture of 70% bronze matrix CuSn10Pb10-C + 30% X102CrMo17 steel with a dispersion of 60–80 μm. Powder consumption during laser surfacing is G_p = 2–5 cm^3/min.

The use of copper-iron composite significantly reduces the coefficient of friction and reduces the wear of the friction pair by 50%. The soft material is present on

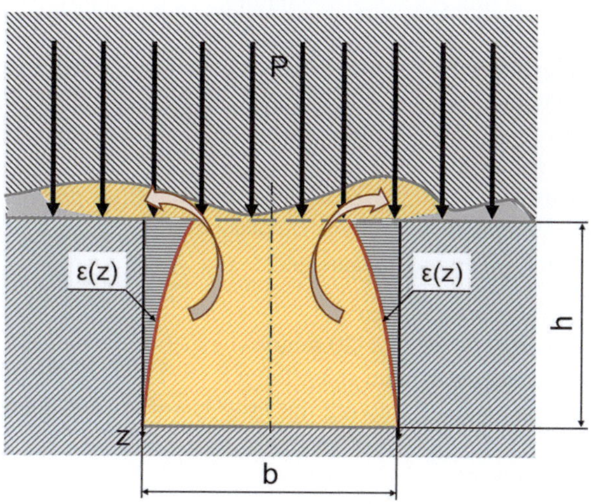

Fig. 6.97 Diagram of loading and deformation of the walls of a rectangular groove (cross section)

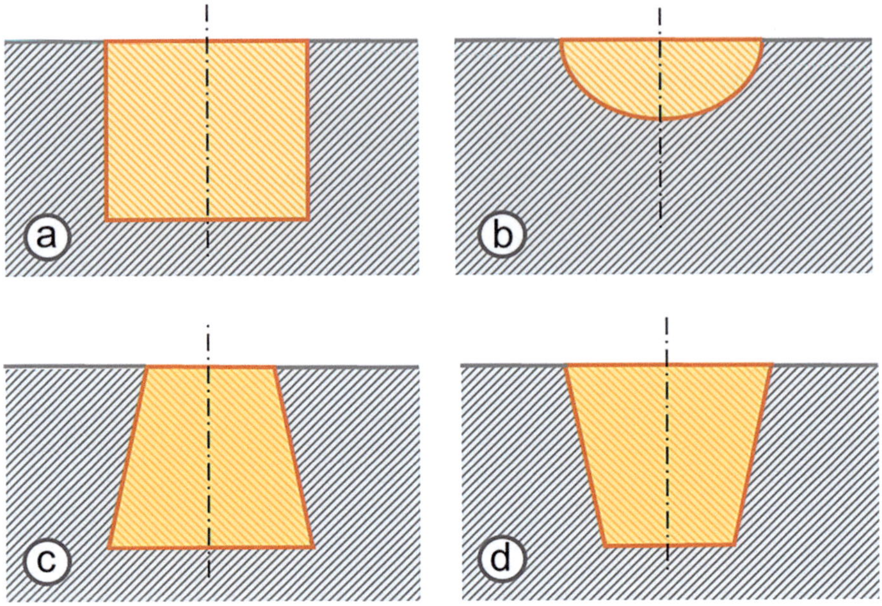

Fig. 6.98 Cross-sectional groove shapes: **a** rectangular; **b** arc; **c** trapezoidal; **d** inverted trapezoidal

both friction surfaces in the form of a light film. The film appears when the soft material (bronze) is rubbed on the friction surfaces and is constantly renewed during the wear process. An important factor is the thickness of the film. A film that is too thin wears out quickly, and a thick film does not provide the necessary resistance to normal loads. At the optimum concentration of soft metal, the film thickness is 5–15 μmicrons and the process is self-regulating.

The ***technology*** of laser-powder surfacing of antifriction material for solid lubrication involves milling intermittent grooves 2 on the surface of the interhcangeable bore 1 along a different contour (Fig. 6.99). The helical intermittent grooves (Fig. 6.99c) provide the best removal of the antifriction material above the interhcangeable bore surface under external load.

The grooves are filled with a laser beam 3 with the filler powder 4 supplied by the transport gas 5 (Fig. 6.100).

An example of the *surfacing parameters*:

- power density $W_p = 0.5$ kW/mm^2;
- laser beam travel speed—$V = 2$–20 mm/s;
- consumption of transporting and shielding gas (Ar)—$G_g = 20$–30 l/h;
- powder feeding scheme—following the laser beam at an angle $\alpha = 45°$ to the laser beam axis;
- width of deposited layers—1–2.5 mm;
- height of the deposited layers—0.5–1.5 mm.

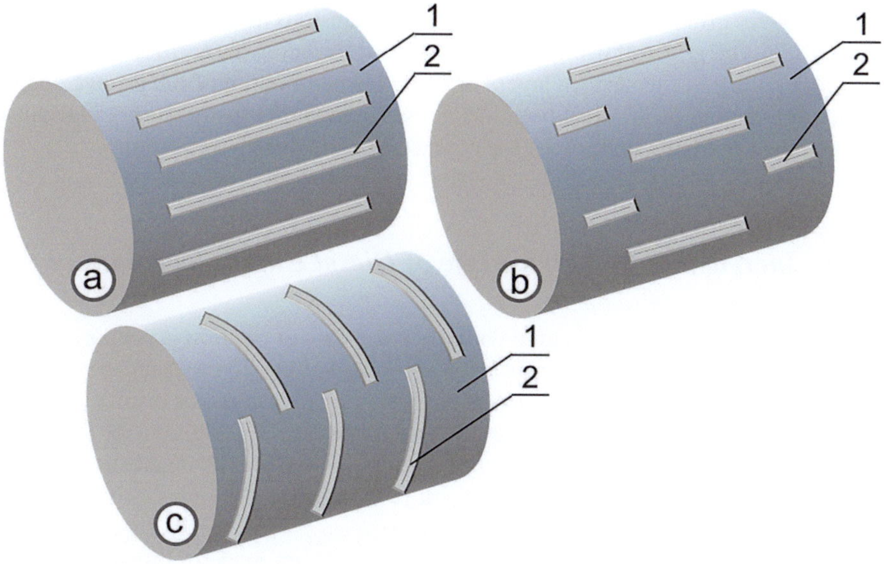

Fig. 6.99 Possible contour of solid lubricant grooves (1—interchangeable bore; 2—grooves): **a** linear; **b** linear intermittent; **c** helical intermittent

Fig. 6.100 Scheme of laser-powder surfacing of antifriction material for solid lubrication: 1—interchangeable bore; 2—groove; 3—laser beam; 4—filler powder; 5—transporting and protective gas; 6—surface of the shoulder

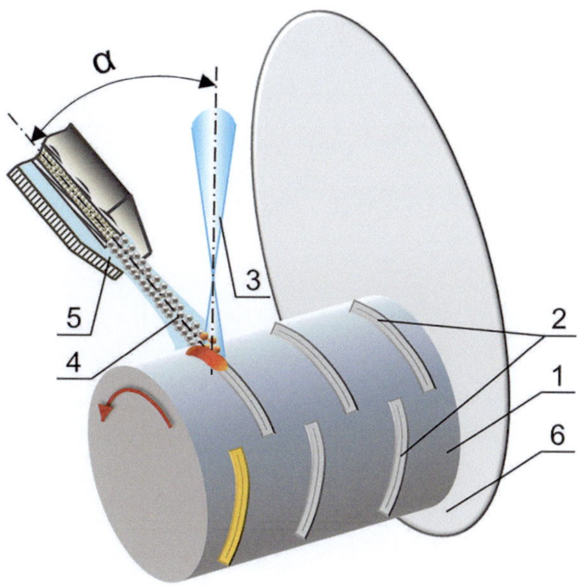

Subsequent grinding provides the required dimensions and surface roughness.

An ***additional effect*** of using laser radiation for surfacing the antifriction material is ***strengthening*** of the surface layers of the walls and bottom of the ***groove*** due to the auto-hardening of the steel base. As a result, the antifriction material is located between the superhard surfaces of the contacting parts during operation.

6.7.6 Increasing the Service Life of Diesel Engine Turbines

The turbine shaft of a diesel engine with a vertical axis of rotation is supported by a plain bearing—a seat. The turbine rotates at a speed of 60–70 thousand revolutions per minute. Wear of the working surface of the seat plate occurs as a result of friction of the contacting surfaces and heating to high temperatures. In addition, the shaft and the seat are subjected to shock loads when the rotation speed is changed. To lubricate and cool the bearing's working surfaces, oil is supplied to the contact zone under a pressure of 2–3 atm.

To ensure high antifriction properties, the bearings are made of bimetallic materials. A highly efficient method of manufacturing the seats is laser surfacing of bronze onto a steel base.

The *advantages* of laser surfacing are ensured by the creation of a *multilayer coating* on the working surfaces of the *seat* (Fig. 6.101):

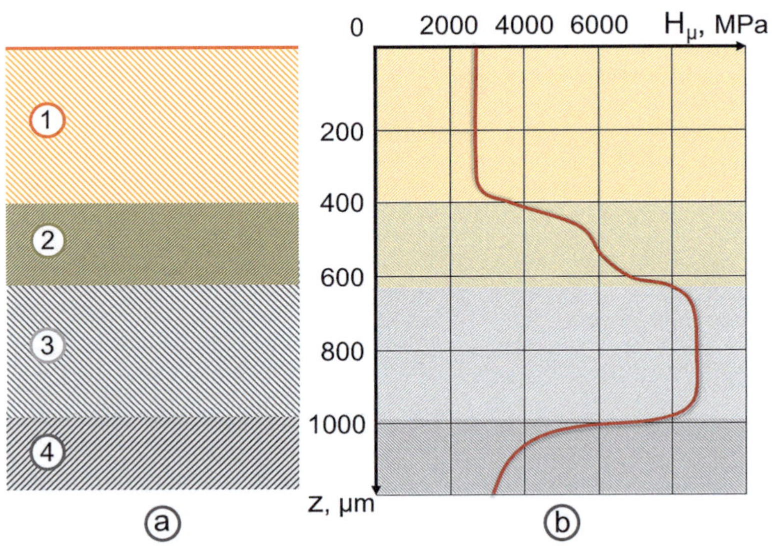

Fig. 6.101 Scheme of formation of a multilayer coating of the base plate by laser surfacing (**a**) and distribution of microhardness H_μ along the depth z (**b**): 1—surface layer of bronze; 2—transition layer of bronze and steel fusion; 3—layer of steel quenched from the solid state; 4—steel base

(1) The *surface layer* of CuSn10Pb10-C *bronze* with a thickness of 400 μm serves as a solid lubricant and has high antifriction properties with a microhardness of 2800–3000 MPa.
(2) A transition *layer* of bronze and steel *fusion* with a thickness of about 80–100 μm ensure their metallurgical connection. Microhardness increases slowly.
(3) The *hardening layer* of steel, where structural and phase transformations occur in the solid state. The high thermal conductivity of the bronze surface layer provides:

- good heat transfer and heating of the steel layer above the A_3 point;
- ultra-high cooling rate, which is much higher than the critical rate.

As a result of auto-hardening, a highly dispersed martensitic structure with a microhardness of 8000–8500 MPa is formed.

(4) The *steel base* ensures the strength of the sliding bearing.

The **parameters of laser surfacing** of CuSn10Pb10-C bronze powder (dispersion 70–80 μm) on a C45 steel base should ensure the temperature range of the melt in the pool:

- the minimum temperature should be higher than the melting point of steel to ensure the fusion of bronze with steel—$T_{min} = 1350\ ^{\circ}C$
- the maximum temperature should not exceed the boiling point of bronze—$T_{max} = 1700\ ^{\circ}C$.

The laser beam speed and filler powder consumption are selected based on the required height of the deposited layer (usually 0.4–0.5 mm)—Figs. 6.102 and 6.103.

Before laser surfacing, circular grooves of triangular shape are milled on the surface of the base plate in cross section. Circular beads with an optimum overlap factor of 0.7 are deposited into the grooves. The surfacing powder is fed in the wake of the laser beam at an angle of $\alpha = 45°$ at a distance to the surface of $L = 10$ mm.

The sequence of surfacing of the circular beads is from smaller to larger diameter.

To reduce the roughness of the deposited surface, it is advisable to re-melt the surface without feeding powder at an increased speed of 20–25%.

After surfacing, the end surfaces of the seats are machined to size and grooves are made to hold the lubricant.

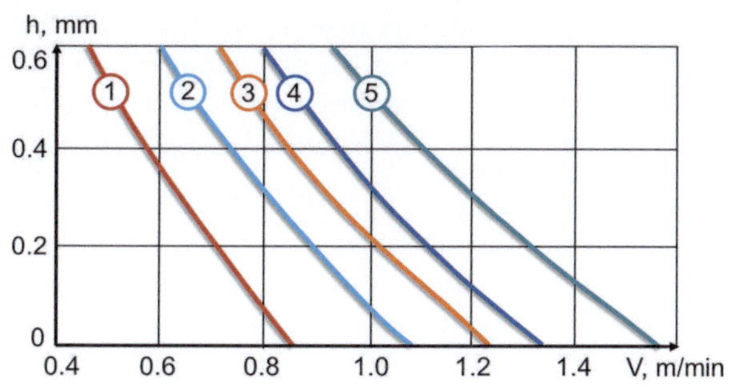

Fig. 6.102 Dependence of the height h of the deposited layer on the speed V of the laser beam at different diameters d_0 of the focusing zone ($P_0 = 2800$ W): 1—at $d_0 = 2$ mm; 2—at $d_0 = 2.5$ mm; 3—at $d_0 = 2.8$ mm; 4—at $d_0 = 3$ mm; 5—at $d_0 = 3.5$ mm

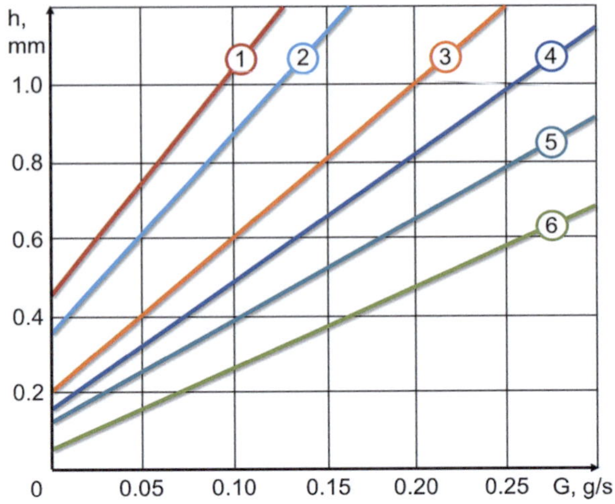

Fig. 6.103 Dependence of the height h of the deposited layer on the filler powder consumption G at different laser beam speeds V (P_0 = 2800 W; d_0 = 3.5 mm): 1—at V = 0.5 m/min; 2—at V = 0.6 m/min; 3—at V = 0.8 m/min; 4—at V = 0.9 m/min; 5—at V = 1.0 m/min; 6—at V = 1.2 m/min

6.8 Laser Technologies for the Production of Bimetals

Components used in the metallurgical, mining, oil and gas production and processing industries, as well as in the chemical, light and food industries, are subject to various types of wear during operation:

- friction in the absence or limited amount of lubricant and the presence of abrasive particles in the contact zone;
- impact-abrasive destruction under the action of pulsed mechanical contact loads under the simultaneous influence of abrasive particles;
- gas-abrasive and hydro-abrasive wear under the influence of high-speed (over 40 m/s) flows of gases or liquids with dispersed abrasive particles at temperatures up to 500 °C;
- corrosion-mechanical wear under the action of an aggressive working environment in the presence of suspended abrasive particles and, as a rule, elevated temperatures.

The service life of parts is determined by the physical and mechanical properties *of the surface layers*. At the same time, the thickness of the surface layer subject to wear usually does not exceed 4–5 mm. Therefore, the use of bimetals is an effective way to increase the service life of equipment.

The main technologies for the production of bimetals are as follows:

(a) Rolling cladding—obtaining bimetallic sheets, plates, strips, pipes, rods by joint deformation during hot rolling of strips (the main method) or cold rolling (for highly ductile materials).
(b) Extrusion cladding—obtaining bimetallic round and shaped profiles by jointly pressing the materials to be joined through an opening in the tool (die). The connection is made by preheating and intensive plastic deformation.

(c) Explosion cladding—obtaining bimetallic plates, sheets, pipes, and parts by plastic deformation during high-speed movement of the workpiece (plate, sheet) thrown towards a stationary workpiece or by mutual counter-movement of the workpieces.
(d) Plasma cladding—multi-pass plasma surfacing with filler materials (solid wires; flux-cored wires; powders; stationary filler in the form of cast or sintered rings, rods, paste).
(e) Electric arc cladding—automated submerged arc surfacing with one or two paired solid or flux-cored electrode strips.
(f) Electroslag cladding—surfacing with one or two electrode strips with the surface to be processed in a horizontal position, with free formation of the surfaced metal layer without the use of crystallizers.
(g) Resistance cladding—welding of steel powders or strips to the surfaces of flat or cylindrical parts by heating with powerful current pulses and simultaneous plastic deformation.
(h) Friction cladding is performed by heating to a plastic state due to the friction of the filler material against the surface of the part, which is achieved by high-speed rotation and pressing.

Laser technologies for the production of bimetals allow for effective control of the aggregate state and surface quality characteristics while intensifying mass transfer processes in the joint area.

Two laser technologies for the production of bimetals are proposed:

(1) laser-casting;
(2) projection-casting.

The final procedure for both technologies is rolling to obtain the specified geometric dimensions of the bimetal.

6.8.1 Laser-Casting Technology

Laser-casting technology involves melting a rectangular section of the surface layer of the bimetal base to a depth of 30–50 μm with a laser beam, while simultaneously feeding the molten functional component of the bimetal from a foundry induction ladle onto the melted surface. Laser radiation is either scanned along a line segment or focused on a line segment with a Top-Hat intensity distribution (rectangular distribution along one coordinate and Gaussian distribution along the other).

There are *two* possible *schemes* for forming bimetals using laser casting technology (Fig. 6.104): with a scanning laser beam and with a linearly focused laser beam [24, 25].

The scanning laser beam method (Fig. 6.104a) involves scanning a rectangular section of the bimetal base surface with a focused laser beam at a given frequency f, trajectory, and amplitude. The workpiece moves at a constant speed V. A laser

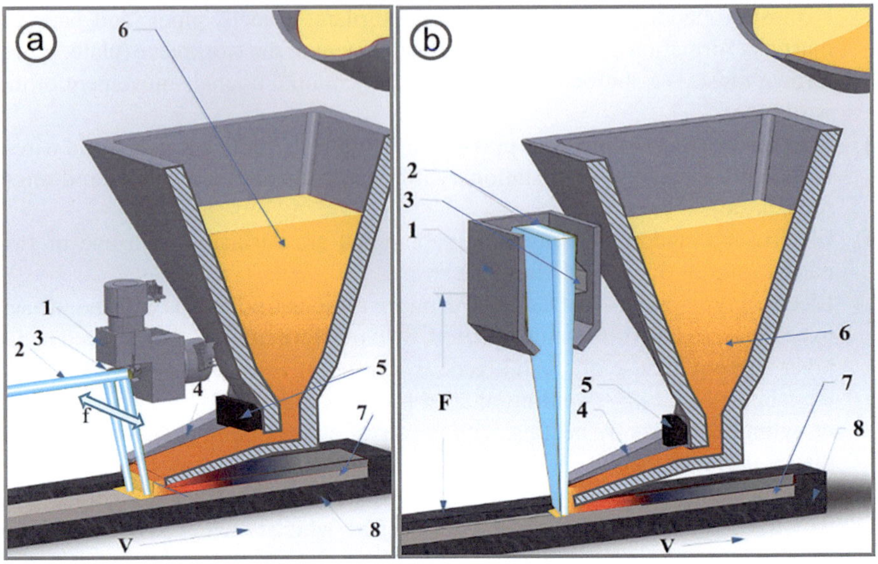

Fig. 6.104 Schemes for manufacturing bimetals using laser-casting technology: **a** scanning laser beam; **b** linearly focused laser beam (1—beam shaping device; 2—laser radiation; 3—scanning mirror (**a**), radiation focusing lens (**b**); 4—foundry system; 5—shutter; 6—melt; 7—workpiece being cast; 8—limiter; f—scanning frequency; F—focal length of the lens; V—workpiece movement speed)

source with a power of P melts the surface area to a depth of Z. Molten metal of the functional component of the bimetal is fed to the melted area at a given flow rate Q. The width B of the molten metal flow is equal to the scanning amplitude of the laser beam.

The advantage is a simpler optical system that forms a quasi-continuous heat source on the surface of the material in the form of a line segment with a relatively low laser power.

The scheme with a linearly focused laser beam (Fig. 6.104b) differs in that the surface of the bimetal base is melted by a laser beam focused into a line segment. The width of the molten metal flow is equal to the width of the laser beam focusing zone.

The bimetal forming process is influenced by four groups of factors that determine its performance characteristics (Fig. 6.105):

(1) laser beam parameters;
(2) properties of the bimetal base;
(3) properties of the molten metal of the functional component of the bimetal;
(4) geometric parameters of the location and kinematics of the relative movement of the bimetal components.

The *control criteria* that have the greatest influence on the surfacing process are:

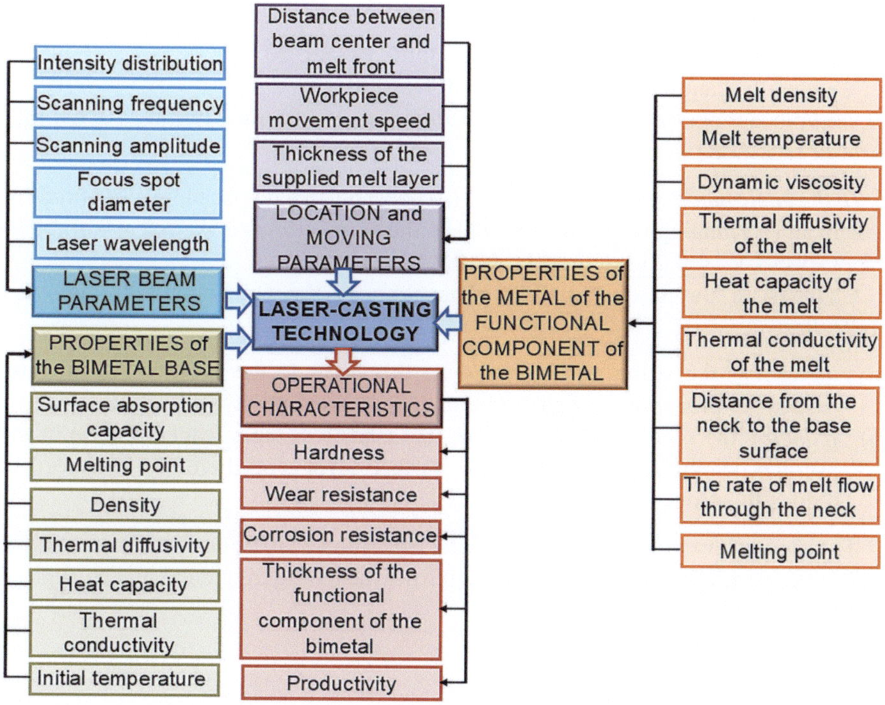

Fig. 6.105 Factors and parameters of laser-casting technology for the production of bimetals

- laser radiation power;
- speed of movement of the bimetal base relative to the laser beam;
- scanning amplitude;
- scanning frequency;
- initial temperature of the bimetal base metal;
- configuration of the opening through which the melt is delivered (length, width, depth);
- thickness of the deposited metal layer;
- height of the metal pressure column.

Modeling of the process of forming a surface molten zone with a scanning laser beam

Numerical analysis of the three-dimensional temperature distribution in a solid body when irradiated by a moving laser beam with a Gaussian distribution of heat flux density in the focus spot was performed using the COMSOL Multiphysics 5.2 software package.

Initial parameters for calculations:

- base material of the bimetal—AISI 420 steel;

- workpiece dimensions—50 × 50 × 2 mm;
- laser beam scanning amplitude A = 20 mm, 30 mm, 50 mm;
- laser beam scanning frequency f = 100 Hz, 200 Hz;
- workpiece movement speed V_x = 0.1 m/min, 0.25 m/min, 1 m/min, 10 m/min;
- laser radiation power P = 2–14 kW.

In the basic mode of scanning with a 2 kW laser beam with a scanning amplitude of 20 mm and a scanning frequency of 200 Hz, the *surface layer* is *melted* at workpiece travel speeds $V_x \leq 0.25$ m/min (Fig. 6.106b).

Changes in the laser radiation parameters in the following directions are *ineffective*:

(a) *increasing* the scanning *amplitude*, as this leads to the need to significantly reduce the processing speed, and thus reduce productivity;
(b) *reducing* the scanning *frequency*, as this leads to a significant decrease in the average heating temperature of the surface layer, and therefore to a need to reduce the processing speed.

The *optimal* laser radiation *power* at which guaranteed melting of the surface layer is ensured, depending on the scanning amplitude, scanning frequency, and workpiece movement speed, is:

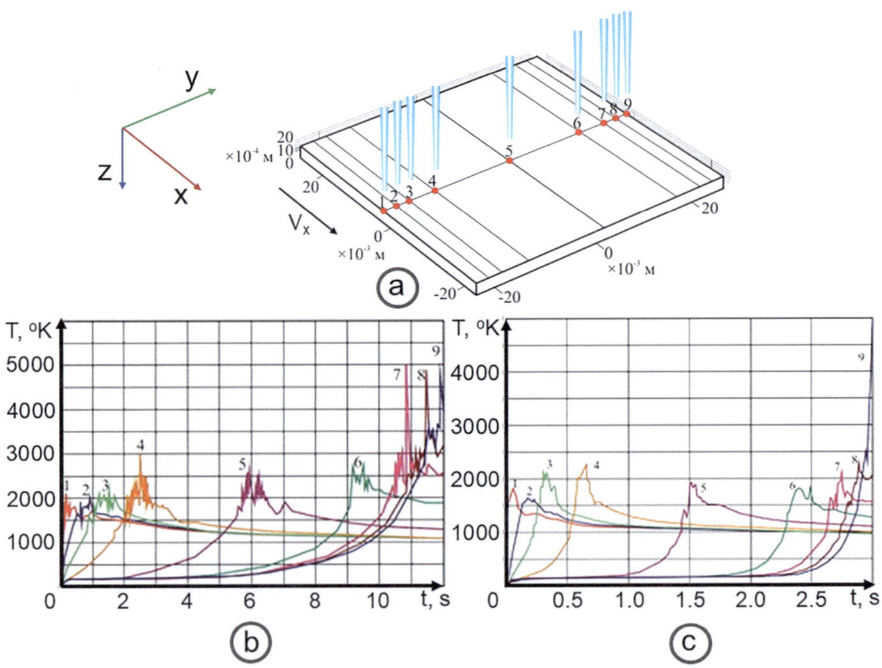

Fig. 6.106 Thermal cycles at points along the laser scanning line under processing conditions: **a** point locations; **b** at P = 2 kW, A = 20 mm, f = 200 Hz, V_X = 0.25 m/min (**a**); **c** at P = 5.25 kW, A = 30 mm, f = 200 Hz, V_X = 1 m/min

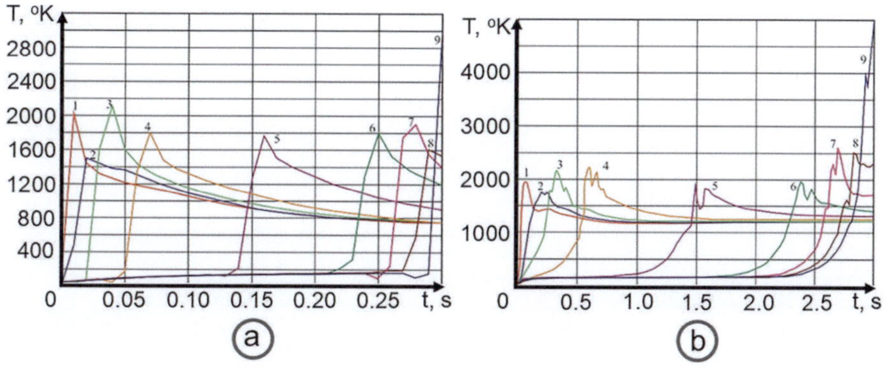

Fig. 6.107 Thermal cycles at points on the central axis of the workpiece when scanning with a laser beam under processing conditions: **a** at P = 14 kW, A = 20 mm, f = 200 Hz, V_x = 10 m/min; **b** at P = 8.5 kV, A = 50 mm, f = 200 Hz, V_x = 1 m/min

P = 5.25 kW—at A = 30 mm, f = 200 Hz, V_x = 1 m/min (Fig. 6.106c);
P = 4 kW—at A = 20 mm, f = 200 Hz, V_x = 1 m/min;
P = 14 kW—when the workpiece movement speed is increased to V_x = 1 m/min (surface melting condition—Fig. 6.107).

The temperature distribution along the depth of the bimetal base (Fig. 6.108) shows that the metal surface is melted in the y direction (scanning direction) in the area less than the scanning amplitude (15–16 mm). To ensure melting of the surface over the entire scanning amplitude, it is necessary to increase the scanning frequency and reduce the speed of movement of the workpiece.

To increase the surface melting zone, it is advisable to use a scanning scheme with two laser beams.

When the laser radiation power is increased to P = 8.5 kW (at A = 50 mm, f = 200 Hz, V_x = 1 m/min), the surface layer is heated across the entire scanning amplitude, which is due to the absence of intensive heat removal from the extreme points of scanning in the unheated areas of the bimetal base (Fig. 6.109).

The width of the stable melting zone in the x direction (workpiece movement direction) is L_x = 1–1.5 mm (Fig. 6.110).

The productivity of bimetal production can be increased by heating the base to temperatures of 800–1000 °C, which is slightly lower than the melting point, but ensures fusion with the melt of the functional component of the bimetal. In this case, the width of the processing zone increases to L_x = 2–2.5 mm under the same processing conditions (P = 8.5 kW, A = 50 mm, V_x = 1 m/min).

A distinctive feature of the method is the possibility of using two surfacing schemes:

(a) the melt of the functional component is applied to the solid base of the bimetal—the main scheme);
(b) a melt of the bimetal base is applied to the solid functional component (additional scheme for preserving the properties of the functional component).

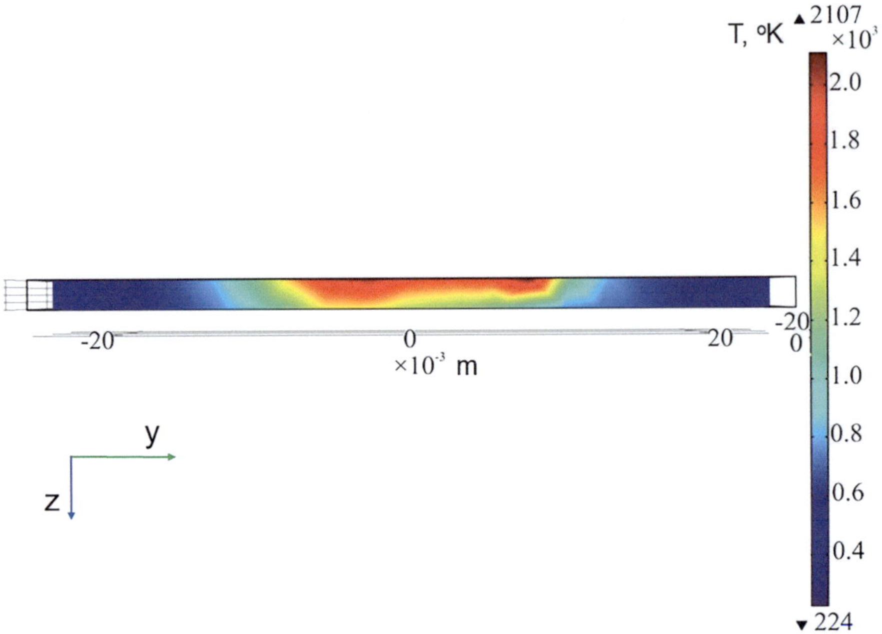

Fig. 6.108 Temperature distribution along the depth of the bimetal base in the y direction (scanning direction) at a distance of 2 mm from the laser beam 6 s after the start of processing at P = 2 kW, A = 20 mm, f = 200 Hz, $V_x = 0.25$ m/min

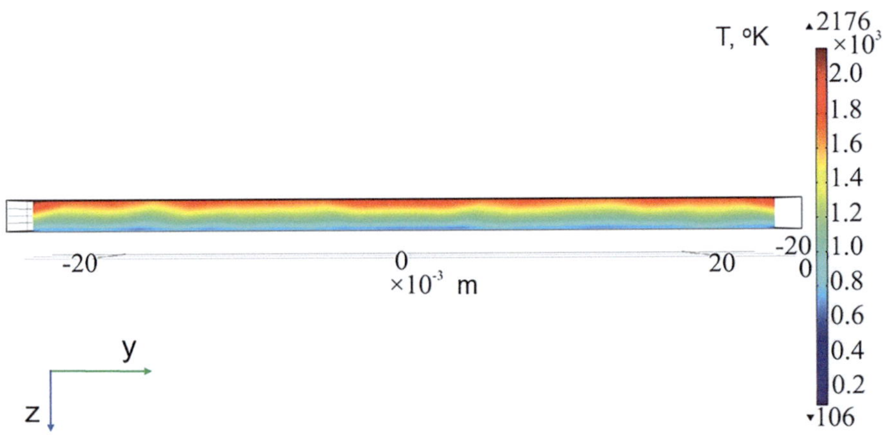

Fig. 6.109 Temperature distribution along the depth of the bimetal base in the y direction (scanning direction) at a distance of 2 mm from the laser beam 6 s after the start of processing at P = 8.5 kW, A = 50 mm, $V_x = 1$ m/min

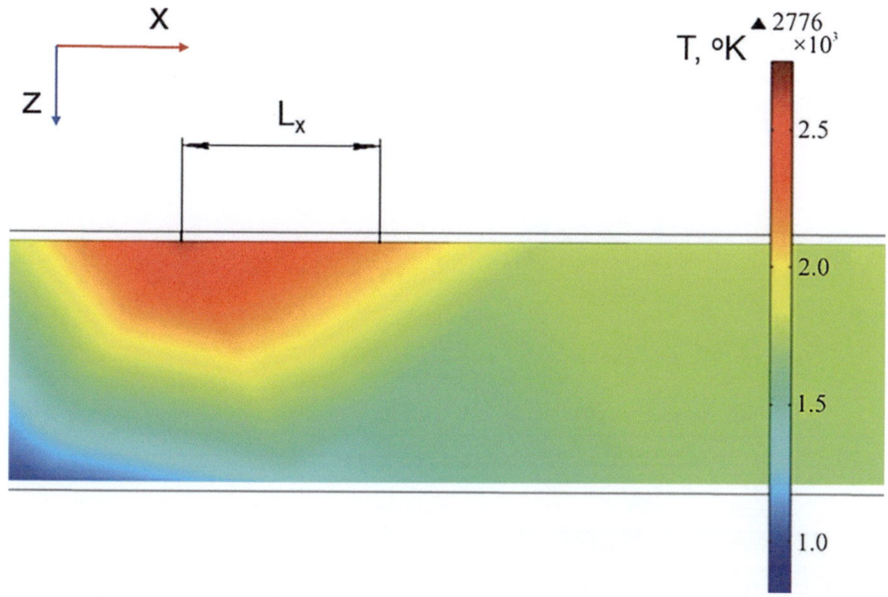

Fig. 6.110 Temperature distribution along the depth of the bimetal base in the x direction (direction of workpiece movement) at P = 8.5 kW, A = 50 mm, V_x = 1 m/min

Determination of the parameters for feeding the molten functional component of the bimetal

One of the technological tasks in the manufacture of bimetal is to ensure a *uniform, specified* flow rate of the molten functional component of the bimetal into the laser radiation zone.

The calculation scheme for supplying the melt of the functional component of the bimetal under pressure p_h through the mold-forming opening of rectangular cross-section (a × b) in the foundry induction ladle is shown in Fig. 6.111.

The foundry induction ladle provides heating and maintains a stable temperature of the liquid metal.

The metal flow rate is regulated by changing the height h of the liquid metal pressure column. The height of the pressure column determines the pressure p_h, which determines the required flow rate and consumption of molten metal through the molding opening.

The *flow rate V_m of liquid metal* through the mold opening is determined from the mass flow balance.

The volumetric flow rate Q of liquid metal flowing through the mold opening:

$$Q = \rho ba V_m \tag{6.50}$$

where ρ is the density of the melt;

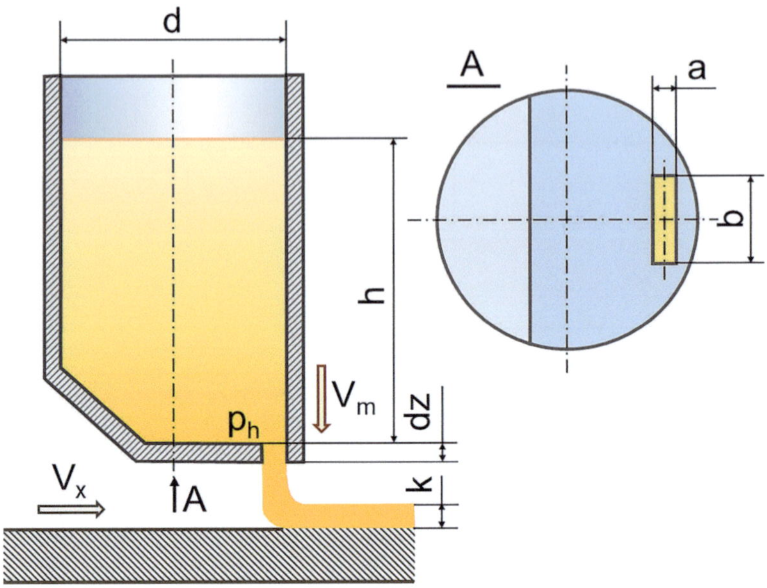

Fig. 6.111 Calculation scheme for feeding the melt of the functional component of the bimetal onto the surface of the base

a, b are the dimensions of the mold opening.

On the other hand, the volumetric flow rate of liquid metal Q, which forms the functional component layer of the bimetal:

$$Q = \rho bkV_x \tag{6.51}$$

where b is the width of the functional component layer of the bimetal, which is equal to the length of the mold opening;

k is the thickness of the functional component layer of the bimetal;

V_x is the speed of movement of the workpiece.

From Eqs. (6.50) and (6.51):

$$V_m = V_x\frac{k}{a} \tag{6.52}$$

The flow rate from the hole at a depth h is equal to:

$$V_m = \sqrt{2gh_v} \tag{6.53}$$

where g = 9.81 m/s^2 is acceleration due to gravity.

Hence, the *height h_v of the molten column*:

$$h_v = \frac{V_x^2 k^2}{2ga^2} \tag{6.54}$$

To *compensate for the reduction in flow* due to hydraulic friction of the liquid metal in the opening, it is necessary to create additional pressure p_L:

$$p_L = \rho g h_L \tag{6.55}$$

The *height of the melt column* h_L to compensate for hydraulic friction losses:

$$h_L = \frac{36(a^2 + b^2)\mu V_m \cdot \mathrm{dz}}{-10a^2 b^2 \rho g} \tag{6.56}$$

where μ is the dynamic viscosity of the melt.

The total height H of the melt column, which ensures the formation of a layer of the functional component of the bimetal with the specified dimensions:

$$\mathrm{H} = \mathrm{h_v} + \mathrm{h_L} \tag{6.57}$$

Example of determining the parameters of melt feed in laser casting of bimetals

Initial data: dimensions of the mold cavity: a = 0.2 mm, b = 50 mm; properties of the steel functional component of the bimetal at a temperature of 1500 °C: ρ = 7070 kg/ m^3, μ = 6 MPa s.

Let's set the required thickness of the functional component of the bimetal k = 3 mm. The required total height of the melt column therefore is H = 12 mm (Fig. 6.112a). The height of the melt column is ensured at a workpiece travel speed V_x = 1 m/min (Fig. 6.112b).

Productivity can be increased by increasing the speed of the workpiece, for example, to V_x = 10 m/min. The total height of the column to ensure the corresponding flow rate of liquid metal must be increased to H = 193 mm (Fig. 6.112b).

6.8.2 *Laser-Projection Technology*

Projection casting technology for the production of bimetals consists of two operations (Fig. 6.113)—*basic diagram*:

(1) *Forming a macro-relief* on the surface of the bimetal base 1 by milling (Fig. 6.113a), or laser cutting (Fig. 6.113b), or plastic deformation. The macro-relief consists of projections 2 of round or square shape in cross-section. The projections have specific dimensions (height and width) and configuration (pattern of arrangement, for example, along mutually perpendicular axes

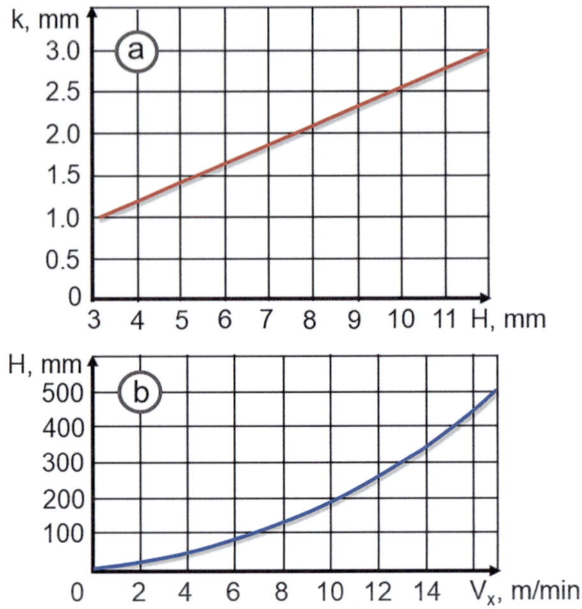

Fig. 6.112 Melt feed parameters in laser casting of bimetals: **a** dependence of the thickness of the functional component of the bimetal k on the height of the melt column h; **b** dependence of the height of the melt column h on the speed of movement of the workpiece V_x

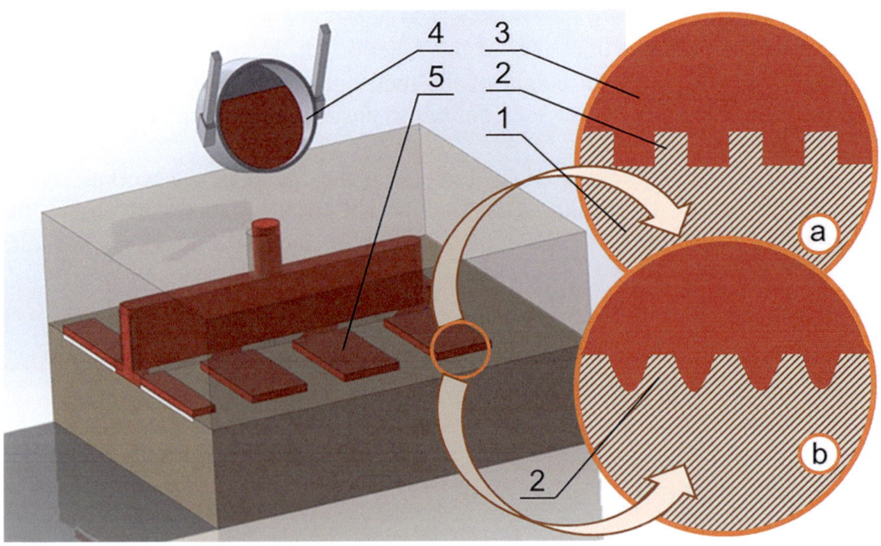

Fig. 6.113 Diagram of bimetal production using projection casting technology (**a** forming macro-relief by milling; **b** forming macro-relief by laser cutting): 1—bimetal base; 2—macro-relief projections; 3–functional component of the bimetal; 4—foundry induction ladle; 5—workpieces

with a certain pitch). Controlling the configuration of the projections allows to control the temperature distribution, the strength of adhesion to the bimetal base, and productivity.

(2) *Pouring the melt* of the functional component of the bimetal 3 from the foundry induction ladle 4 over the entire length of the workpiece 5 with a surface macrorelief. The molten metal of the functional component of the bimetal is heated above the melting point.

Depending on the metallurgical properties of the functional component (for example, when using heat-resistant alloys) and the thickness ratio, an *additional scheme* can be applied:

(1) *Macrorelief* is formed on the surface of the *functional* component of the bimetal.
(2) The molten base of the bimetal *is poured.*

The advantage of the additional scheme is the preservation of the physical and mechanical properties of the functional component of the bimetal.

Modeling of the interphase boundary of liquid metal and the relief of the bimetal base

Objective is to calculate the temperature distribution in the fusion zone and to establish the characteristics of changes in the position and shape of the interphase boundary over time depending on the temperature at different stages of bimetal formation.

The modeling was performed using the moving boundary method in accordance with Stefan's problem in the Comsol Multiphysics v.5.5 software package [26].

Bimetal components: solid base with relief—nickel alloy UNS 6230 (nickel-chromium-tungsten-molybdenum); liquid metal of the functional component—structural carbon steel Fe37-3FN.

The Deform Geometry interface was used to calculate the position of the interphase boundary. The dependence of liquid density on temperature using the Boussinesq linear approximation:

$$\rho(T) = \rho_0 \cdot (1 - \alpha \Delta T) \tag{6.58}$$

where α is the coefficient of volumetric expansion;

$\Delta T = T - T_0$ is the deviation of the temperature from the equilibrium state;

ρ_0 is the density of the molten liquid at a certain equilibrium temperature T_0.

Since the coefficient of volumetric expansion and temperature deviation are usually small, the linear approximation has acceptable accuracy in the problems under study.

The phase boundary velocity is determined as the ratio of the heat flux density at the boundary to the specific heat of phase transition, which follows from the compatibility condition or Stefan's condition:

$$V_{1\to2} = {}^{T_{im}}/_{\Delta H_{1\to2}} \tag{6.59}$$

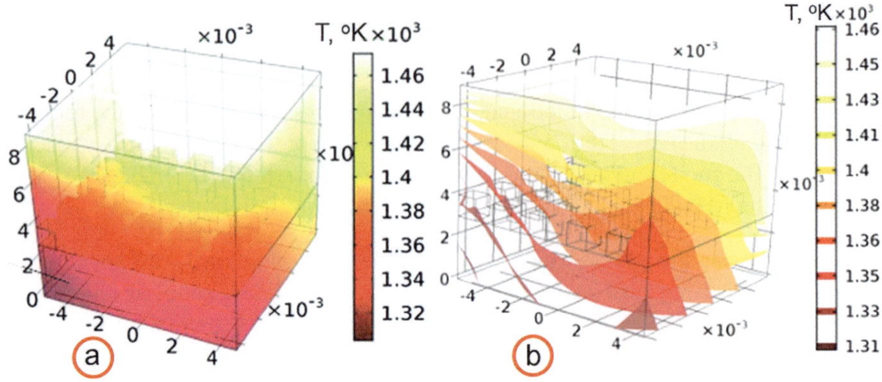

Fig. 6.114 Temperature field in the relief projections of the base under the action of the functional component melt: **a** temperature distribution; **b** isothermal surfaces

where $V_{1\to2}$ is the phase boundary velocity (m/s);

T_{im} is the heat flux density;

$\Delta H_{1\to2}$ is latent heat of fusion (for selected metals $\Delta H_{1\to2} = 305$ kJ/kg).

The liquid and solid phases are separated by a moving crystallization front of the melt and are resolved separately (Fig. 6.114). The position of the interphase boundary in time is calculated according to Stefan's energy balance condition.

Assumptions made in modeling of thermal processes:

- mass transfer is not taken into account;
- thermal conductivity coefficients of materials are considered constant;
- heat capacity coefficients are approximated by functions that have a spike in the phase transition interval.

The basic heat transfer equation in the studied area:

$$C(U, x, y, z)\frac{\partial U}{\partial t} = \lambda(x, y, z)\left(\frac{\partial^2 U}{\partial x^2} + \frac{\partial^2 U}{\partial y^2} + \frac{\partial^2 U}{\partial z^2}\right) \tag{6.60}$$

At the initial moment of time, the temperature of the studied area is determined by the initial conditions:

$$U(x, y, z, 0) = U_1 \quad at \quad (x, y, t) \in D_1$$
$$U(x, y, z, 0) = U_2 \quad at \quad (x, y, t) \in D_2$$

The boundary conditions that model the interaction of the sample with the environment are as follows:

$$\lambda\frac{\partial U}{\partial n}\bigg|_{\Omega} = \alpha(U)(U - U_C) = 0 \quad at \quad (x, y, t) \in \Omega$$

At optimal values of height, area, and configuration of relief projections, ***heat accumulates at the tops of the projections*** due to simultaneous heating of the upper and lower parts and, consequently, the absence of a temperature gradient and heat dissipation from the top (Fig. 6.114b). Thus, the heating temperature of the top by the melt of the functional component may exceed the melting temperature, which ensures the formation of a strong metallurgical bond between the functional component and the base of the bimetal.

In the melt, metal movement occurs on the projections of the functional component of the bimetal, caused by natural convection. This movement affects the displacement of the front between the liquid and solid phases.

The features of the changes in the crystallization front 1 between the liquid 2 and solid 3 phases are considered in a two-dimensional area (Fig. 6.115). The temperature of the upper boundary of the region is maintained at $T_h = 1872$ K, and the lower boundary at $T_l = 293$ K. The side boundaries of the region are thermally insulated 4. The temperature distribution varies linearly in the vertical direction from the hot to the cold boundary (Fig. 6.115b).

The liquid phase of the metal is described by the Navier–Stokes equation in the Boussinesq approximation:

$$T_{ref} = T_f$$

where T_{ref}—temperature at the boundary;

T_f—melting temperature.

The density of metal depends on temperature. A change in density in a closed environment can lead to a significant change in pressure in the calculation area, which will negatively affect the accuracy of the calculation. For the same reason, in

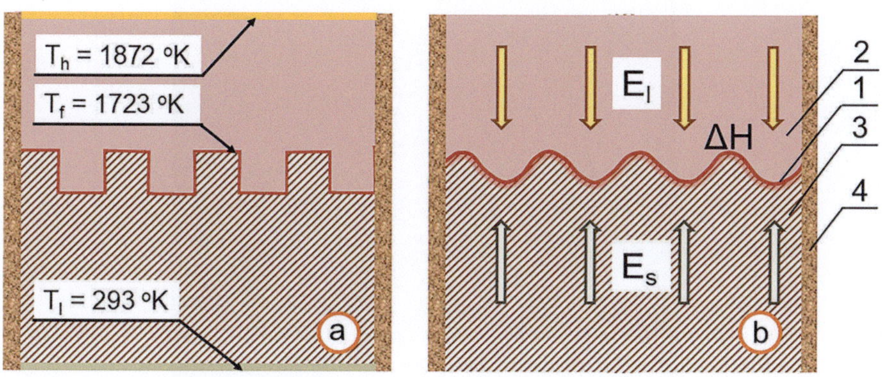

Fig. 6.115 Geometry and boundary conditions at the crystallization front between the liquid and solid phases (**a**) and heat flows around the crystallization front at the moment of melt action (**b**): 1—crystallization front; 2—liquid phase (Fe37-3FN steel); 3—solid phase (nickel alloy UNS 6230); 4—thermal insulation

convection problems involving incompressible fluids, it is advisable (where possible) to disregard the hydrostatic component.

As melting proceeds, the interphase boundary moves toward the solid metal. The energy balance at the interphase boundary is described by the equation:

$$\rho_0 = \Delta H v n = (E_l - E_s)n \tag{6.61}$$

where n is the normal vector to the interphase boundary;

E_l is the heat flux coming from the liquid phase (Fig. 6.115b);

E_s is the heat flux coming from the solid phase.

The expression $(E_l - E_s)n$ in Eq. (6.61) is more accurately estimated using the Lagrange multiplier. This variable is available when Weak Constraint is enabled in the COMSOL Multiphysics calculation domain, as in the case of fixed temperature constraints at the front of the interphase boundary. To construct the mesh and obtain a more accurate solution, a mesh adaptation algorithm was selected (Fig. 6.116).

The temperature in the projection zone is within the range of 1740–1750°K. From 12 to 20 s, the process slows down slightly. The interphase boundary front begins to move slowly at a speed of 1–1.5 m/s relative to the initial temperature. At 30 s of simulation, the melting front no longer moves, as a balance is achieved between the upper and lower adjacent heat flows.

To control the movement of the interphase boundary front, mesh deformation is required. During this transformation, a certain amount of the solid phase of the substance decreases, and the corresponding amount of liquid metal increases. The

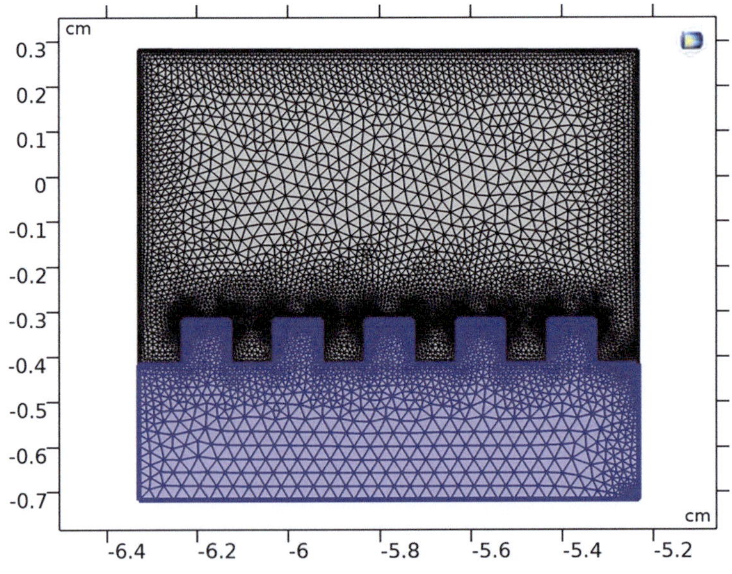

Fig. 6.116 Adaptive mesh at the phase transition boundary

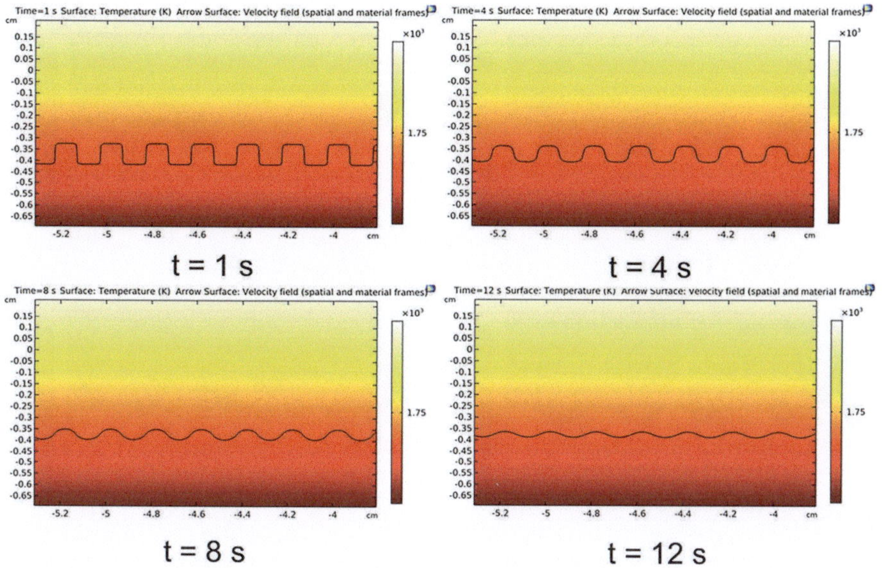

Fig. 6.117 Temperature field and interphase boundary shape at different time intervals t after pouring the melt of the functional component of the bimetal

tool for deforming the mesh (without reflecting any expansion or compression effects in the material properties) is the Deformed Geometry module setting.

Figure 6.117 shows the change in the temperature field and the shape of the interphase boundary during the interaction of the liquid and solid phases at different time intervals after pouring the melt of the functional component of the bimetal. Starting from 4 s, the temperature on the surface of the projections exceeds the melting temperature of the solid component (1728 K) and active smoothing of the projection profile occurs, which lasts up to 12 s.

Figure 6.118 shows the distribution of the interphase boundary front velocity at different time intervals after pouring the melt of the functional component of the bimetal. Already at the beginning of the process at t = 1–4 s, the melting rate is high due to the large temperature gradient. Later, at 12 s, the melting rate slows down due to the stability of the interphase boundary and the thermal field. At 30 s, the melting rate drops to almost zero as the system reaches a steady state.

The peaks of the interphase boundary front velocity are represented by lighter areas above the projections. At the same time, in the recessions, the velocity is directed in the opposite direction. This leads to the formation of an area with minimum velocity. Starting from 4 to 7 s, the velocities in the depressions equalize with the velocities in the molten projections.

An example of projection casting technology

Projection casting technology was tested during the manufacture of bimetal according to an additional scheme:

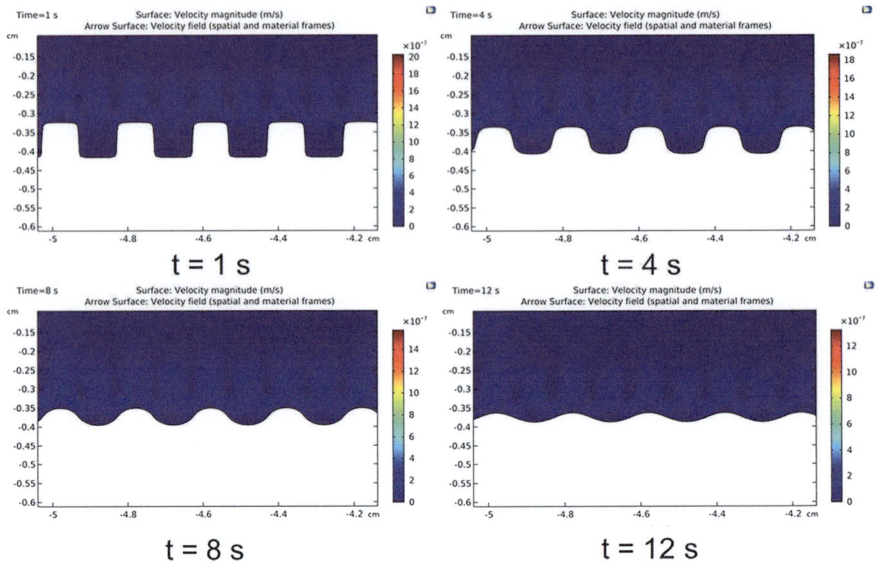

Fig. 6.118 Field of the interphase boundary front velocity at different time intervals t after pouring the melt of the functional component of the bimetal

(1) A relief was applied to the functional component in the form of a plate made of UNS 6230 alloy (nickel-chromium-tungsten-molybdenum). The UNS 6230 alloy is characterized by high heat resistance, heat strength at temperatures up to 1150 °C, low thermal expansion coefficient, and good machinability at room temperature.
(2) The base made of S235JRG2 structural steel was formed as a melt casting.

A regular relief in the form of square projections with cross dimensions of 1 × 1 mm was created on the surface of UNS 6230 alloy plates 100 × 50 × 4 mm in size by evaporation with laser radiation in combination with mechanical processing. Laser radiation parameters: laser radiation power P = 400–500 W; workpiece travel speed V = 1–2 m/min; focusing spot diameter d_0 = 0.4–0.5 mm. The height of the projections on the samples was h = 0.5; 1.0; 1.5 mm, the distance between the projections in two coordinates s = 1.0; 1.5; 2.0 mm (Table 6.6).

Plates made of UNS 6230 nickel alloy with relief (Fig. 6.119b) are placed in a sand form (Fig. 6.119a), heated to a temperature of 300 °C, and kept in the furnace for 3 h. The molten base of the S235JRG2 steel bimetal is preheated to a temperature of 1600 °C and poured (Fig. 6.119d) onto the surface of the plates through a pouring system (Fig. 6.119c). The ***pouring temperature of the melt of 1600 °C*** is the first significant parameter of projection casting technology. After casting is complete, the samples are *slowly cooled* together with the mold *for* 4 h. Slow cooling minimizes residual stresses, reduces the risk of hot and cold cracks, and ensures the dimensional stability of the resulting bimetal workpieces.

Table 6.6 Schematic representation of the macro-relief in the form of square projections 1 × 1 mm on the surface of the plates (h—height of projections; s—pitch of projections)

	s = 1.0 mm	s = 1.5 mm	s = 2.0 mm
h = 0.5 mm	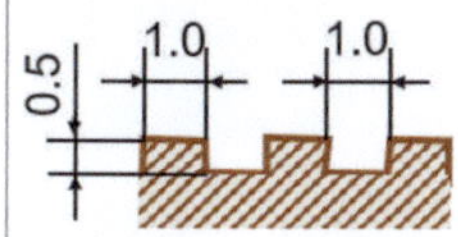	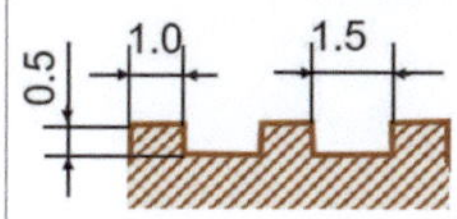	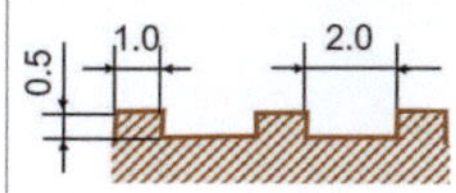
h = 1.0 mm	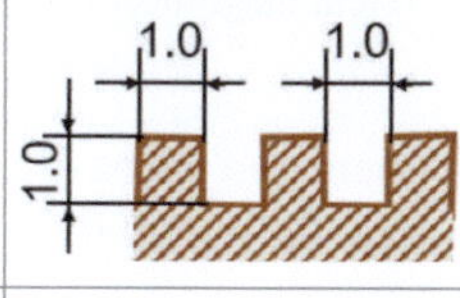	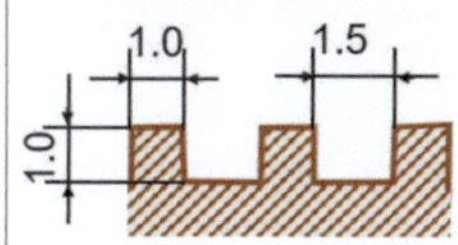	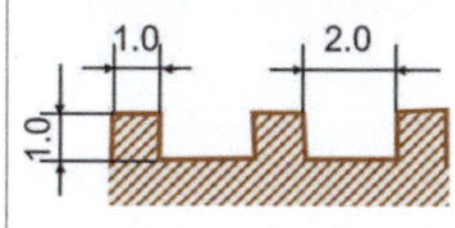
h = 1.5 mm	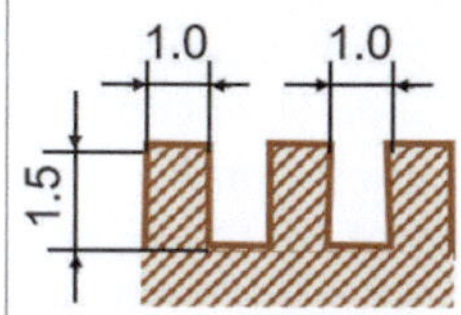	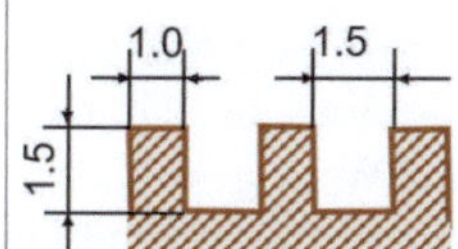	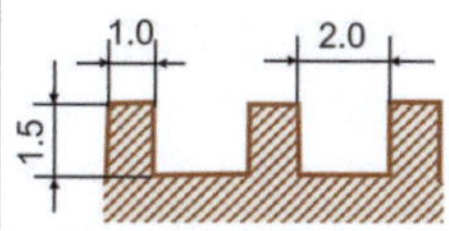

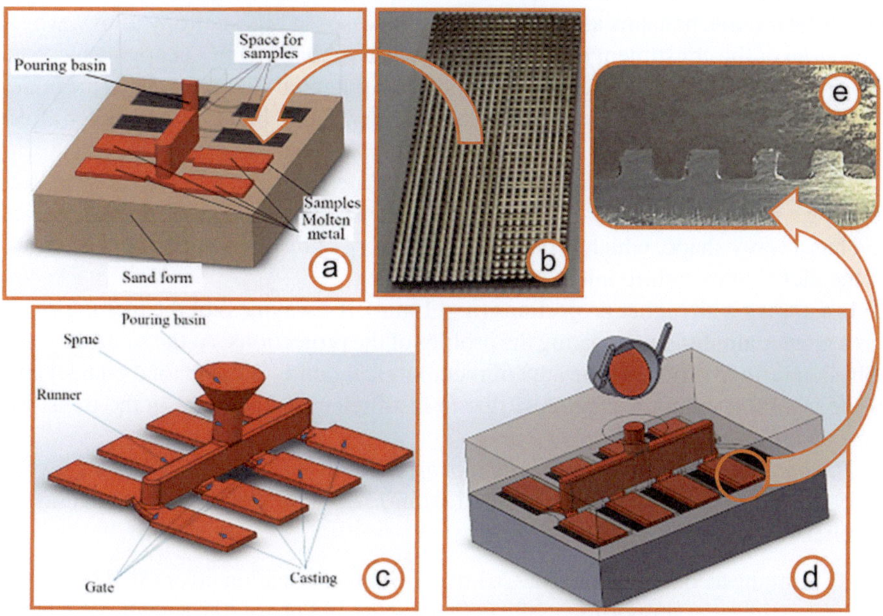

Fig. 6.119 Diagram of the bimetal base casting: **a** sand form for casting; **b** plate made of UNS 6230 nickel alloy (relief parameters—h = 0.5 mm, s = 1.5 mm); **c** design of the pouring system; **d** pouring of the molten base of the bimetal made of S235JRG2 steel; **e** cross-section of the bimetal being formed

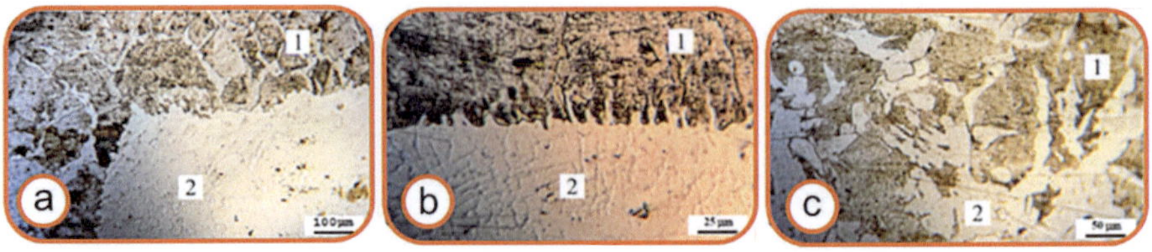

Fig. 6.120 Microstructure of bimetallic castings (1—S235JRG2 steel; 2—UNS 6230 nickel alloy): **a** general view of the fusion line; **b**, **c** transition layer with different magnification factors

The ***transition layer*** between the functional component and the base largely determines the microstructure and mechanical properties of the bimetal (primarily, the strength of the bond).

The S235JRG2 carbon steel base has a ferritic-pearlitic structure (Fig. 6.120). Clusters of white grains of various sizes in the transition layer contain an increased concentration of Cr, Ni, and W (Fig. 6.120c). The transition layer is formed as a result of mutual diffusion across the fusion boundary of Ni, Cr, W, Fe, and C atoms and is 80–170 μm thick. Diffusion occurs through vacancies and between positions in the crystal lattice nodes and depends on temperature and concentration gradients. The diffusion mechanisms are explained by processes in high-entropy alloys, where each atom of the constituent elements has an equal probability of occupying a crystal lattice site. Due to the heterogeneity of atoms, the potential energy between the nodes varies. Fluctuations in the potential energy of interatomic bonds significantly affect the kinetics of diffusion and its activation energy.

The projections on the functional component are partially or completely melted, forming a solution of nickel alloy in carbon steel. The separation boundary is blurred and has a wavy shape, which is due to the geometry of the projections on the surface of the nickel alloy before interaction with the melt.

In samples with higher projections (h $\geq$ 1 mm), the waviness of the melting zones has a greater amplitude, repeating the profiles of the projections. At the same time, the adhesion ratio (the ratio of the area of metallurgical bonding to the total contact area of the layers) does not exceed 20% due to insufficient temperature in the transition zone and high cooling rates.

The optimal dimensions of the relief projections h = 0.5 mm, s = 1.5 mm (with a cross-section of the projections of 1.0 × 1.0 mm) are the second significant parameter of relief casting technology and ensure:

(a) The highest content of the Ni–Cr-Fe phase in the transition layer (86.4%). At the same time, the proportion of the Fe0.507–Ni0.493 phase (tetrataenite) increases to 13.6%. The solid phase in the melt leads to the formation of a *homogeneous microstructure, a reduction in grain size*, and a decrease in the number of defects. This has a positive effect on mechanical properties, increasing the hardness and strength of the transition layer of the bimetal.

(b) *Uniform distribution of microhardness* across the thickness of the bimetal with a smooth decrease from 5 to 7 GPa in the nickel alloy to 4–5 GPa in the transition layer and further to 3–4 GPa in the carbon steel base (Fig. 6.6.2.17). A smooth decrease in microhardness indicates a uniform distribution and mixing of bimetal components, which has a positive effect on mechanical properties (Figs. 6.121 and 6.122).

c) The highest (97–98%) ratio of contact areas with a *metallurgical bond.*

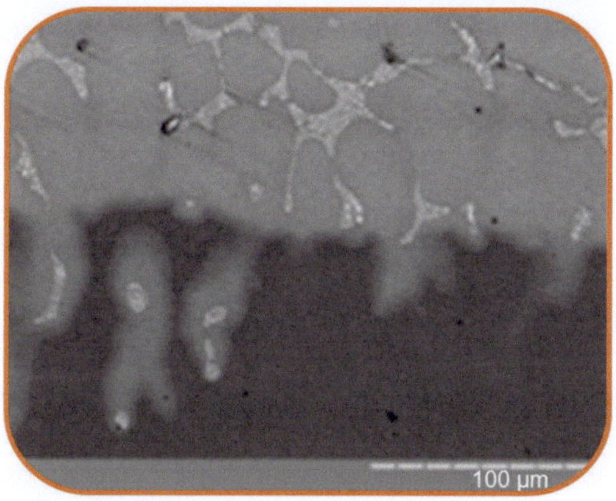

Fig. 6.121 Microstructure of bimetal with optimal relief dimensions (h = 0.5 mm, s = 1.5 mm)

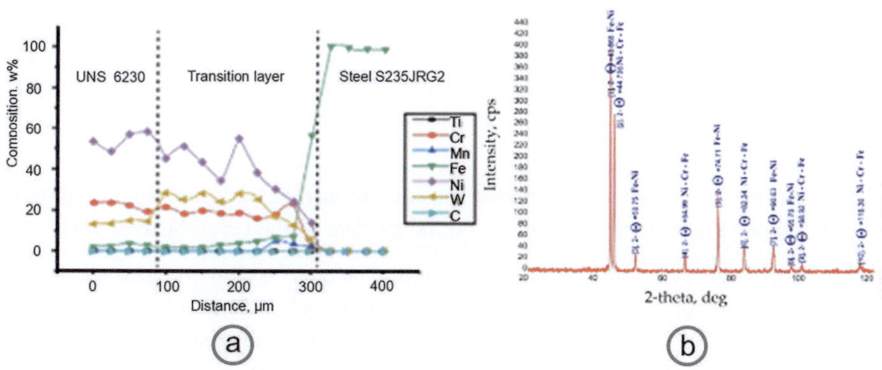

Fig. 6.122 Characteristics of bimetal with optimal relief dimensions (h = 0.5 mm, s = 1.5 mm): **a** EDS analysis; **b** diffraction pattern of the transition layer

6.9 Laser Cleaning of the Surface of Welding Wire

The fundamental principles, research, and technologies of laser cleaning are presented in [27, 28]. In this section, the authors describe the features of laser cleaning with subsequent possible amorphization of the surface using the example of welding wire processing.

For MIG/MAG welding, ER 70–6 steel welding wire with a diameter of 0.8–1.6 mm is most commonly used.

The surface condition of the welding wire significantly affects the quality of the welded joint, because it:

(a) *Determines the level and stability of the contact resistance* of the current lead. The contact resistance of the current lead is connected in series to the welding circuit. The presence of oxides and dirt on the surface of the welding wire increases the contact resistance (up to 10 times), which leads to a jump in the welding current and arc voltage, and consequently to the formation of defects in the welded joint (uneven weld width and incomplete fusion at the root of the weld).

(b) *Determines* the friction coefficient, and therefore the *level of wear* of the working channel of the contact tip. The contact tip is designed to conduct current to the welding wire and is made of copper. Wear of the contact tip increases the gap between the channel surface and the wire, which leads to a significant increase and fluctuation in the contact resistance of the current conductor.

(c) *Determines the risk of oxides entering the weld pool*, which leads to the formation of defects in the welded joint (gas pores).

To ensure better electrical contact and reduce the friction coefficient, a copper electroplated coating with a thickness of 0.15–0.6 μm is applied to the welding wire. However, the copper coating has a significant number of defects, which are due to two reasons:

- a large number of non-metallic inclusions on the surface of the original steel wire;
- significant roughness and scratches on the surface of the original steel wire.

An alternative to copper plating is grinding and polishing the surface of the welding wire, which increases the cost of the wire by 5–8% and leads to intense atmospheric corrosion in the period between manufacture and use in welding.

Laser treatment of the welding wire surface involves two stages:

- *Stage 1*—*laser cleaning* of the welding wire surface from oxides.
- *Stage 2*—use of one of two alternative technologies: application of a copper electroplated coating to the cleaned surface, or *laser modification* with the formation of a surface amorphous structure (amorphization).

The mechanism of laser cleaning of the welding wire surface from oxides (Fig. 6.123) is based on the fact that laser radiation of a certain wavelength is well absorbed by oxides and other dielectrics (without melting the metal surface), resulting in the following phenomena:

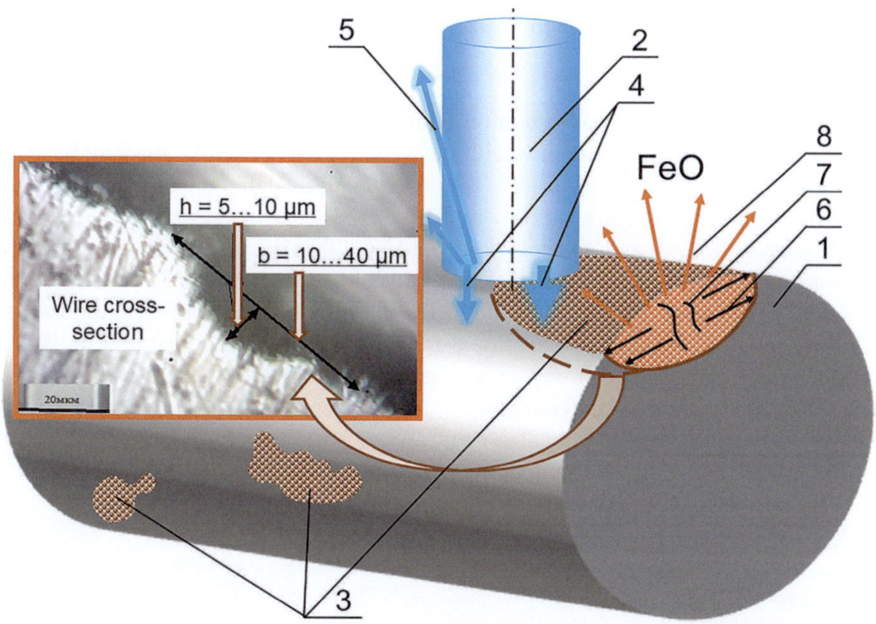

Fig. 6.123 Diagram of laser cleaning of the welding wire surface: 1—welding wire; 2—laser beam; 3—oxide spots; 4—absorbed laser radiation; 5—reflected laser radiation; 6—increase in oxide volume during heating; 7—oxide destruction; 8—oxide evaporation

(a) sharp *increase in the volume* of oxides during heating;
(b) partial *destruction* of oxides due to thermal deformation;
(c) *evaporation* of oxides due to ultra-fast laser heating.

Most of the volumetric defects in the raw wire are iron oxides, mainly FeO + Fe_2O_3 scale. Iron oxides are located at different distances from the surface. When wire is manufactured by drawing, oxides located near the surface of the blank are brought to the surface. The concentration of iron oxides on the surface increases.

Polishing the wire significantly reduces surface roughness and has virtually no effect on the concentration of surface oxides.

Iron oxides evaporate at a temperature of 950 °C (the temperature at which FeO, a component of wustite, exists). This forms the basis of laser cleaning technologies for welding wire surfaces.

Technology and equipment for laser cleaning of the welding wire surface shall ensure:

(a) use of laser radiation with a ***wavelength of*** $\boldsymbol{\lambda = 10.6\ \mu m}$, at which oxide absorption is maximum;
(b) ring cleaning of the entire surface of the welding wire, for which the optical system includes a ***toroidal mirror***.

In the laser optical system for surface treatment of the welding wire [29, 30] (Fig. 6.124), the laser radiation hits the central conical mirror 3 and is directed to the ring conical mirror 2. The laser beam is converted into a beam with a ring-shaped intensity distribution in the cross section. Reflection from the flat ring mirror 8 rotates the beam so that its axis coincides with the axis of the wire. Next, the ring beam is reflected from the toroidal mirror 11, which concentrates all the energy of the laser beam on the surface of the welding wire 31.

All mirrors 2, 3, 8, and 11 must be cooled by a flowing liquid. Therefore, these parts are hollow and hermetically sealed with covers 1, 9, and 12. An additional cooler 4 is installed for the mirror cone 3. The cooling fluid is supplied and removed using fittings 30. The flat ring mirror 8 is equipped with an adjustment unit that provides three adjustment movements: rotation of the mirror around two axes and movement along the axis of symmetry.

The design of the laser optical system provides the supply of protective gas to the processing area and the removal of erosion products from the processed surface from the optical system volume.

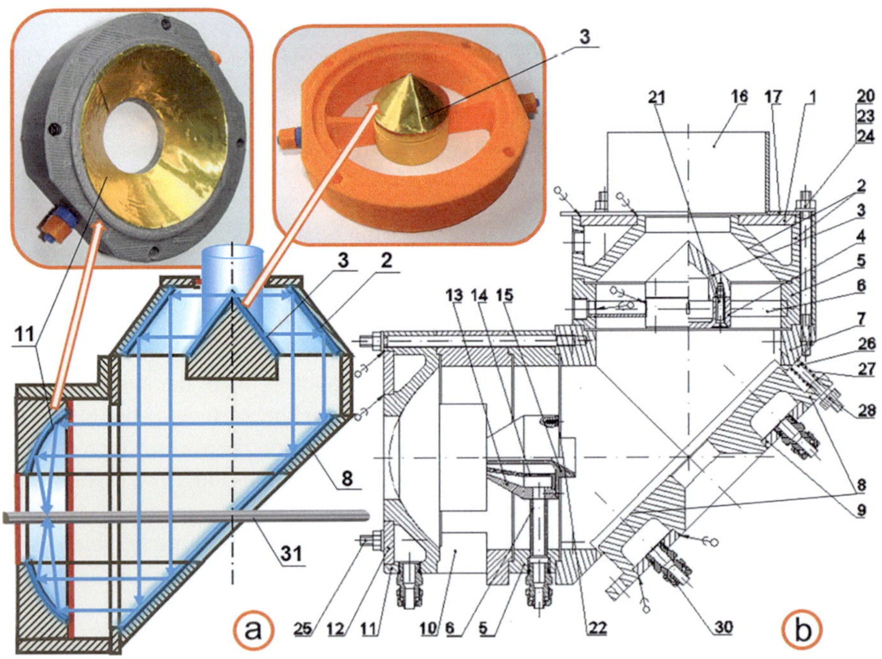

Fig. 6.124 Laser optical system for surface treatment of the welding wire (**a** optical system diagram; **b** assembly drawing): 1—conical mirror cover; 2—ring conical mirror; 3—central conical mirror; 4—mirror cone cooler; 5—mirror cone and nozzle system housing; 6—diaphragms; 7—optical system housing; 8—flat ring mirror; 9—flat mirror cover; 10—insert; 11—toroidal mirror; 12—toroidal mirror cover; 13—nozzle system cover; 14—nozzle system insert; 15—nozzle system cover; 24 and 25—pins; 26—adjustment screws; 30—fittings; 31—welding wire

The laser unit [29, 30] for processing the surface of welding wire (Fig. 6.125) includes a stand 1 on four supports. The stand is equipped with an electric motor with a gearbox and a welding wire receiving cassette 2, a cassette fixing mechanism with a length counter and a wire speed sensor 3. The welding wire 4 is forcibly unwound at a specified speed from the output wire coil 5 and, guided by rollers 6, is fed through the laser optical system for cleaning the wire surface 7. Laser irradiation of the wire surface occurs in a laser beam focused into a ring with a wavelength of 10.6 μm. The wire surface is cleaned of oxides (Fig. 6.125b). An example of the cleaning process parameters: the radiation power was P = 700 W; the diameter of the concentrated beam on the treated wire surface was d = 2 mm; the wire rewinding speed was V = 1.5–3.5 m/min.

Next, the wire passes through the laser optical system for amorphization of the wire surface 8. Heating to the melting temperature and ultra-fast cooling of the surface (108°/s) occur. The designs of the laser optical systems for cleaning and amorphization of the wire surface are the same.

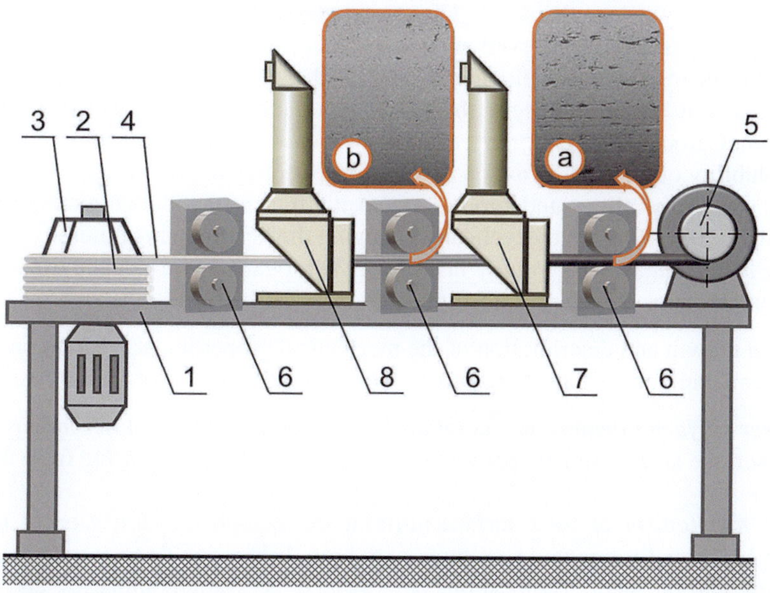

Fig. 6.125 Laser unit for surface treatment of the welding wire (**a** wire surface in its initial state; **b** wire surface after laser cleaning): 1—stand; 2—welding wire cassette; 3—cassette fixing mechanism; 4—welding wire; 5—initial wire coil; 6—guide rollers; 7—laser optical system for cleaning the wire surface; 8—laser optical system for amorphization of the wire surface

6.10 Laser Surface Nitriding

Chemical-thermal treatment (nitriding, carburizing, nitrocarburizing, boriding, silicizing)—changing the chemical composition of the surface layers of metals by diffusion at high temperatures without melting chemical elements (N, C, C + N, B, Si) from a gaseous environment and the formation of functional chemical compounds.

Nitriding is a type of chemical-thermal treatment of metals alloyed with elements that form nitrides with nitrogen (ferritic, martensitic, and austenitic steels and titanium and molybdenum-based alloys). Nitriding provides surface hardening and a significant increase in hardness (up to HV 800–1200 for steels [31]), wear resistance, corrosion resistance, and fatigue strength of parts.

The *thickness of the nitrided layer* is a fraction of a millimeter and depends on the heating temperature, holding time, nature, and degree of alloying.

Ferritic and austenitic high-chromium steels are nitrided at temperatures of 560–600 °C. When the temperature rises above 700–800 °C, nitride particles coagulate, which significantly reduces the thickness of the layer with a high nitrogen content and significantly reduces hardness. The greatest decrease in hardness with increasing nitriding temperature is observed in ferritic steels.

As the degree of steel alloying increases, the thickness of the nitrided layer decreases. For austenitic steels, the maximum thickness of the nitrided layer is 0.15 mm. Carbon steels are practically not strengthened during nitriding due to the low solubility of nitrogen in α-iron at the corresponding temperatures.

For metals and alloys undergoing martensitic transformation, an increase in the diffusion rate is achieved by cyclically changing the heating temperature.

The *disadvantages of complete heating of the product* and prolonged exposure to high temperatures during nitriding are:

- grain growth and deterioration of the mechanical properties of the material;
- a significant increase in energy costs, which affects the cost of the process.

Laser surface nitriding is a combined process that includes: laser *modification* of the surface layers of metal parts (Sect. 2.2) and subsequent *nitriding* from the gas phase.

The ***mechanism*** of laser surface nitriding consists in the formation of highly dispersed, spatially oriented structures with high dislocation density in the surface layer of carbon steel. Dislocations are located at grain boundaries and form "*dislocation tubes*"—paths through which atoms of various chemical elements (N, C, C + N, B, Si) penetrate deep into the metal. Chemical elements form corresponding compounds (nitrides, carbides, borides) with metal ions and are evenly distributed in the surface layer where structural and phase transformations took place.

Laser surface nitriding eliminates the disadvantages of complete heating of the product and prolonged exposure to high temperatures and provides a significant increase in the depth of hardening while increasing the productivity and economic efficiency of the process.

Mathematical modeling of the laser surface nitriding process

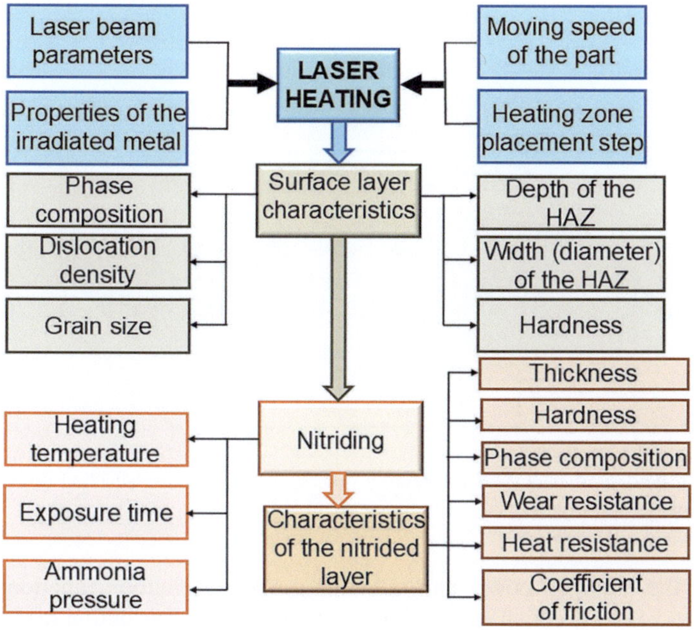

Fig. 6.126 Technological diagram of the laser surface nitriding

The technological diagram of the laser surface nitriding process with processing factors and parameters and operational characteristics of products is shown in Fig. 6.126.

The main *parameters of a laser beam* that significantly affect the processes of structure formation are:

- wavelength of laser radiation;
- area and shape of the laser beam focus zone;
- spatial distribution of radiation intensity across the focus zone;
- for continuous laser radiation—power;
- for pulsed periodic laser radiation—energy, shape, duration, and frequency of pulses.

The initial properties of metals that must be taken into account during laser surface nitriding:

- absorption capacity of laser radiation with the wavelength used;
- melting point;
- density;
- coefficients of thermal conductivity, heat conductivity, and heat capacity;
- initial temperature.

The *temperature* in the heating zone must be higher than the lower limit of structural and phase transformations and lower than the melting point. It is necessary to

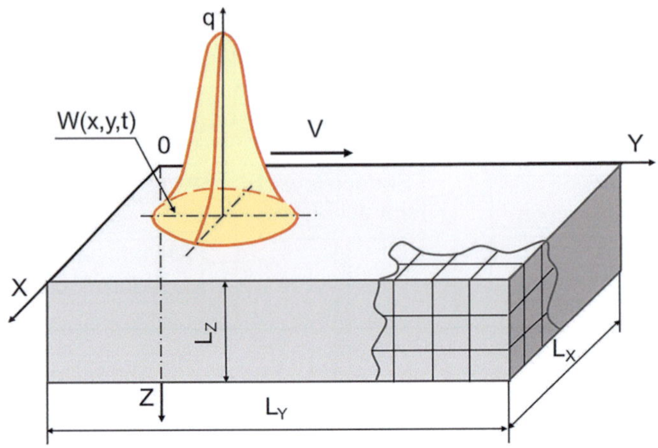

Fig. 6.127 Calculation scheme of the heating process during surface nitriding with continuous laser radiation

determine the radiation power and radiation power distribution functions, as well as the beam velocity, which ensure uniform temperature distribution in the working zone.

The *modeling task* is to determine the temperature distribution in the working zone, which has the shape of a rectangular parallelepiped with sides L_x, L_y, L_z, and an initial temperature T_0 (Fig. 6.127). The surface of the workpiece is irradiated with a laser beam with arbitrary power distribution. The laser beam moves along a free trajectory with a variable speed V.

The distribution of radiation power in the focusing plane is described by the law W(x, y, t). Its action causes the formation of a heat flux on the surface of the part with an intensity distribution q(x, y, t).

The mathematical model is based on a three-dimensional heat conduction equation:

$$c\gamma \frac{\partial T}{\partial t} = \frac{\partial}{\partial x}\left(\lambda \frac{\partial T}{\partial x}\right) + \frac{\partial}{\partial y}\left(\lambda \frac{\partial T}{\partial y}\right) + \frac{\partial}{\partial z}\left(\lambda \frac{\partial T}{\partial z}\right) \tag{6.62}$$

relatively to the unknown temperature T = T(x, y, z, t),

with the initial condition: $T(x, y, z, 0) = T0$,

Boundary conditions on the surface being processed in the laser irradiation zone:

$$\lambda \frac{\partial T(x, y, 0, t)}{\partial z} + q(x, y, t) = 0 \tag{6.63}$$

Outside the laser irradiation zone:

$$\lambda \frac{\partial T(x, y, 0, t)}{\partial z} + \alpha(T(x, y, 0, t) - T_c) = 0 \tag{6.64}$$

Boundary conditions at other boundaries of the calculation domain simulate heat exchange with the surrounding environment:

$$\lambda \frac{\partial T(0, y, z, t)}{\partial x} + \alpha(T(0, y, z, t) - T_c) = 0 \tag{6.65}$$

$$\lambda \frac{\partial T(x, 0, z, t)}{\partial y} + \alpha(T(x, 0, z, t) - T_c) = 0 \tag{6.66}$$

$$\lambda \frac{\partial T(L_x, y, z, t)}{\partial x} + \alpha(T(L_x, y, z, t) - T_c) = 0 \tag{6.67}$$

$$\lambda \frac{\partial T\left(x, L_y, z, t\right)}{\partial y} + \alpha\left(T\left(x, L_y, z, t\right) - T_c\right) = 0 \tag{6.68}$$

$$\lambda \frac{\partial T(x, y, L_z, t)}{\partial z} + \alpha(T(x, y, L_z, t) - T_c) = 0 \tag{6.69}$$

where x, y, z are spatial coordinates;

t is time;

c is heat capacity coefficient of the material of the part;

ρ is metal density;

λ—thermal conductivity coefficient;

q is radiation power density;

α is heat transfer coefficient;

T_0 is initial temperature of the part;

T_c—ambient temperature.

The temperature of each point of the workpiece at any given time is calculated using the finite difference method. To do this, the area under investigation is divided into parallelepipeds with sides L_x, L_y, L_z parallel to the axes Ox, Oy, Oz.

The calculation takes into account the actual spatial–temporal distribution of laser radiation power on the treated surface, measured using an LBA device. The radiation power was assumed to be constant, equal to 1000 W. The absorption capacity of the irradiated surface with a ZnO-based technological coating was 80%. The relative speed of the laser beam and the part varied in the range (0.4–2) m/min, the diameter of the focusing zone $d_0 = 4; 5; 6$ mm. The material of the product is CT80 steel after volume quenching and tempering.

The *thermal state* of the surface layer of the product can be *controlled* over a wide range by changing the speed of the laser beam and the diameter of the focus

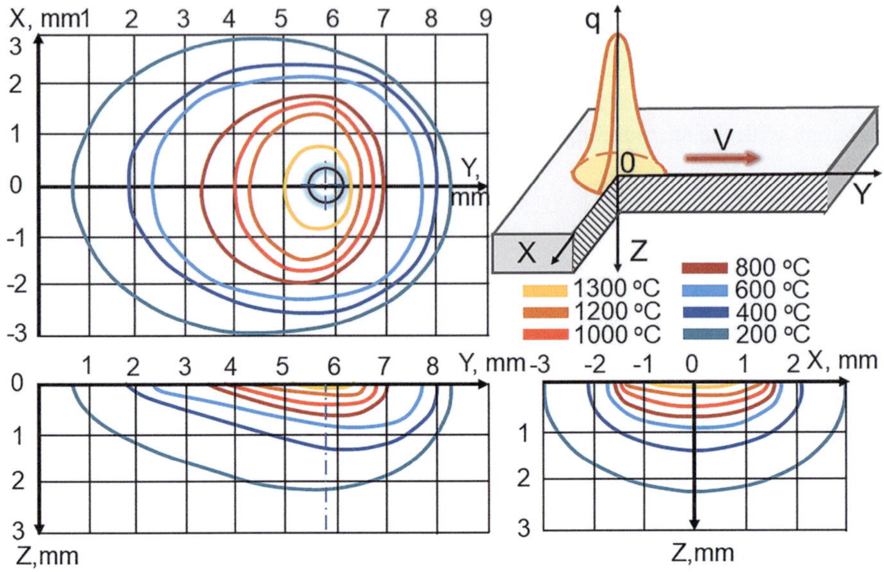

Fig. 6.128 Calculated isotherms during heating with continuous laser radiation of CT80 steel (P = 1000 W; d_0 = 5 mm; V = 0.8 m/min)

zone. When the processing speed is increased from 0.4 to 1.4 m/min, the maximum surface temperature on the laser beam axis decreases from 2084 to 834 °C. At the same time, reducing the laser radiation exposure time from 0.75 to 0.2 s leads to a decrease in the depth and width of the heat-affected zone (limited by an isotherm of 2000 °C) and an increase in the cooling rate from 2000 to 5000 °C/s.

The heating zone, limited by the isotherm 780–800 °C, varies in depth from 0.2 to 0.9 mm and in width from 5 to 8 mm under all conditions of continuous laser irradiation of CT80 steel (Figs. 6.128 and 6.129). The cooling rate in all modes significantly exceeds the critical rate for martensite formation.

Preliminary laser surface hardening

Samples measuring 40 × 40 × 5 mm were made from CT80 carbon steel, austenitic (X10CrNiTi18-10) and martensitic (X40Cr13) stainless steels.

The CT80 steel samples were subjected to standard quenching, low tempering (Fig. 6.130) and grinding. Further chemical oxidation with a 3 μm thick Fe_3O_4 coating at a temperature of 1000–1200 °C provides a surface absorption capacity of 80% for laser radiation with a wavelength of 10.6 μm [29].

Samples made of X10CrNiTi18-10 and X40Cr13 steels were coated with a 20 μm thick ZnO-based coating to increase the absorption capacity to 80%.

Parameters of laser radiation (Fig. 6.131):

- modes—continuous and pulsed;
- laser radiation power P = 1200 W;

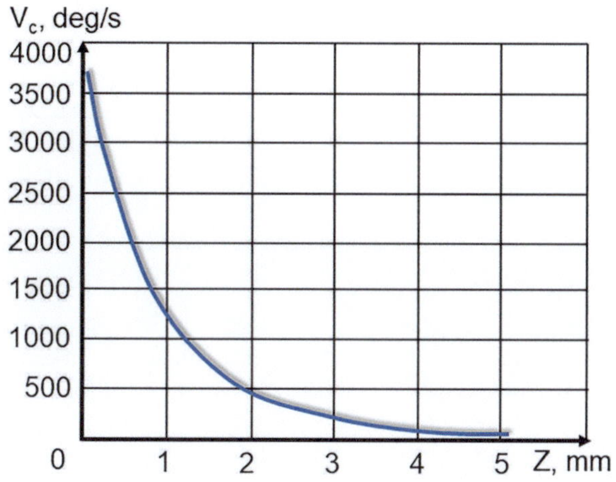

Fig. 6.129 Distribution of cooling rates V_c across depth in CT80 steel

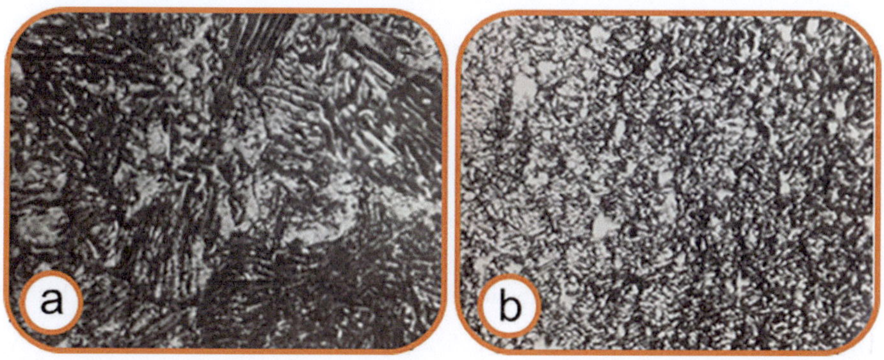

Fig. 6.130 Microstructure of CT80 steel (×500): **a** as supplied; **b** after standard quenching and low tempering

- focal length F = 300 mm;
- beam diameter in the focal plane d = 5 mm;
- laser beam speed V = 0.5; 0.8; 1.2; 1.4 m/min;
- pulse duration $\tau = 0.1$–0.01 s, processing zone step along the X and Y axes is $S_x = S_y = 7$ mm (when processing with pulsed radiation).

The *heat-affected zone* of CT80 steel, hardened by continuous laser irradiation with surface melting ($W_p = 4 \times 10^3$ W/cm^2; $V \leq 0.6$ m/min), has *three* layers:

(a) The *surface molten* layer consists of dispersed rounded grains with a martensitic-austenitic structure and a microhardness of 9500–10,500 MPa. The presence of residual austenite is clearly evident. As the speed decreases, the amount of residual austenite increases and reaches 35% at V = 0.2 m/min.

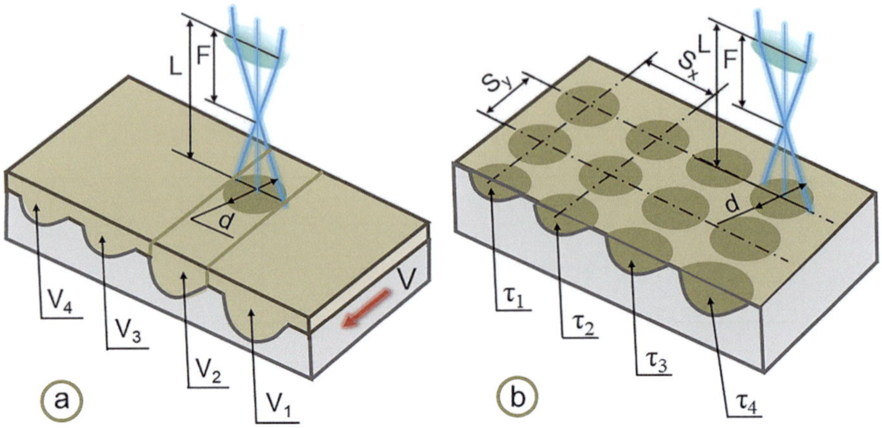

Fig. 6.131 Scheme of laser surface nitriding: **a** continuous laser radiation ($V_1 = 0.5$ m/min; $V_2 = 0.8$ m/min; $V_3 = 1.2$ m/min; $V_4 = 1.4$ m/min); **b** pulsed laser radiation ($\tau_1 = 0.01$ s; $\tau_2 = 0.04$ s; $\tau_3 = 0.07$ s; $\tau_4 = 0.1$ s)

(b) The *near-surface* layer, where the heating temperature does not exceed the critical point A_1, is characterized by high structural dispersion and an increase in microhardness to 11,000–11,500 Mpa.

(c) The *transition* layer, where austenite homogenization does not have time to occur and, along with high-carbon martensite, martensite with a low carbon content is observed. Microhardness decreases to the initial value.

The *heat-affected zone* of CT80 steel with an initial lamellar pearlite structure, hardened by continuous laser irradiation without melting the surface, has *two* layers:

(a) The *surface* layer is formed in the heating zone at temperatures above the A_1 point, but not higher than the melting point. Structural and phase transformations occur in the solid state. Highly dispersed martensite, residual austenite, and indistinct troostite are formed along the former boundaries of austenite grains (Fig. 6.132a). Microhardness is 11000–12,000 MPa. As the laser irradiation rate increases, the dispersion of the structure increases and the content of residual austenite increases.

(b) The *transition* layer is located at the boundary with the base. The microstructure changes smoothly from martensitic-troostitic to the initial pearlitic (Fig. 6.132b).

The *heat-affected zone* of CT80 steel with an initial martensitic-troostitic structure, hardened by continuous laser irradiation *without melting* the surface, has *two* layers:

(a) The *surface* layer is formed in the heating zone at temperatures above the A_1 point, but not higher than the melting point. Structural and phase transformations occur in the solid state. A microstructure of weakly etched medium-needle martensite is formed (Fig. 6.133a) with a gradual increase in the content of residual austenite and refinement of martensite needles in depth.

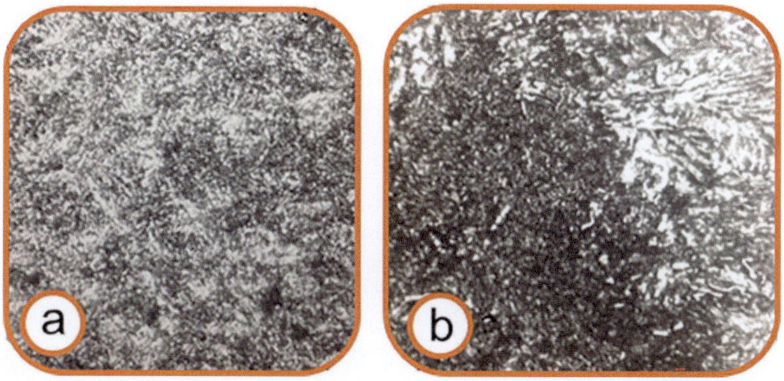

Fig. 6.132 Microstructure of the heat-affected zone of CT80 steel with an initial lamellar pearlite structure, hardened by continuous laser irradiation in modes that do not cause surface melting (× 500): **a** surface layer; **b** transition layer

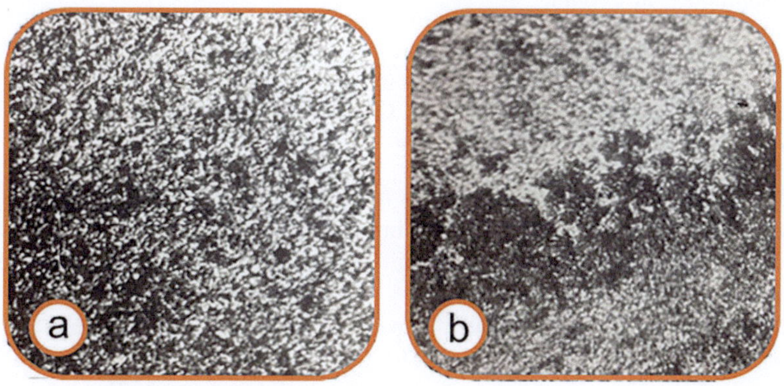

Fig. 6.133 Microstructure of the heat-affected zone of CT80 steel with an initial martensitic-troostitic structure, hardened by continuous laser irradiation in modes that do not cause surface melting (×500): **a** surface layer; **b** transition layer

(b) The *transition* layer at the boundary with the base is a tempering zone and has a troostite structure with cementite-type carbides. The microstructure changes smoothly from a troostite-cementite structure to the martensite-troostite structure of the base (Fig. 6.133b).

Preliminary laser surface hardening creates conditions for laser surface nitriding by forming microstructures with high dislocation density and an increased number of barriers to their movement. Dislocations connect with each other to form "channels" through which nitrogen can diffuse.

Laser surface nitriding

Gas nitriding *parameters* after preliminary laser surface hardening:

Fig. 6.134 Microstructure of CT80 steel after laser hardening at a speed of 0.8 m/min and subsequent gas nitriding (×500)

- environment—ammonia;
- temperature T = 800–860 K;
- exposure time t = 15 h.

Features of the *heat-affected zone* of CT80 steel after laser hardening without melting and subsequent gas *nitriding* (Fig. 6.134):

(a) The *surface* layer is characterized by a significantly fragmented structure, which causes a significant increase in the length of grain boundaries and their surface area, thus accelerating the diffusion of nitrogen atoms and increasing microhardness to 8800 MPa.
(b) The *transition* layer has lower dispersion and microhardness and a clearly defined boundary with the base. The thickness of the layer is uneven in the center of the laser heating zone and in its peripheral areas. The use of either focused laser beams in scanning mode or laser beams with a "Top Had" type radiation intensity distribution allows the formation of a transition layer with a rectangular cross-section and thus increases the efficiency of the laser nitriding process.

Table 6.7 shows the characteristics of the surface layer of CT80 steel after nitriding compared to the combined process of preliminary laser treatment followed by nitriding.

Features of the phase composition of the nitrided layer:

(a) In case of nitriding without laser processing, oxides and carbides such as FeC, Fe_3C (orthorhombic lattice), Fe_2C (hexagonal lattice) and a small amount of nitrides Fe_3N, Fe_2N, Fe_4N, α-Fe ($Fe_{2\text{-}3}N$) are fixed in the surface layer.
(b) Laser processing and subsequent nitriding practically do not change the qualitative phase composition of the surface layer. At the same time, the intensity of the formation of iron nitrides Fe_3N, Fe_2N, Fe_4N, α-Fe ($Fe_{2\text{-}3}N$) increases, which indicates an increase in nitrogen concentration. The ε-phase component

Table 6.7 Characteristics of the nitrided surface layer depending on the type of treatment and the speed of laser beam movement

Characteristics of the nitrided layer	Nitriding without laser processing	Nitriding after laser processing at different laser beam speeds V, m/min			
		V = 0.5	V = 0.85	V = 1.2	V = 1.4
Layer thickness, mm	0.115	0.461	0.200	0.190	0.120
Microhardness, MPa	3500	7300	8700	7300	7000
Nitrogen content, %	0.071	0.158	0.150	0.143	0.143
Phase composition of the diffusion layer	α-Fe_2O_3; γ-Fe_2O_3; Fe_2O_3;FeC; Fe_3C; Fe_2C; FeO; Fe_3N; Fe_2N; Fe_4N; ε-Fe	Fe_3O_4; Fe_2O_3; FeO; α-Fe_2O_3; γ-Fe_2O_3; FeO; Fe_2C; Fe_3C; FeC; Fe_2N; Fe_3N; Fe_4N; α- Fe; ε-Fe	FeO; Fe_2O_3; α-Fe_2O_3; γ-Fe_2O_3; Fe_2O_3; FeC; Fe_2N; Fe_3C; Fe_5C_2; α-Fe; ε-Fe	α-Fe_2O_3; γ-Fe_2O_3; Fe_2O_3; FeC; Fe_3C; Fe_2C; FeO; Fe_3N; Fe_2N; Fe_4N; ε-Fe	Fe_3O_4; Fe_2O_3; FeO; α-Fe_2O_3; γ-Fe_2O_3; FeO; Fe_2C; Fe_3C; FeC; Fe_2N; Fe_3N; Fe_4N; α- Fe; ε-Fe

is a solid solution based on $Fe_{2\text{-}3}N$ nitride and is present in a wide range of nitrogen concentrations (4.55–11.0%). The basis of the ε-phase crystal structure is a dense hexagonal packing of iron atoms. Nitrogen atoms are located in pores that form a graphite-type lattice. The ε-phase has increased wear resistance and high corrosion resistance. As the distance from the surface increases, the hardness of the layer gradually decreases to the level of the initial matrix. The effective depth of the layer after processing at a laser beam travel speed of 0.8 m/min is 0.45 mm (Fig. 6.135).

Preliminary *laser processing* followed by *nitriding* allows:

(a) increase the depth of nitriding up to 5 times;

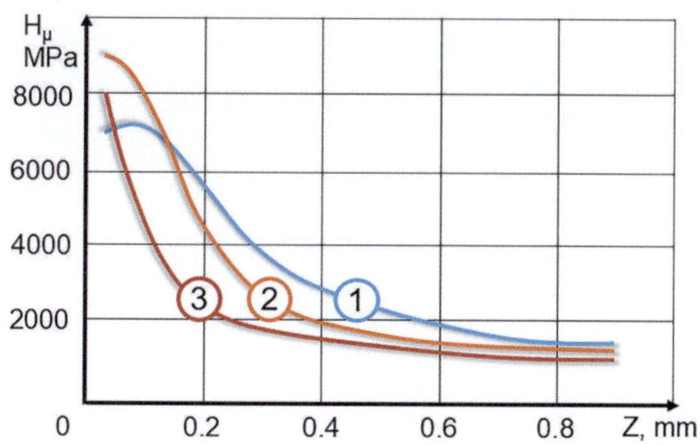

Fig. 6.135 Distribution of microhardness H_μ along the depth z of the nitrided layer after laser treatment at different laser beam speeds: 1—V = 0.5 m/min; 2—V = 0.81 m/min; 3—V = 1.41 m/min

(b) purposefully change the depth of nitriding within a wide range from 0.12 to 0.46 mm;
(c) obtain high microhardness of the nitrided layer (up to 7000–8700 MPa—Fig. 6.135), which is maintained when heated to temperatures of 550–600 °C;
(d) control the wear process of parts by changing the ratio of areas and the topography of nitrided areas on friction surfaces.

Nitriding with preliminary laser processing of *X10CrNiTi18-10* and *X40Cr13* steels does not change the phase composition of the nitrided layer compared to nitriding without laser processing. The difference lies only in the percentage content of the main phases. The basis consists of two structural phases: α-Fe (BCC) and $Fe_{24}N_{10}$ (hexagonal).

Without preliminary laser processing, the ε-phase (hexagonal nitride Fe_3N) predominates in the nitride layer.

After laser processing, the basis of the nitride layer is the γ'-phase (cubic nitride Fe_4N). In addition, there is a tendency for the interference lines of the α-phase to expand. Such changes in the structural state of the surface not only reduce the amount of brittle and porous ε-phase and increase the concentration of nitrogen in the α-N and γ'- phases, but also significantly accelerate diffusion processes.

The microhardness of the nitrided layer with a depth of 0.1 mm of X10CrNiTi18-10 steel was 12,000 MPa.

The microhardness of the nitrided layer with a depth of 0.21 mm of X40Cr13 steel was 3700 MPa.

References

1. O. Kapustynskyi, N. Visniakov, Laser Treatment for strengthening of thin sheet steel. Adv. Mater. Eng. **43**, 1–13. https://doi.org/10.1155/2020/59630/2
2. O. Kapustynskyi, N. Visniakov, O. Iernasejus, L. Golovko, V. Chayeuski, Optimization of the parameters of local laser treatment for the creation of reinforcing ribs in thin metal sheets, in *Proceedings of 24th International Conference "MECHANIKA 2019"* (2019), pp. 71–75
3. O. Kagljak, B. Romanov, K. Romanova, *Repeatability of Sheet Material Formation Results and Interchangeability of Processing Modes at Multi-Pass Laser Formation*, vol. 1038 (Trans Tech Publications Ltd., Switzerland, 2021), pp. 15–24. ISSN: 1662-9752. https://doi.org/10.4028/www.scientific.net/MSF.1038.15
4. O. Kagljak, B. Romanov, K. Romanova, *Stress-Strain State and Structural-Phase Transformations of Metal. Articles Obtained by Laser Shaping*, vol. 1038 (Trans Tech Publications Ltd., Switzerland, 2021), pp. 25–32. ISSN: 1662-9752. https://doi.org/10.4028/www.scientific.net/MSF.1038.25
5. E.-S. Salah, R. Mostafa, M.M. Tawfik, M. Dewidar, Laser forming technology: a comprehensive review of mechanisms, process optimization, and industrial applications. Int. J. Mater. Forming **18**, article number 82 (2025)
6. T. Pretorius, Laser forming, in *The Theory of Laser Materials Processing* (SSMATERIALS, т. 119, 2017), pp. 307–340
7. M. Safari, R.A. de Sousa, J. Joudaki, Fabrication of saddle-shaped surfaces by a laser forming process: an experimental and statistical investigation. Met (Basel) **10**, 1–13 (2020). https://doi.org/10.3390/met10070883

8. Z. Hu, H. Wang, Theoretical and Experimental investigation of sheet metal bending. Int J Mach Tools Manuf **21**, 332–347
9. J. Magee, K. Watkins, W.M. Steen, Advances in laser forming. J. Laser Appl. **10**, 235–246 (1998)
10. J. Chen, *Modelling of simultaneous transformations in steels*, PhD thesis, Department of Materials Science and Metallurgy University of Cambridge England (2009)
11. O. Kapustynskyi, L. Golovko, Effects of laser cladding and treatment methods on wear resistance in heavy-loaded units. Int. J. Multidiscip. Res. **7**(1), 17–26 (2024). eISSN 2783-0179. https://doi.org/10.53430/ijmru.2024.7.1.0029
12. D. Lesyk, M. Hruska, V. Dzhemelinskyi, O. Danyleiko, M. Honner, Selective surface modification of complexly shaped steel parts by robot-assisted 3D scanning laser hardening system, in *Lecture Notes in Networks and System. New Technologies, Development and Application V*, vol. 472 (Springer International Publishing, Switzerland, 2022), pp. 30–36. https://doi.org/10.1007/978-3-031-05230-9_3
13. Q. Yuan, L. Chai, J. Shen, H. Wang, H. Guan, N. Guo, Y. Li, Microstructural characteristics, hardness and wear resistance of a typical ferritic/martensitic steel surface-treated by pulsed laser. Surf. Coat. Technol. **418**, 127261 (2021). https://doi.org/10.1016/j.surfcoat.2021.127261
14. S., Li, X. Dong, S. Guo, X. Liu, H. Cao, Status and perspective of high-energy beam surface strengthening: high-speed steel. Materials **15**. https://doi.org/10.3390/ma15176129
15. V. Kovalenko, L. Golovko, N. Fialko, Laser hardening of components and tools with complicated working elements shaped for heat transfer, in *ICALEO'93 Orlando, Florida*, October 28, 1993
16. L. Golovko, O. Kaglyak, V. Romanenko, M. Bloshchytsyn, Application of laser thermocycling to improve the quality of plasma wear-resistant coatings. Mech. Adv. Technol. **3**, 328–341. https://doi.org/10.20535/2521-1943.2022.6.3.268173
17. M. Lu, H. Wang, X. Song et al., Fabrication and grinding performance of CVD diamond abrasive tool. Proc IMechE, Part B: J. Eng. Manuf. **237**, 192–202 (2023)
18. J. Wu, S. Zhang, F. Qu et al., Matrix material for a new 3D-printed diamond-impregnated bit with grid-shaped matrix. Int. J. Refract. Met. Hard. Mater. **82**, 199–207 (2019)
19. O. Goncharuk, O. Kaglyak, D. Lesyk, L. Golovko, S. Voloshko, O. Savchenko, Laser synthesis of CBN-contained abrasive layers, in *Welding and Related Technologies: Proceedings of the VIIth International Conference on Welding and Related Technologies*, vol. 1 (2025), pp. 88–91. https://doi.org/10.1201/9781003518518
20. O. Kaglyak, L. Golovko, O. Goncharuk, S. Voloshko, O. Kapustynskyi, N. Visniakov, Application of laser thermal deformation sintering in the manufacture of drum-type diamond tools. J. Manuf. Mater. Process. **9**(8), 1–22 (2025). eISSN 2504-4494. https://doi.org/10.3390/jmmp9080251
21. O. Kagljak, O. Goncharuk, R. Zhuk, V. Dzhemelinskyi, D. Lesyk, Laser sintering of abrasive layers with inclusions of cubic boron nitride grains, *in Lasers in Manufacturing and Materials Processing* (2018). ISSN 2196-7229. https://doi.org/10.1007/s40516-018-0068-0
22. O. Goncharuk, L. Golovko, S. Voloshko, O. Kaglyak, A. Burmak, Y. Kliuchnikov, Mathematical modeling of laser synthesis of the working layer of an abrasive tool. Tekhnolohiia I Tekhnika Druk **3**, 51–68 (2024)
23. I. Ryabtsev, S. Fomichov, V. Kuznetsov, Y. Chvertko, et al., Surfacing and additive technologies in welded fabrication, in *International Welding Engineers Textbook Series*, ed. by B. Paton (Springer, 2023), 229p. ISBN 978-3-031-34389-6. https://doi.org/10.1007/978-3-031-34390-2
24. L. Golovko, S. Salii, M. Bloshchytsyn, W. Alnusirat, Development of the laserfoundry process for manufacture of bymetalls. East.-Europ. J. Enterprise Technol. **4/1** (94), 47–54 (2018)
25. S. Salii, L. Golovko, V. Romanenko, A. Golovko, Combined laser-foundry manufacturing process of bimetals. Mech. Adv. Technol. **88**(1), 93–107
26. S. Salii, L. Golovko, O. Kaglyak, O. Kapustynskyi, N. Visniakov, The formation of a low-carbon steel/Ni-Cr-W alloy bimetallic material via liquid–solid compound casting with a laser assisted solid surface. Machines **12**, 522. https://doi.org/10.3390/machines12080522

27. S. Wang, *Laser Cleaning. Key Technology of Substrate Surface Morphology and Performance Control* (Springer, Switzerland, 2025), 351p. ISBN 978-981-95-2934-6
28. F. Song, X. Lin, *Laser Cleaning. Fundamentals and Applications* (Springer, Switzerland, 2024), 337p. ISBN 978-981-97-6630-7
29. L. Golovko, V. Roman, M. Bloshchytsyn, E. Omidvary, Possibilities of controlling the formation of a reinforced layer in wedge-shaped products by changing the conditions of laser irradiation. Probl. Tribol. **2**, 105–110
30. V. Kotlyarov, V. Dubnyuk, Design algorithm and technological support for laser surface hardening operations of cutting tools. Bull. NTUU "KPI". Mech. Eng. Ser. **3**(72), 161–168 (2014)
31. L. Golovko, S. Lukyanenko, *Laser Technologies and Computer Modeling*. "Vistka", Kyiv (2009), 296p

Chapter 7
Laser Welding

Abstract The scheme and physical phenomena are presented for the processes in laser welding that lead to a reduction in the width of the heat-affected zone (structural heterogeneity decreases), a reduction in the width of the elastic–plastic zone (shrinkage force decreases), minimization of the time the metal remains at a temperature above A_1 (conditions for diffusion transformations disappear), an increase in the cooling rate in the weld and HAZ (the grain in the weld becomes finer, conditions for carbide formation disappear), improvement of the microstructure and mechanical properties of the welded joint, reduction of residual stresses and deformations, disappearance of conditions for the formation of hot and cold cracks, disappearance of imperfections in geometry of welded joints, increase in corrosion resistance, including resistance to intergranular corrosion. The energy coefficients of absorption and reflection of electromagnetic radiation have been calculated, ensuring conditions for maximum absorption of laser radiation by the weld metal and, consequently, maximum efficiency of laser welding, taking into account the influence of the plasma torch. Technological recommendations are provided for determining the parameters of the laser welding: laser power, laser radiation intensity, wavelength, welding speed, focal length, focal plane depth, focused beam diameter, laser beam deflection angle, protective gas composition and consumption. A diagram is provided, the features of the process are analyzed, and general technological recommendations for hybrid laser-plasma surfacing are given.

7.1 Features of Physical Processes

Works [1, 2] present the fundamental principles, research, and technologies of laser cleaning. This section considers the formation of the heat affected zone, develops a mathematical model of the process, and provides examples of laser welding technology.

Laser welding is carried out with lasers with a power of 1.0–25.0 kW and higher in the environment of shielding gases He, Ar, Ar + He mixture, CO_2, N_2 (Fig. 7.1).

I. Krivtsun et al., *Laser and Hybrid Technologies for Surface Engineering and Welding*, Springer Series in Materials Science 358, https://doi.org/10.1007/978-3-032-15038-7_7

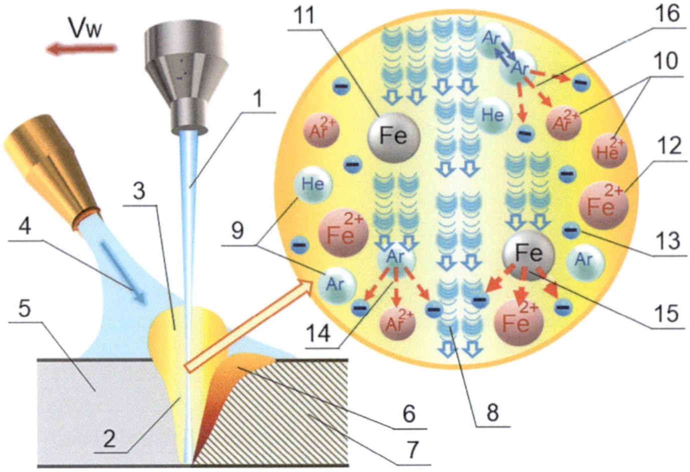

Fig. 7.1 Scheme of laser welding: 1—laser beam; 2—vapor–gas channel; 3—plasma torch; 4—shielding gas; 5—base metal; 6—welding pool; 7—weld; 8—photons; 9—shielding gas atoms (He + Ar); 10—positive ions of shielding gas (He + Ar); 11—metal atom (vapor); 12—metal ion; 13—free electrons; 14—photoionization of argon atoms by a laser beam; 15—photoionization of metal atoms by a laser beam; 16—thermal ionization of argon atoms

When a metal is melted by a laser beam, a vapor–gas channel 2 is formed from photons 8, shielding gas (argon) atoms 9, and metal vapors (atoms) 11.

As the laser intensity increases (above 10^6 W/cm^2), shielding gas atoms and metal atoms are ionized by photons 14 and 15. In addition, due to the high temperature, argon atoms 16 are thermally ionized. The vapor–gas channel is converted into a hybrid laser-plasma process.

The formation of plasma in metal vapors leads to an increase in volume and pressure in the vapor–gas channel, which can initiate a metal splash. To prevent the formation of plasma during welding with high-power lasers (e.g., pulsed deep penetration welding), shielding gases with high ionization energy (helium and mixtures) are used.

Above the welding pool, the vapor–gas channel turns into a plasma torch 3, which is characterized by a lower density of constituent particles and temperature. The presence of a plasma torch reduces efficiency because it absorbs laser radiation outside the welding zone.

Laser welding is characterized by a *"keyhole" penetration* with a small width and a large depth (Fig. 7.2). Thus, the volume of the welding pool is several times smaller, which reduces shrinkage and, consequently, contraction deformations by up to 10 times and eliminates angular deformations.

The *distribution and change of temperatures* during welding is characterized by the following (Fig. 7.3):

(a) The ***width of the heating zone***, where the maximum temperatures are higher than A_1, decreases, which means that the *width of the heat-affected zone decreases*

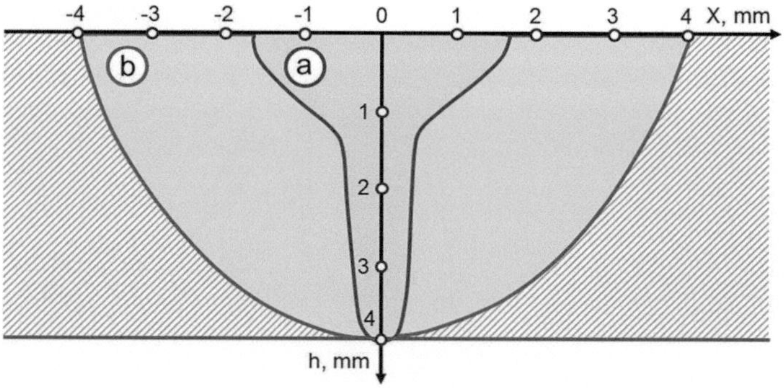

Fig. 7.2 Weld penetration pattern for 4 mm thick steel: **a** laser welding (power $Q^0 = 5.0$ kW); **b** TIG electric arc welding

(structural heterogeneity decreases) and the *width of the elastic–plastic zone decreases* (shrinkage force decreases).

(b) The ***thermal cycles of welding*** are "flattened" (due to the increased cooling rate), therefore:

- the *time* the metal stays at a temperature above A_1 *is minimized*, thus eliminating the conditions for diffusion transformations;
- the *cooling rate* in the weld and the weld zone *increases*, resulting in a *finer grain* in the weld, and the conditions for carbide formation, which is the main cause of *intergranular corrosion* of stainless steels, disappear [3].

(c) As a result:

- the *microstructure* of the weld and the heat-affected zone improve, and thus the mechanical properties of the welded joint improve;
- residual *stresses and deformations* are significantly reduced;
- conditions for the formation of *hot cracks* in welded joints of austenitic steels and heat-resistant nickel alloys are eliminated, and corrosion resistance is increased;
- susceptibility for *cold crack* formation in hardened steels reduces;
- *imperfections in the shape* of welded joints practically disappear [4];
- resistance to *mechanical and chemical corrosion* increases [3].

7.2 Mathematical Modeling of Processes in Laser Welding

The ***object*** of mathematical modeling is the conditions for maximum absorption of laser radiation by the metal being welded, and hence maximum efficiency of laser welding, taking into account the influence of the plasma torch.

The ***purpose*** of the mathematical model is:

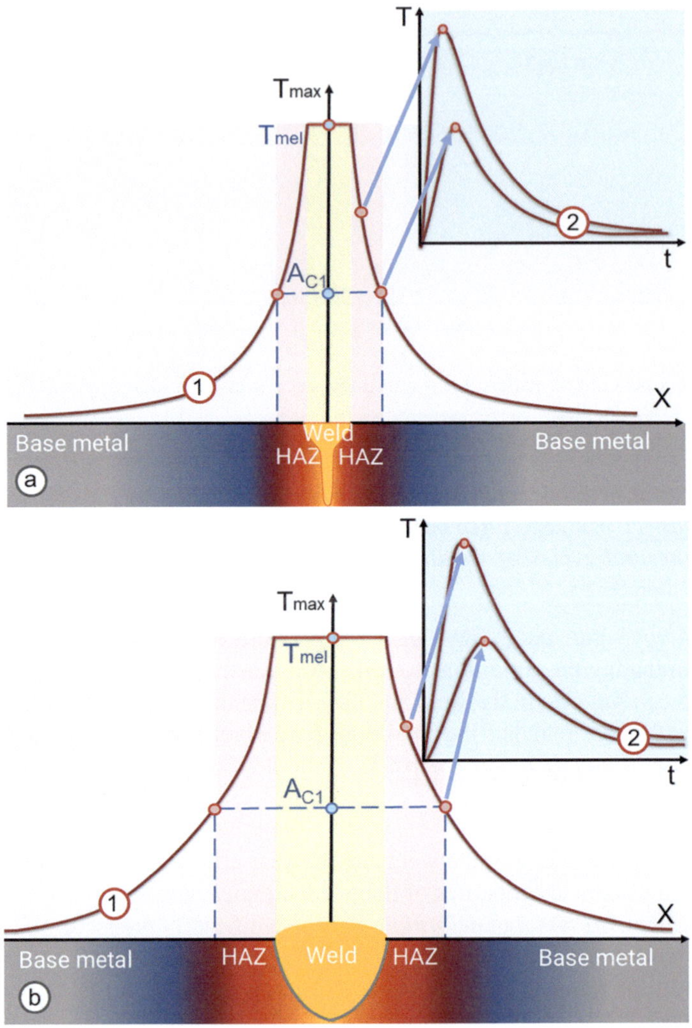

Fig. 7.3 Distribution of maximum heating temperatures T_{max} depending on the distance X to the weld axis (1) and examples of thermal cycles of welding two points of the heat-affected zone (2): **a** for laser welding; **b** for TIG welding with filler wire

to *obtain* the desired ***output*** parameters: energy coefficients of absorption and reflection of electromagnetic radiation;

with the required ***input*** parameters: physical properties of the laser beam, polarization and spatial distribution of laser intensity.

The mathematical model is based on [5].

The *following assumption* are made: the electromagnetic field of the laser beam is transverse and homogeneously polarized (linearly or circularly).

In the normal incidence of a Gaussian beam on a semi-infinite metal with a near-surface layer (– L < z < 0) of inhomogeneous plasma, the electric field strength Ē of the incident wave is:

$$\overline{E}^{inc}(\bar{r},t) = \overline{E}^{inc}_{0\omega} \cdot \left(\frac{w_0}{w(z)}\right) \cdot e^{-(x^2+y^2)\left(\frac{1}{w^2(z)} - \frac{i\cdot\bar{k}}{2R(z)}\right)} \cdot e^{i[\cdot\bar{k}\cdot(z-F)-\eta(z)-\omega t]} \tag{7.1}$$

where:

$w^2(z) = w_0^2\left(1 + \frac{2(z-F)}{\bar{k}w_0^2}\right);$

$R(z) = (z-F)\left(1 + \left(\frac{\bar{k}w_0^2}{2(z-F)}\right)^2\right);$

$\eta(z) = \text{arctg}\left(\frac{2(z-F)}{\bar{k}w_0^2}\right);$

$\bar{k} = \frac{\omega\sqrt{\varepsilon}}{c};$

$w_0 = D_F/2$ – effective radius of the beam in the focal plane (z = F);

$\overline{E}^{inc}_{0\omega}$ is the amplitude of the field intensity of the incident wave;

D_F is the diameter of the focused laser beam;

ε is the dielectric constant of the external environment (z < – L).

Expression (7.1) of the electric field is a superposition of plane electromagnetic waves incident on the boundary of a layer-by-layer inhomogeneous plasma-metal system. Thus, the equations for calculating the energy coefficients of *absorption* $\Gamma(w_0)$ and *reflection* $R(w_0)$ of electromagnetic radiation ***by metal*** are as follows:

$$\Gamma(w_0) = \frac{\int_0^{\bar{k}2} dp^2\left(\frac{\Gamma_p(p^2)\cdot\bar{k}}{\bar{k}_z(p^2)} + \frac{\Gamma_s(p^2)\bar{k}_z(p^2)}{\bar{k}}\right)\cdot e^{-\psi}}{\int_0^{\bar{k}2} dp^2\left(\frac{\bar{k}}{\bar{k}_z(p^2)} + \frac{\bar{k}_z(p^2)}{\bar{k}}\right)\cdot e^{-\psi}} \tag{7.2}$$

$$R(w_0) = \frac{\int_0^{\bar{k}2} dp^2\left(\frac{R_p(p^2)\cdot\bar{k}}{\bar{k}_z(p^2)} + \frac{R_s(p^2)\bar{k}_z(p^2)}{\bar{k}}\right)\cdot e^{-\psi}}{\int_0^{\bar{k}2} dp^2\left(\frac{\bar{k}}{\bar{k}_z(p^2)} + \frac{\bar{k}_z(p^2)}{\bar{k}}\right)\cdot e^{-\psi}} \tag{7.3}$$

where:

$\psi = p^2 w_0^2/2$

$\bar{k}_z(p^2) = \sqrt{\frac{\omega^2\varepsilon}{c^2} - p^2}$

Γ_p is the absorption coefficient of p-polarized plane electromagnetic waves for a layer-by-layer inhomogeneous plasma-metal system;

Γ_s is the absorption coefficient of s-polarized plane electromagnetic waves for a layer-by-layer inhomogeneous plasma-metal system;

R_p is the reflection coefficient of p-polarized plane electromagnetic waves for a layer-by-layer inhomogeneous plasma-metal system;

R_s is the reflection coefficient of s-polarized plane electromagnetic waves for a layer-by-layer inhomogeneous plasma-metal system.

The energy *absorption* coefficient *$D(w_0)$* of electromagnetic radiation ***by a plasma layer*** is determined from the following condition:

$$R(w_0) + D(w_0) + \Gamma(w_0) = 1 \tag{7.4}$$

The analysis of Eqs. (7.2), (7.3), (7.4) shows that:

(1) The appearance of a plasma torch above the surface of the metal being welded leads to a significant reduction in the energy absorbed by the metal, and therefore to a significant decrease in efficiency.
(2) A plasma torch is a much more efficient energy absorber than metal. For example, during laser welding of aluminum in argon ($T_{me} = 933.5$ K, $T_{pl} = 15{,}000$ K, $L = 10$ mm), the plasma torch absorbs 75.3% of the energy, while the metal absorbs 7.0%. Even in the absence of a plasma torch, the metal absorbs only 15.5% of the laser beam energy. Thus, the way to increase the efficiency of laser welding is to use the radiation energy absorbed by the plasma. This is utilized in hybrid laser-plasma welding.

7.3 Technology of Laser Welding

Parameters of laser welding (Fig. 7.4):

- laser power Q^0 (kW);
- laser radiation intensity P (kW/mm^2);
- wavelength λ (μm);
- welding speed V_w (m/h);
- focal length F (mm);
- focal plane depression ΔF (mm);
- diameter of the focused beam D_F (mm);
- laser beam deflection angle Θ (angular degrees);
- composition and flow of shielding gas C_P (l/min).

Power Q^0 is the amount of energy produced by the laser per unit of time.

The laser power is selected depending on the:

- composition and properties of the metal being welded—the power required increases in proportion to the boiling point (or melting point) of the metal;

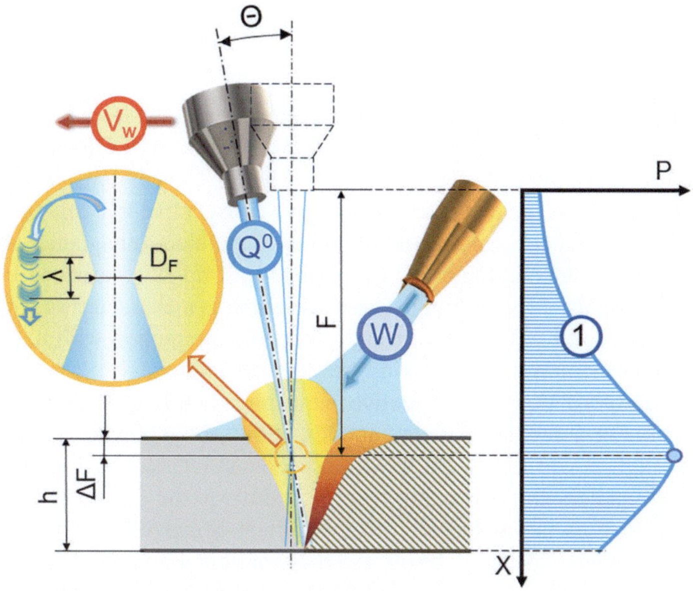

Fig. 7.4 Laser welding parameters (1—distribution of laser radiation intensity P along the longitudinal axis X of the laser beam)

- *thickness* of the metal to be welded *h (mm)*—the required power increases with increasing thickness according to a dependence close to linear (Fig. 7.5).

The range of thicknesses for laser welding is quite wide—from several micrometers to tens of millimeters.

Fig. 7.5 Dependence of the required laser power on the thickness of the metal to be welded in the welding speed range 60 m/h (17 mm/s) to 300 m/h (84 mm/s)

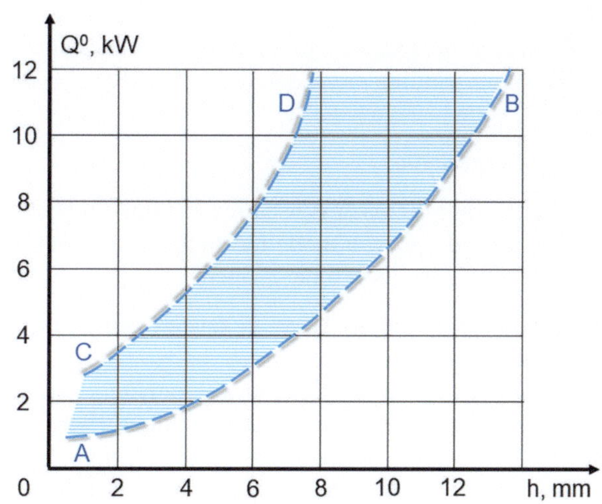

Increasing the laser *power* leads to deepening of the vapor–gas channel, reducing losses due to beam energy reflection and, consequently, increasing the maximum *efficiency*.

Intensity P is the power per unit cross-sectional area of the laser beam ($P = 4Q^0/(\pi D^2)$). The maximum intensity of laser radiation occurs in the focal plane of the laser beam (curve 1 in Fig. 7.4).

The ***wavelength λ*** is the distance between two consecutive maxima or minima of the intensity of the electric and magnetic fields of the electromagnetic wave of the laser beam. Wavelengths in the range from 1 μm in solid-state lasers to 10.6 μm in CO_2 gas lasers are used for welding. The wavelength is a characteristic of the laser, it determines the color of the laser beam, affects the interaction with materials, but is not taken into account when selecting welding parameters.

The ***welding speed*** V_w is most effective in the range of 60 m/h (17 mm/s) to 160 m/h (45 mm/s). In Fig. 7.5, curve AB corresponds to low (about 60 m/h) welding speeds. When using high-power lasers, the welding speed can increase to 300 m/h (84 mm/s), while the thickness-to-power ratio (Fig. 7.5) shifts to the CD position.

Focal length F is the distance from the optical center of the focusing lens to the focal plane where the laser beam has the smallest diameter. The focal length range for laser welding is from 30 to 500 mm. The most widely used focal length is 100–250 mm.

Maximum radiation absorption and, therefore, maximum efficiency is achieved when the ***focal plane is buried*** by $\Delta F = -1$ mm below the surface of the workpiece. This explains the expansion of the upper part of the penetration zone (Fig. 7.2a).

The ***diameter of the focused beam*** D_F is determined in the range of 0.5–1.0 mm. A smaller diameter of the focused laser beam is achieved by reducing the focal length.

The ***laser beam deflection angle*** $\Theta = 10°$ when welding in the "*backward angle*" position increases efficiency by 30%. The effect of the beam deflection is explained by a decrease in the shielding effect of the vapor–gas torch, a more even distribution of beam energy in the weld pool channel, and a decrease in the effect of reflected radiation on the operation of the quantum generator.

The laser welding ***mode*** can be:

(a) continuous;
(b) pulse-periodic.

The *continuous mode* is used for welding thicknesses over 1 mm at *high speeds*.

The *pulsed-periodic* mode has better heat input control capabilities and higher melting efficiency (3–4 times), since:

- the power of pulsed lasers is several orders of magnitude higher than that of continuous lasers;
- pulse-periodic mode has more parameters; during the pause between pulses, cooling occurs, which reduces heat input when welding small thicknesses;
- during the pause between pulses, the plasma torch is partially dissipated (which has a negative effect because it absorbs laser radiation outside the welding zone).

Additional welding *parameters* in relation to the *pulse-periodic* mode are:

- ***pulse frequency*** f (Hz)—the number of pulses emitted per unit time;
- ***pulse duration τ*** (s) is the time interval during which the laser emits a pulse $\tau = 1/f$;
- ***pulse energy*** E (J) – $E = Q^0 \tau$.

The pulsed-periodic mode is used for welding in a wide range of thicknesses (from several micrometers to tens of millimeters) *at low speeds* (about 60 m/h).

The focal length is reduced to F = 30–100 mm.

Pulse frequency is f = 20–200 Hz.

Pulse duration: for steel $\tau = 5 \times 10^{-3}$ to 8×10^{-3} s, for copper $\tau = 10^{-4}$ to 5×10^{-4} s, for aluminum $\tau = 5 \times 10^{-4}$ to 2×10^{-3} s.

The ***shielding gas*** for laser welding can be: He, a mixture (He + Ar), Ar, CO_2, N_2.

The *plasma torch* (3 in Fig. 7.1) *negatively* affects the efficiency of laser welding due to the absorption of part of the energy and the deterioration of the laser beam focusing, so it is advisable to:

- Use shielding gases with *high ionization energy*. The most effective is He or a *mixture of He + Ar* (40–50% He). *He increases the penetration depth*. The presence of *Ar* in the mixture increases the density, helps to retain the gas above the weld, and thus *improves protection*.
- "*Blow off*' the torch by increasing gas flow rate.
- Use the *pulsed* mode.

Shielding gas flow rate is about 5–10 L per hour.

Filler wires with a diameter of up to 1.5 mm are mainly used in pulsed laser welding.

In laser welding, the width of the weld (Fig. 7.2) is several times smaller than that in arc and plasma welding. Therefore, the requirements for assembly accuracy and joint tracking are increasing, which is ensured by the use of ***auxiliary welding equipment***: robots, manipulators, centralizers, welding stands, and clamps.

Advantages of laser welding:

- the ability to *transmit* a laser beam using a fiber optic system over long distances to hard-to-reach places without energy loss;
- easy *adjustment of the energy* characteristics of the laser beam;
- easy *control of* laser welding *processes*;
- *easy protection* in inert gases, carbon dioxide, etc., and the ability to weld large structures;
- no *influence of magnetic fields* on the laser beam, which ensures stable weld formation along its entire length.

Disadvantages of laser welding:

low efficiency of laser industrial systems;
complexity and *high cost* of equipment and maintenance;
the need for high *operator qualifications*.

7.4 Hybrid Laser-Plasma Surfacing

The scheme of hybrid laser-plasma surfacing is shown in Fig. 7.6, a. A direct current electric arc burns between the refractory tubular cathode 1 and the workpiece (anode) 2. In the initial section, the arc is compressed and stabilized by the plasma nozzle 3. Filler powder 4 is fed through the nozzle 5. The plasma nozzle and powder feed nozzle are made of copper and are cooled by water 6. A focused laser beam is injected into the plasma-forming channel through the opening of the tubular cathode 7. The plasma-forming gas is injected into the plasma-forming channel through the opening of the tubular cathode 8 and through the gap between the tubular cathode and the plasma-forming nozzle 9. The flow rates of the two plasma gas streams are independently controlled. The filler powder is supplied by a stream of transport gas 10 (usually argon), which also acts as a shielding gas. To ensure the alignment of the laser beam and plasma flow, cathode position correctors must be provided in the plasmatron.

The absorption of laser radiation by plasma leads to:

- a significant increase in the *temperature* of the central region of the plasma (the maximum temperature value increases with the growth of laser radiation power);
- increase in *electrical conductivity* and, consequently, *current density* in the central region of the plasma;

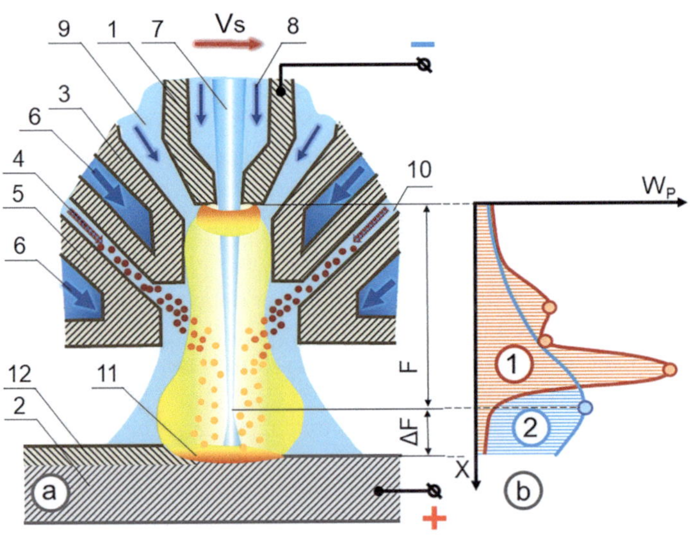

Fig. 7.6 Hybrid laser-plasma surfacing: **a** scheme of the plasmatron (1—refractory tubular cathode; 2—product (anode); 3—plasma-forming nozzle; 4—filler powder; 5—nozzle for distributed filler powder supply; 6—cooling water; 7—laser beam; 8—flow of plasma-forming gas through the refractory tubular cathode; 9—flow of plasma-forming gas through the plasma-forming nozzle; 10—flow of transporting and protective gas; 11—deposited pool; 12—deposited layer); **b** distribution of power density W_P along the X axis of the plasmatron (1—hybrid discharge; 2—laser radiation)

- increase in the *power density* of the central plasma region;
- improvement of the *spatial stability* of the hybrid pole;
- improvement of the *stability* of electric arc *burning at low currents*;
- improvement of the *stability* of electric arc *burning at high* deposition *rates*, which allows to increase the deposition rate by 2–3 times;
- increase of the *absorption coefficient* of laser radiation, which allows the use of low-power lasers.
- reducing the plasma arc voltage, which reduces the likelihood of double arc formation;
- reduced plasma density, hence *reduced gas-dynamic pressure* of the hybrid flow on the surface of the product, which can significantly *reduce* the *penetration depth* and the *proportion of base metal in the deposited metal* by up to 5%;
- a significant increase in the axial ***speed*** of the plasma in the central region.

Approximate *parameters* of the surfacing [6]:

- current $100 < I < 300$ A;
- laser power $1 < W^0 < 4$ kW;
- focal length F ~ 14–22 mm;
- length of the open hybrid discharge column (distance from the tip of the plasmatron to the surface of the product) d ~ 8–12 mm;
- displacement of the focal plane relative to the surface ΔF ~ 1–2 mm;
- deposition rate $12 < Vs < 30$ m/h.
- plasma gas flow rate through the refractory tubular cathode C_{p1} ~ 0.5 l/min;
- plasma gas flow rate through the plasma nozzle C_{p2} ~ 2.5 l/min;
- transport and shielding gas flow rate C_t ~ 5.0 l/min.

General recommendations for hybrid laser-plasma surfacing:

(a) use of a *coaxial arrangement* of the laser beam and plasma flow;
(b) *commensurability of laser* radiation *energy* and *plasma energy* due to the flow of electric current when selecting a laser and calculating the surfacing parameters (laser power must exceed 0.3 kW for laser-plasma discharge to occur);
(c) displacement of the *focal plane above the surface of the product* (ΔF), which leads to maximum heat release above the surface (Fig. 7.6b), and thus minimizes the penetration depth with maximum filler powder melting.

References

1. S. Katayama, *Fundamentals and Details of Laser Welding* (Springer, Singapore, 2020), p. 198. ISBN 978-981-15-7932-5. https://doi.org/10.1007/978-981-15-7933-2
2. A.B. Pereira, E.S.V. Marques, F.J.G. Silva, *A Guide to Laser Welding* (Nova Science Publishers, 2024), p. 369. ISBN: 979-8-88697-613-7. https://doi.org/10.52305/MCBI4234
3. S. Fomichov, O. Linyucheva, G. Vasyliev, E. Chvertko, *Corrosion of Welded Structures* (Springer, Switzerland, 2024), p. 222. ISBN 978-3-031-68992-5. https://doi.org/10.1007/978-3-031-68993-2

4. S. Fomichov, E. Chvertko, I. Skachkov, S. Minakov, A. Banin, *Quality Management in Welded Fabrication* (*Transactions on Intelligent Welding Manufacturing*) (Springer, Switzerland, 2023), p. 219. ISBN 978-3-031–34799-3. https://doi.org/10.1007/978-3-031-34800-6
5. V. Gvozdetsky, G. Korchinsky, I. Krivtsun, A. Zagorodniy, I. Jakimenko, Calculation of energy absorption and reflection coefficients of electromagnetic radiation during laser welding, in *Physical Processes During Welding and Processing of Metals. Collection of Articles* (DIA, Kyiv, 2018), pp. 293–299
6. A. Som, I. Krivtsun, Laser + plasma: searching for new possibilities in surfacing, in *Physical Processes During Welding and Processing of Metals. Collection of Articles* (DIA, Kyiv, 2018), pp. 591–596

MIX
Papier aus verantwortungsvollen Quellen
Paper from responsible sources
FSC® C105338

If you have any concerns about our products,
you can contact us on
ProductSafety@springernature.com

In case Publisher is established outside the EU,
the EU authorized representative is:
Springer Nature Customer Service Center GmbH
Europaplatz 3, 69115 Heidelberg, Germany

Printed by Libri Plureos GmbH
in Hamburg, Germany